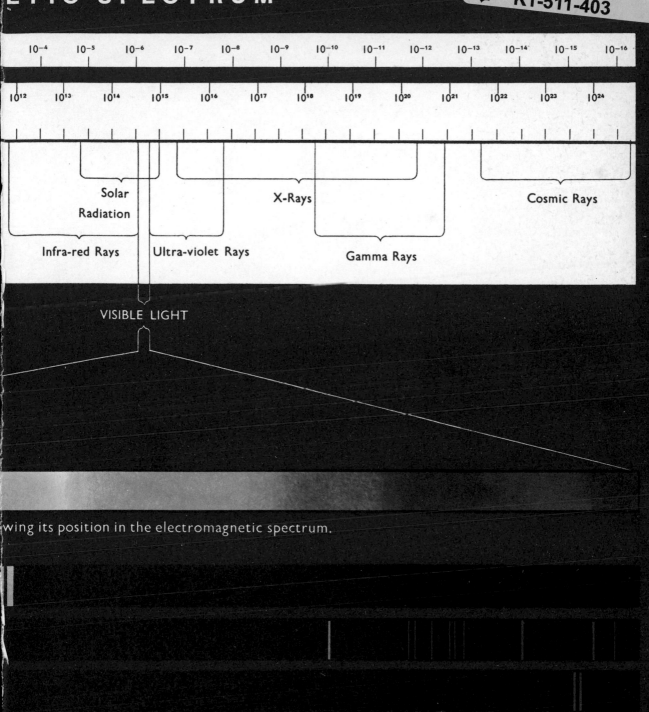

10^{-4}	10^{-5}	10^{-6}	10^{-7}	10^{-8}	10^{-9}	10^{-10}	10^{-11}	10^{-12}	10^{-13}	10^{-14}	10^{-15}	10^{-16}

10^{12}	10^{13}	10^{14}	10^{15}	10^{16}	10^{17}	10^{18}	10^{19}	10^{20}	10^{21}	10^{22}	10^{23}	10^{24}

Solar
Radiation

X-Rays

Cosmic Rays

Infra-red Rays

Ultra-violet Rays

Gamma Rays

VISIBLE LIGHT

wing its position in the electromagnetic spectrum.

ome elements.

A MODERN APPROACH TO CHEMISTRY Second Edition

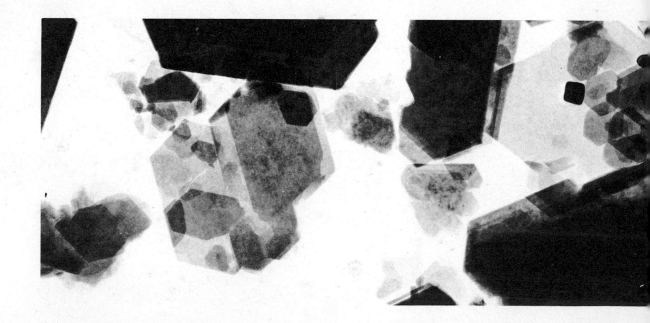

A MODERN APPROACH TO CHEMISTRY Second Edition

J. D. STOVE · K. A. PHILLIPS

Illustrations by Florence Caddy

HEINEMANN EDUCATIONAL BOOKS LTD

HEINEMANN EDUCATIONAL BOOKS LTD
London Edinburgh Melbourne Auckland Toronto
Singapore Hong Kong Kuala Lumpur
Ibadan Nairobi Johannesburg
Lusaka New Delhi

ISBN 0 435 64856 X

First published 1963
Reprinted six times
Second Edition 1971
Reprinted 1972, 1974

THE AUTHORS

J. D. Stove is Principal of the Koonung High School, Box
Hill, Victoria

K. A. Phillips is Lecturer-in-charge of the Faculty of Inorganic Chemistry, Royal Melbourne Institute of
Technology

Published by Heinemann Educational Books Ltd.,
48 Charles Street, London W1X 8AH

Printed in Great Britain by
Fletcher & Son Ltd, Norwich

Contents

Preface to the First Edition

THIS book has been planned on the basis of several years' experience in teaching Chemistry along the lines of the new Victorian course. The approach used is based on this experience. It has been fully tested and found thoroughly satisfactory.

We would like to drawn the reader's attention to one point. The development of understanding of scientific concepts during the course of the Leaving year is very rapid. As a result, a class which at the beginning of the year has very little systematized understanding of the scientific approach and of the basic concepts of Chemistry, should by the end not only have a thorough understanding of scientific method but should also be able to grasp new concepts easily and fluently.

The pace of this book takes account of this change. In the early chapters the foundations are laid slowly and carefully and classes which have had a really good grounding in Chemistry in earlier years should be able to move through these chapters fairly quickly. In later chapters, however, the pace quickens considerably to allow for the expected development of the students' ability.

The book adopts the internationally accepted basis for atomic weights determined by mass spectrometric measurements using the carbon-12 isotope as standard. Furthermore, ionic equations are used very widely after their introduction in chapter 14.

An important objective of the new course is to develop an appreciation of scientific method and give experience in critical, logical thought. This has been a significant factor in the design of the book. Thus, wherever possible, experiments have been described or discussed in order to present the experimental facts before introducing the relevant theories, and not to demonstrate or illustrate theories already propounded. It is hoped that the students will see that the theories or explanations simply explain the facts, and that new facts may lead to modification of the theories.

We would like to thank Mr. D. R. Driscoll, of the Secondary Teachers' College, for his very careful reading of the manuscript. A great many of his suggestions have been incorporated in the text. We would further like to thank Mr. L. Ackland, of Brighton Grammar School, Mr. N. Henry, of Oakleigh High School, Mr. D. Lugg, of University High School, Mr. G. Miller, of Upwey High School, Mr. A. Smith, of Carey Baptist Grammar School and Mr. M. E. W. Stump, of Scotch College, for their assistance in reading and checking the proofs, and for their many valuable suggestions.

We are also indebted to the commercial firms and Government agencies which have assisted us with the illustrations. A full list of these will be found on page 267.

We hope that this book will prove useful to both teachers and students.

J.D.S.
K.A.P.
Melbourne, January, 1963.

Preface to the Second Edition

THE second edition of this text is closely based on the successful first edition. The general scope and approach to the subject has been retained, but some major changes, and additions, as well as many minor alterations have been made. The introductory material to many of the topics is now covered in the junior secondary school, so the pace of development of many topics, particularly in the earlier chapters, has been increased.

Some major rearrangements of the sequence of presentation have been made. The calculations of empirical formulae have been brought forward and now follow immediately after the determination of relative atomic masses. The electrolysis of ionic liquids has been combined with the section on electrovalency and the writing of empirical formulae of ionic compounds. The periodic table has been introduced with the chapter on electronic configurations of atoms and is then frequently referred to throughout the remainder of the text. The sequence in which the nonmetals appear is the sequence of the periodic table but formal group studies have not been attempted because we feel that too many exceptions arise in this approach.

Some major alterations and additions have been made. All calculations in stoichiometry are now based on the mole. The electronic structures of atoms are now described in terms of regions occupied by electrons. Protonic definitions of acids and bases have been introduced and are subsequently referred to in the text. A new chapter introduces structural principles of organic compounds using a non-traditional approach, so that it is possible immediately to introduce many examples of organic materials in common use. The study questions have been increased in number for several of the chapters.

Most physical quantities are now given in the appropriate multiples and sub-multiples of SI units. The main effect of this is in measurements of volume: the litre and its subdivisions are not used in SI, but are replaced by volumes expressed in (length)^3, i.e. cm^3, dm^3 or m^3 as appropriate. It was also considered useful to retain the following non-SI units: the mmHg for pressure and the angstrom unit of length. Chemical names of inorganic compounds have also been altered to the current IUPAC conventions.

J. D. STOVE
K. A. PHILLIPS
January 1970.

1: Atoms and their Relative Masses

One of the foundations of modern science is that matter consists of particles which are extremely small and light. Much of the evidence for the existence of particles in matter will emerge in a natural sequence during the course of this book.

As a starting point, it is useful to focus attention on the *size* and *mass* of particles comprising matter. An experiment which can help to develop the notion of the extreme smallness and lightness of particles uses methylene blue.

An Experiment with Methylene Blue

Methylene blue is an intensely coloured dye. If 0.1 gramme of methylene blue powder is dropped into one dm³ of water in a glass beaker, the blue colour is seen to spread through the water as the powder dissolves. When all the dye has dissolved, 100 cm³ of the solution may be placed in another beaker and diluted to one dm³ by the addition of more water. This solution contains 0.01 gramme of the dye. If 100 cm³ of this solution is diluted to one dm³, the blue colour can still be seen, though this solution contains only 0.001 gramme of the dye. Finally, if a drop of this solution is placed on a white glazed tile, the blue colour can just be seen. A drop is about 0.1 cm³, and therefore contains 0.000 000 1 gramme of the dye (see fig. 1.1).

Since at least one particle of methylene blue must be present in the drop of solution to cause the blue colour,

Fig. 1.1. Dilution of a coloured solution.

After two dilutions the beaker on the right contains one thousandth of a gramme of methylene blue. Each drop of this solution shows a blue tint even though one drop contains only one ten-millionth of a gram of methylene blue.

each particle of methylene blue must weigh no more than one ten-millionth of a gramme.

The mass of one ten-millionth of a gramme is clearly the upper limit for the mass of one particle of methylene blue. At first, this mass seems surprisingly small, but the real surprise lies in the fact that this figure is too great by a factor of many billions (10^{14}).

If 100 cm³ of the most dilute solution prepared above were used as a further starting point and diluted to 1 dm³, the blue colour would be barely visible. Nevertheless, this process of diluting the solution to one-tenth of its previous concentration, would have to be repeated eleven more times if a single drop of liquid were to contain only one particle of methylene blue on the average.

Particles of methylene blue consist of many atoms combined together. Atoms are therefore much smaller and lighter than particles of methylene blue.

Atoms

The existence of atoms is so widely accepted that it is taken as a fact. However, our evidence for the existence of atoms is entirely indirect because no instrument has been built which can magnify sufficiently to allow us to see atoms, and what is more, no such instrument can ever be built.

Optical microscopes can magnify very small objects, such as an amoeba, and thus make them visible. However, there is a limit to the magnification possible with optical instruments because their resolving power is limited by the wavelength used to view the object. An image is formed by an optical microscope because light waves are differentially absorbed, transmitted or reflected by the material on which they fall or through which they pass. But if an object is very much smaller than the wavelength of light, the object causes negligible interference with the light waves, and hence cannot be 'seen'. An analogy can be made with ocean waves. A large rock standing above water will interrupt the wave pattern but a thin post will scarcely affect it. In general, no wave pattern can be disturbed by an object whose diameter is substantially smaller than the wave length. Atoms have diameters of about 10^{-10} m and the wavelengths of visible light range from about 10^{-7} to 10^{-6} m. Thus atoms are very much smaller than the wavelengths of visible light, and no optical microscope can ever enable anyone to 'see' an atom.

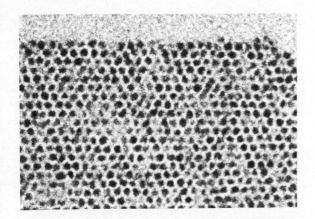

Fig. 1.2. Crystal of the protein *ferritin*, magnified ×
350 000.

The dark spots show iron(III) hydroxide clusters which are
contained in the centres of the hollow particles of protein.
In this case, the particles are large enough to be resolved
by an electron microscope.

There are other forms of radiation of shorter wavelength than visible light. Electron microscopes, for
example, use a beam of high energy electrons instead
of light, and the wavelength of the electron beam is
short enough to show some of the larger molecules
(see fig. 1.2). However, these massive molecules each
contain several hundred-thousand atoms. If a higher-
energy, shorter-wavelength beam of electrons is
generated, another complication arises. Such a high-
energy beam is extremely destructive and when it is
focused on the target, the target is vaporized. It
would be like trying to study a heap of feathers in a
wind tunnel.

Although it is thus impossible to *see* atoms in any
normal sense, a technique has been developed which
gets quite near to it. Whereas both optical and
electron microscopes are based on the *absorption* of
incident waves, this technique is based on the *emission*
of beams of charged particles by the substance itself.
A charged particle is called an *ion*, and the technique
is called *field ion emission* (see fig. 1.3). It can be
used to show the arrangements and approximate sizes
of atoms in metals of high melting point. For this
technique, a very finely pointed needle, made of a
metal of high melting point, is mounted inside a
container which is almost completely evacuated. The
container contains some helium gas at very low
pressure and when atoms of this gas collide with the
metal of the needle they become electrically charged,
i.e. ions. Because the metal and the helium ions are
both positively charged, the helium ions are repelled.
Some of these ions strike the photographic plate, and
when this is developed, it shows a pattern of spots
which corresponds to the pattern of atoms in the
metal. In the diagram, the atoms in the positively
charged metal are represented by circles marked
with plus signs and the helium ions being repelled are
represented by the smaller circles marked with plus
signs, the arrows showing the direction of their
motion.

The magnification factor for this technique is the
ratio of the radius of curvature of the photographic
plate to the radius of curvature of the tip of the metal
needle. Magnification factors of several million times
can be obtained. If this factor is known then the actual

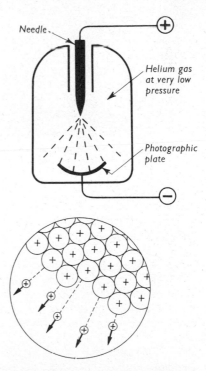

*Highly magnified representation of positive helium ions
being repelled from atoms in the tip of the metal needle*

Fig. 1.3. Field emission.

diameter of the atoms in the metal can be calculated from the distance apart of the spots on the photographic plate. Thus:

$$\text{Diameter of atoms} = \frac{\text{distance between spots}}{\text{magnification factor}}$$

For the metal rhenium, a field emission pattern at 5×10^6 times magnification shows a pattern of spots in which the closest spacing is about 0.5 mm or 5×10^{-4} m. Thus, the approximate diameter of the atoms in the metal can be calculated:

$$\text{Diameter} \approx \frac{5 \times 10^{-4} \text{ m}}{5 \times 10^6} = 10^{-10} \text{ m}$$

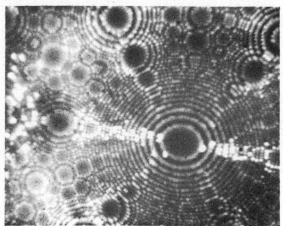

Crystal of rhenium, magnified $\times 5\,000\,000$.

Sub-atomic Particles

It is now generally accepted that atoms are composed of a number of smaller particles. The sub-atomic particles which are important to a study of chemistry are *electrons*, *protons* and *neutrons*.

Other particles have been detected when atoms are split, but because these particles become apparent only under conditions which are not encountered in carrying out chemical reactions, they need not be considered in developing theories concerning chemical behaviour.

The Electron

The existence of electrons can be demonstrated by experiments involving the passage of electricity through gases at low pressures. All gases glow brightly

under these conditions. A convenient arrangement is shown in fig. 1.4 (a).

Two metal electrodes are sealed into the tube and these are connected to a source of electricity capable of supplying a very high voltage. An induction coil is suitable for this purpose. It will be noticed that the cathode used in this experiment has a hole drilled through its centre.

If the glass tube contains a gas at very low pressure and the induction coil is switched on, a number of regions will appear to light up when the tube is observed in a dark room. These glowing regions are *not* obtained if the tube is completely evacuated and they are *not* obtained unless the induction coil is switched on. The appearance of the glowing regions may be attributed to the effect of the electricity on the molecules of the gas in the tube.

Two glowing regions appear to originate from the vicinity of the cathode and these are labelled on the diagram. The green glow on the glass seems to be caused by something moving from the cathode towards the left hand end of the tube and the faint red glow seems to be caused by something moving to the right. The nature of these phenomena can be better understood by bringing a magnet near the tube. It is known that a magnet will deflect a moving charged particle in a direction that depends on the sign of the charge of the particle. The position of the green glow is greatly affected by a magnet, and the direction in which it is deflected suggests that it is caused by negatively charged particles streaming away from the cathode (see fig. 1.4 (b)).

The red streamer in the gas on the other side of the cathode is not apparently affected when an ordinary magnet is placed near it. However, if a very powerful magnet is used, the beam is deflected in a direction which suggests that it is composed of positively charged particles moving towards the right in the diagram.

These observations suggest that the result of the high electrical potential on the molecules of the gas is to produce two streams of oppositely charged particles and this suggests that the molecules can be separated into negative particles and positive particles. Consider one molecule of the gas. This is an electrically neutral particle. Suppose that it is separated by the electricity into two charged particles. It follows that these two particles will carry charges of the same magnitude

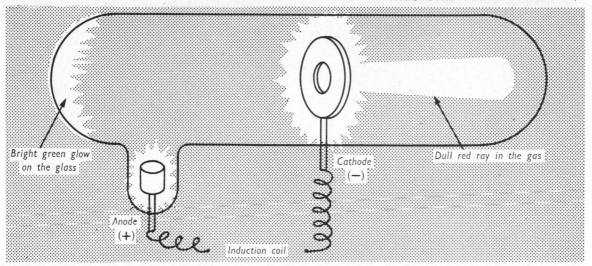

Fig. 1.4 (a). Electrical discharge in a gas at low pressure.

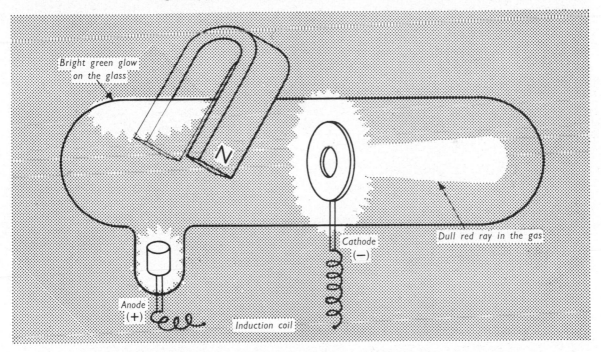

Fig. 1.4 (b). Deflection caused by a magnetic field.

but of opposite sign. Since they carry equal charges, the relative ease with which the negative particles are deflected suggests that they are much lighter than the positive particles. The heavy positive particles are called *ions*. In general, single uncharged particles of elements are called *atoms*; uncharged groups of atoms are called *molecules*; electrically charged atoms or molecules are called *ions* and these may be positively or negatively charged.

The light negative particles are called electrons and it has been found that no matter what gas is used in the discharge tube, electrons always have the same mass and the same charge.

mass of an electron =
$(9.101\ 9 \pm 0.000\ 4) \times 10^{-31}\,\mathrm{kg}$

charge of an electron =
$-(1.602\ 10 \pm 0.000\ 07) \times 10^{-19}\,\mathrm{C}$

The Structure of the Atom

The way in which the heavy positive particles and the electrons are arranged within the atom was shown by an experiment devised by H. Geiger and E. Marsden in 1909. The experiment made use of alpha particles emitted by some radioactive substances. Alpha particles, which are positively charged and about four times as heavy as hydrogen atoms, penetrate a thin metal foil without any apparent effect on the foil. Most of the alpha particles pass straight through the foil but about 1 in every 20 000 is deflected through an angle of 90° or more. An alpha particle can be detected because it causes a bright spot of light called a scintillation when it strikes a screen coated with zinc sulphide (fig. 1.5).

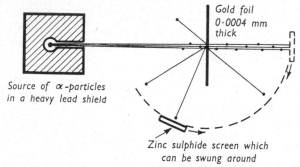

Fig. 1.5. Experiment of Geiger and Marsden.

These observations were explained by E. Rutherford in 1911. He suggested that most of the mass of an atom was concentrated in a positively charged nucleus at the centre of the atom, and that the electrons surround the nucleus. The way in which this theory explains the above experiment is shown in the diagram (fig. 1.6).

Calculations based on the scattering of alpha particles and on other experiments give the picture of the atom shown in the diagram (fig. 1.7). It should be noted that the nucleus and the electrons are very small compared to the size of the atom. The atom is mostly empty space.

Before any details of the structure of the nucleus can be discussed it is necessary to explain how the mass and charge of the nuclei of various atoms can be measured.

The most accurate way of measuring the mass of an atom is to change it into a positive ion by removing one or more electrons, and then determine the mass of the positive ion from its deflection in a magnetic field. Instruments designed to make such measurements are called *mass spectrometers*.

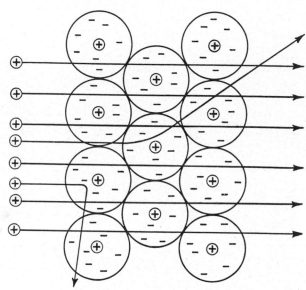

Most α-particles pass straight through the atoms but a few approach a nucleus and are deflected.

Fig. 1.6. Rutherford's explanation.

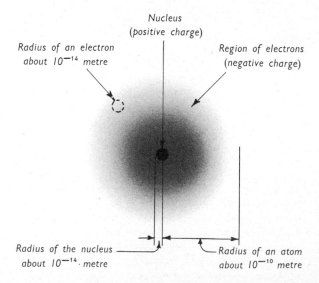

Fig. 1.7. The nuclear atom.

RELATIVE ATOMIC MASSES

Mass Spectrometers

Two famous designs which illustrate the principles used are represented diagrammatically in fig. 1.8.

In these instruments, positive ions are usually produced by bombarding a sample in the form of a gas or vapour, with electrons from a heated filament.

The positive ions are allowed to enter the space between the two charged plates where they are accelerated to a high velocity in the direction of the negative plate. A narrow slit in the plate selects a narrow beam of ions for analysis, the method of analysis providing the key difference between the two designs.

1. Dempster's design:

The beam of ions is analysed in the semi-circular chamber at the end of the spectrometer. This chamber is sandwiched between the poles of a powerful electromagnet. As the stream of ions enters the magnetic field, the ions travel in semi-circular paths. Light particles will be deflected into paths having a smaller radius of curvature than heavy particles having the same speed and charge. Each of these separate beams is made to pass in turn through the final slit by changing the potential between the charged plates.

As each beam of ions passes through the final slit, an electrical device detects the ions and makes a record in the form of a graph which has a series of peaks. Each peak corresponds to one of the separate ion beams, and the relative height of each peak indicates the relative intensity of each ion beam.

2. Bainbridge's design:

The beam is passed through a combined magnetic and electric field which has the effect of allowing only particles with a certain speed to pass on to be analysed. The ion beam is analysed by a magnetic field in the semi-circular chamber. The detecting device is a photographic plate which records all of the separate ion beams simultaneously. The beams are allowed to fall on the photographic plate which, on subsequent development, shows a series of lines corresponding to each beam of ions. The relative intensity of each line indicates the relative intensity of each ion beam.

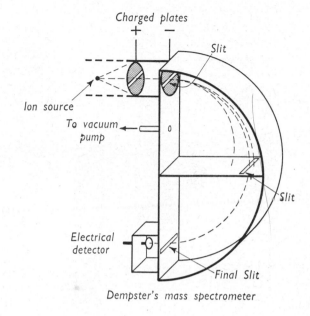

Dempster's mass spectrometer

Fig. 1.8. Mass spectrometers.

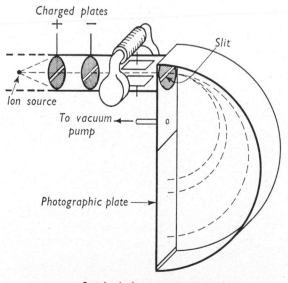

Bainbridge's mass spectrometer

A modern high precision mass spectrometer.
The analysis tube is mounted on the framework at the left. The operator is adjusting the ion source. From the source, the ions are accelerated upwards and deflected through an angle of 60° by the magnet which is in the upper part of the framework. Each selected ion beam passes up the tube to the extreme top left where an electrical device detects the ions. The tubes leading down in front of the operator lead to the vacuum pumps. The paper rolls in the centre of the instrument cabinets record the mass spectrum.

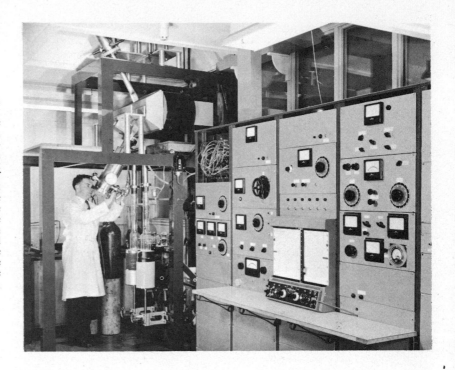

From the position of the peaks on the graph or the lines on the photographic plate it is possible to calculate the masses of the positive ions.

Modern mass spectrometers are capable of giving very precise results. They use the same principles as the instruments described above but have many refinements which give greater accuracy.

Three very important features of nuclear structure become evident from the results obtained from mass spectrometers. These will now be discussed under the headings "*The Proton*", "*The Neutron*" and "*Isotopes*".

The Proton

The lightest positive particle detected in a mass spectrometer is obtained when hydrogen gas is analysed. This particle is called a *proton*. Experiments show that its charge is the same in magnitude as the charge of an electron but of opposite sign, that is, positive whereas the charge of an electron is negative. Since this particle is obtained when an electron is removed from an atom of hydrogen, the nucleus of a hydrogen atom must be a single proton. Thus a hydrogen atom consists of one proton and one electron and the atom may be represented diagrammatically as in fig. 1.9.

It is now well established that all other atoms contain protons in their nuclei. The number of protons in the nucleus of an atom is equal to the number of electrons revolving around the nucleus.

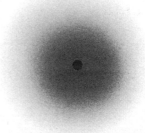

Fig. 1.9. A representation of a hydrogen atom, which consists of one proton and one electron.

Atomic Number

It has been found that all atoms of the same element have the same number of protons in their nuclei.

● *The number of protons in the nucleus is called the atomic number.*

The atomic numbers of elements have been determined in two ways:

(i) The atomic number can be determined by examining the radiations given off when an element is bombarded by X-rays. H. G. Moseley devised a suitable experimental procedure for this in 1913. His method was so reliable that it was used to predict that certain elements were then undiscovered. The missing elements were later discovered.

(ii) The atomic numbers of some elements can be confirmed by measuring the number of alpha particles scattered by a thin sheet of the element. This was carried out by J. Chadwick in 1920.

The atomic numbers of elements determined by these methods are related to the positions of the elements in the periodic table. This adds further support to the results and thus it is possible to state the atomic numbers of the elements with considerable certainty.

Singly and Doubly Charged Ions

When helium is examined in a mass spectrometer a graph with two peaks is obtained (see fig. 1.10).

The atomic number of helium is known to be two and therefore atoms of helium must each contain *two protons* and *two electrons*. Therefore it is possible to obtain two different kinds of positive helium ions: one with a single positive charge represented by the symbol He^{1+} and the other with a double positive charge represented by the symbol He^{2+}.

The He^{1+} ion consists of a helium nucleus with only one electron and the He^{2+} ion is a helium nucleus without any electrons. (It has been found that the He^{2+} ion is identical to the alpha particle mentioned earlier.)

Because the ions differ in structure by only one electron, they have virtually the same mass and therefore the different positions of the peaks on the graph are due to the different charges on the ions. The He^{2+} ions will be deflected into a path having a smaller radius of curvature than the path of the

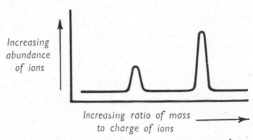

Fig. 1.10. Mass spectrometer recording for helium.

He^{1+} ions. Thus the small peak corresponds to the He^{2+} ion beam and the larger peak corresponds to the He^{1+} ion beam.

The Neutron

It was pointed out above that a helium nucleus contains only *two* protons but the positions of the peaks show that the helium nucleus is about *four* times as heavy as a proton. Similarly, it has been found that for all atoms other than hydrogen, the mass of the nucleus cannot be accounted for by the protons alone. To account for this, E. Rutherford suggested in 1920 that such nuclei contain particles which have about the same mass as a proton but are electrically neutral. These particles were called *neutrons*. This was verified by J. Chadwick in 1932 when he showed that experimental observations made by W. Bothe and H. Becker in 1930 confirmed Rutherford's suggestion. For helium it is apparent that the nucleus contains *two protons* and *two neutrons* (fig. 1.11).

In addition to the atomic number of each element it is now possible to define a *mass number*.

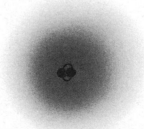

Fig. 1.11. A representation of a helium atom, which consists of two protons, two neutrons and two electrons.

Mass Number

● *The Mass Number is the number of protons and neutrons in the nucleus of each atom.*

This can be determined by the mass spectrometer. A few examples of elements together with their values for atomic number (Z) and mass number (A) are quoted in the table (fig. 1.12).

Element	Symbol of Element	Atomic Number Z	Mass Number A
helium	He	2	4
beryllium	Be	4	9
sodium	Na	11	23
aluminium	Al	13	27
phosphorus	P	15	31

Number of neutrons in the nucleus of an
atom $= A - Z$
Fig. 1.12.

Isotopes

If mercury vapour is analysed in a mass spectrometer a number of peaks are obtained in the recorded graph. Examination of these peaks shows that mercury contains atoms with six different mass numbers (see fig. 1.13). Since mercury has an *atomic number* of 80, it follows that the nuclei of mercury atoms contain

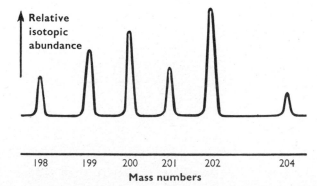

Fig. 1.13. Mass spectrum of mercury.

80 *protons* together with 118, 119, 120, 121, 122 or 124 *neutrons*. These different kinds of atoms of mercury are called *isotopes*.

Isotopes of an element are different atoms of the same element. All isotopes of an element contain the same number of protons but have different numbers of neutrons.

The relative heights of the peaks enable the relative proportion of each isotope to be determined. The relative abundances of the isotopes of a few well known elements are listed in the table (fig. 1.14).

Element	Symbol of Element	Atomic Number	Mass Number	Percent Abundance
hydrogen	H	1	1	99.985
			2	0.015
carbon	C	6	12	98.89
			13	1.11
oxygen	O	8	16	99.759
			17	0.037
			18	0.204
uranium	U	92	234	0.006
			235	0.719
			238	99.274

The elements listed in the previous table do not occur naturally as a mixture of isotopes.
Fig. 1.14. Isotopes of some elements.

Most elements occur naturally as a mixture of isotopes. Thus natural hydrogen always consists of the two isotopes called protium (Z = 1, A = 1) and deuterium (Z = 1, A = 2). The symbols used for these two isotopes are $_1^1H$ for protium and $_1^2H$ for deuterium.

Relative Atomic Mass

Although it is possible to express the masses of atoms in kilogram or in any other unit of mass it has been found more convenient and useful to determine their *relative* masses. The masses of atoms relative to each other are ratios and have no unit. The relative masses of the elements have been called "atomic weights" for many years and this name is still widely used. Hence the terms *relative atomic mass* and *atomic weight* should be regarded as synonymous.

At present, the most precise measurements of relative atomic masses are being obtained from modern mass spectrometers. For purely technical reasons bound up with the operation of these instruments, it has been found that relative atomic masses can be expressed most accurately relative to the most abundant isotope of carbon, namely *carbon-twelve*.

● *The carbon-twelve isotope is now accepted as the standard of relative atomic masses.*

This is written as $^{12}C = 12$. (The twelve is understood to be followed by as many zeros as the precision of any particular determination justifies.)

The method for determining relative atomic masses of elements is to use precise mass spectrometric measurements to obtain two sets of results:

(a) an accurate value for the mass of an atom of each isotope of the element relative to the ^{12}C isotope. These are called the *isotopic masses*;

(b) the relative abundance of each isotope in the sample of the element.

A mean of the isotopic masses for the element can then be calculated, weighted according to the relative abundance of each isotope. This weighted mean is the relative atomic mass of the element.

NUMERICAL ILLUSTRATION

Natural chlorine is composed of the two isotopes ^{35}Cl and ^{37}Cl and the values needed to calculate the relative atomic mass of chlorine are given in the table (fig.1.15).

Isotope of Chlorine	Isotopic Mass	Percent Abundance
^{35}Cl	34.968 9	75.77
^{37}Cl	36.965 9	24.23

Fig. 1.15

The relative atomic mass of chlorine

$$= \frac{34.968\ 9 \times 75.77}{100} + \frac{36.965\ 9 \times 24.23}{100}$$

$$= \frac{2\ 649.6 + 895.7}{100}$$

$$= 35.45(3)$$

Note: The accuracy of the atomic mass is limited by the accuracy of the least accurate figure, viz. the percentage abundance of each isotope.

The Relative Atomic Mass of Carbon

In table 1.14 natural carbon is shown as being composed of two isotopes: ^{12}C and ^{13}C. Thus the atomic mass of natural carbon is slightly greater than twelve (see below). It should be clearly understood that the standard for expressing relative atomic masses is the carbon-twelve isotope and not natural carbon.

RELATIVE ATOMIC MASS OF CARBON

Isotope of Carbon	Number of Protons	Number of Neutrons	Number of Electrons	Isotopic Mass	Percent Abundance
^{12}C	6	6	6	12.000 0	98.893
^{13}C	6	7	6	13.003 4	1.107

The relative atomic mass of natural carbon $= \dfrac{12.000\ 0 \times 98.893}{100} + \dfrac{13.003 \times 1.107}{100}$

$$= \frac{1186.72 + 14.39}{100}$$

$$= 12.011\ 1$$

The currently accepted value for the relative atomic mass of carbon is: $12.011\ 15 \pm 0.000\ 05$.

SUMMARY AND DEFINITIONS

A great deal of experimental evidence exists (a little of which has been outlined in this chapter) which suggests that all atoms are composed of much smaller particles. The **nucleus** of an atom contains all the **positive charge** and almost all the **mass**. It contains **protons** and **neutrons**. **Electrons** occupy the region around the nucleus and they carry all the **negative charge** but very little of the mass. (See table.)

Particle	Mass (kilogram)	Charge (coulomb)
proton	$1.672\,5 \times 10^{-27}$	$+ 1.602 \times 10^{-19}$
neutron	$1.674\,8 \times 10^{-27}$	No charge
electron	9.109×10^{-31}	$- 1.602 \times 10^{-19}$

(i) **An Element** is composed of atoms, all of which have the same atomic number.

(ii) **Isotopes of an Element** are different atoms of the same element that have the same atomic number but different mass numbers.

(iii) **The Atomic Number** of an element is the number of protons in the nucleus of an atom of the element. It is numerically equal to the number of electrons in the neutral atom.

(iv) **The Mass Number** of an isotope of an element is the number of protons and neutrons in the nucleus of an atom of the isotope. It is numerically equal to the whole number nearest to the isotopic mass.

(v) **The Isotopic Mass** is the mass of an atom of the isotope expressed on the scale in which an atom of ^{12}C is taken as 12 exactly.

(vi) **The Relative Atomic Mass** of an element is the weighted mean of the isotopic masses on the scale in which an atom of ^{12}C is taken as 12 exactly.

STUDY QUESTIONS

1. The isotopic masses on the scale $^{12}C = 12$, and the relative abundances of the isotopes of the elements copper and silver are:

(a) copper:
   ^{63}Cu. Isotopic mass 62.93. Abundance 69.0%.
   ^{65}Cu. Isotopic mass 64.93. Abundance 31.0%.

(b) silver:
   ^{107}Ag. Isotopic mass 106.9. Abundance 51.4%.
   ^{109}Ag. Isotopic mass 108.9. Abundance 48.6%.

What are the atomic masses of these elements?

2. Sodium has only one natural isotope. The atomic number of sodium is 11 and an accurate value for the mass of a sodium atom on the scale $^{12}C = 12$ is 22.989 78.
   (The mass of an electron on the ^{12}C scale is 0.000 55.)

(a) What is the mass number of a sodium atom?

(b) How many protons, neutrons and electrons are there in a sodium atom?

(c) What is an accurate value for the mass of a sodium ion with a single positive charge?

3. The distance between the centres of atoms in a solid can be determined by an experimental technique known as X-ray analysis. This method shows that in copper, each atom requires a volume of 1.184×10^{-23} cm³ in the crystal. The density of copper is 8.92 gramme per cm³. Calculate the mass in gramme of an atom of copper.

4. Radium is a radioactive element. Its atoms disintegrate into a number of products. If radium is placed in a closed vessel it is found that each atom that disintegrates produces, amongst other products, one atom of helium. Radium decomposes at a rate of 1.16×10^{18} atoms per gramme per year. In the same time 7.68×10^{-6} gramme of helium is produced. Calculate the mass in gramme of an atom of helium.

Answers to Numerical Problems

1. (a) 63.5(5), (b) 107.9.

2. (a) 23, (b) 11 protons, 12 neutrons, 11 electrons, (c) 22.989 23.

3. 1.056×10^{-22} g.

4. 6.6×10^{-24} g.

2: Empirical Formulae

For almost all chemical compounds, it has been found that the composition by mass is constant from sample to sample. This has been summarized in the *law of definite proportions* which can be stated as follows. A given chemical compound is always composed of the same elements combined in the same proportions by mass. Since the elements are always present in the same proportions by mass, they must always be present in the same ratio of the numbers of atoms. A formula which shows the simplest ratio of numbers of atoms is called an *empirical formula*.

Definition of Empirical Formula

The empirical formula of a compound is the formula which represents the simplest ratio of the numbers of the respective atoms or ions in the compound. Examples of empirical formulae are:

NaCl (ratio 1 : 1)
$MgCl_2$ (ratio 1 : 2)
Na_2O (ratio 2 : 1)

These formulae indicate the simplest ratio of the numbers of atoms or ions in a sample of the compound but in an experimental determination of the composition of a compound, the relative *masses* of the elements are obtained. It is necessary to define a unit which enables the masses of elements to be related to the numbers of atoms. This unit is the *mole*.

The mole can be defined for any element, compound or part of a compound, but for the first example, only the *mole of an element* will be defined.

Definition of a Mole of an Element

A mole of an element is the amount of the element which contains the same number of atoms as 12 g of pure ^{12}C.

The way in which this definition can be used to relate masses of elements to numbers of atoms of the elements can be illustrated by an example. Suppose that samples of two elements contain the same number of atoms. Let this number be N (see fig. 2.1). The total mass of each element in the sample is the product of the mass of each atom multiplied by the number of atoms.

Mass of carbon in the sample = $12.0 \times N$ units.

Mass of magnesium in the sample = $24.3 \times N$ units.

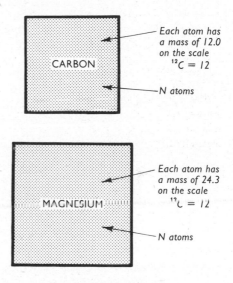

Fig. 2.1. Equal numbers of atoms of carbon and magnesium.

Thus the ratio of the masses of the elements in the two samples is the same as the ratio of the relative atomic masses no matter what value N may have. There is a particular number of atoms of C in 12 g of ^{12}C and this same number of atoms of Mg will be present in 24.3 g of Mg. *Therefore 24.3 g is the mass of one mole of Mg.* Similarly, the mass of one mole of atoms of any element is its relative atomic mass expressed as a mass in gramme.

The definition of the mole has been devised so that you can simply look up the relative atomic mass, add the unit 'g' and you have the mass of a mole of the element.

● *It is not necessary to know the actual number of atoms in one mole of an element to use the definition of a mole for calculating relative numbers of atoms. However, the number of atoms per mole has been determined and it is known as the Avogadro constant and its symbol is N.*
$$N = (6.022\,52 \pm 0.000\,28) \times 10^{23} \text{ mol}^{-1}$$

The definition of the mole makes it very easy to relate the composition by mass of a compound to the empirical formula of the compound. For example, sodium chloride contains sodium and chlorine combined in the ratio of 23.0 to 35.5 by mass. Thus 23.0 g of sodium is combined with 35.5 g of chlorine or

1 mole of Na is combined with 1 mole of Cl. Since a mole is defined as the amount of element containing N atoms, it follows that sodium chloride contains equal numbers of atoms of the two elements. The simplest ratio of the numbers of atoms is therefore 1 : 1 and the empirical formula is NaCl.

When an experiment is performed it is unlikely that exact mole amounts of substances will be used. One other step is usually necessary in the calculation. After determining the mass of each element in the sample of the compound, the number of mole of each element is calculated as follows:

$$\text{number of mole} = \frac{\text{mass of element}}{\text{mass of a mole}}$$

The way in which an empirical formula is calculated can be summarized as follows:

 (i) Determine the relative masses of the elements in a sample of a compound by a suitable experiment.
 (ii) Look up the relative atomic masses of the elements and add the unit 'g'.
(iii) Divide the mass of each element by the mass of the mole amount of the element. This gives the number of mole of each element in the sample.
(iv) Find the ratio of the number of mole of each element and convert the ratio to a simple ratio of whole numbers.

Determination of Empirical Formulae

The method used to analyse a compound to find its composition by mass depends on the properties of the particular compound under consideration. One

example will be given here. It must be understood that different methods may be used for other compounds.

EXPERIMENT:

To determine the empirical formula of magnesium oxide.

A typical procedure is:

 (i) Weigh a crucible and lid.

 (ii) Weigh a strip of clean magnesium ribbon in the crucible fitted with the lid.

(iii) Carefully ignite the magnesium taking care that no magnesium oxide escapes from the crucible. A small amount of magnesium nitride will also be formed during the combustion, because magnesium combines with the nitrogen of the air. The nitride can be converted to the oxide by heating the residue with a little water. All of the magnesium has now been converted to the oxide (see fig. 2.2).

(iv) Weigh the magnesium oxide in the crucible fitted with the lid.

Typical results:

Mass of crucible and lid	= 18.100 g
Mass of crucible and lid and magnesium ribbon	= 18.343 g
Mass of crucible and lid and magnesium oxide	= 18.503 g
∴ mass of magnesium	= 0.243 g
and mass of magnesium oxide	= 0.403 g
∴ mass of oxygen which has combined with magnesium	= 0.160 g

Thus it can be stated that magnesium oxide contains magnesium and oxygen in the ratio of 0.243 : 0.160 by mass. Note that no units are quoted in this statement as the result is a *ratio* of masses. This ratio can be converted to a percentage ratio if desired.

$$\% \text{ magnesium} = \frac{0.243}{0.403} \times \frac{100}{1} = 60.3\% \text{ by mass}$$

$$\% \text{ oxygen} = \frac{0.160}{0.403} \times \frac{100}{1} = 39.7\% \text{ by mass}$$

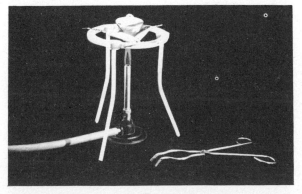

Fig. 2.2. Ignition of magnesium in a crucible supported on a pipe-clay triangle.

CALCULATIONS:

The required formula of magnesium oxide can be calculated from either result. The ratio by mass must be changed to a ratio of numbers of mole by dividing by the appropriate molar amounts.

One mole of Mg has a mass $= 24.3$ g

One mole of O has a mass $= 16.0$ g

$\therefore$ the ratio of the number of mole of Mg to the number of mole of O is:

$$\frac{0.243}{24.3} : \frac{0.160}{16.0} \quad OR \quad \frac{60.3}{24.3} : \frac{39.7}{16.0}$$

$$= 0.0100 : 0.0100 \qquad = 2.48 : 2.48$$

$$= 1 : 1 \qquad\qquad = 1 : 1$$

The empirical formula of magnesium oxide is MgO, i.e., the compound contains equal numbers of atoms or ions of magnesium and oxygen.

● *The ratio of the numbers of atoms is the same as the ratio of the numbers of mole.*

NUMERICAL ILLUSTRATION

A certain compound is found by experiment to have the following composition by mass: 43.4% sodium, 11.3% carbon, 45.3% oxygen.

The ratio of the number of mole of the respective elements is:

$$= \frac{43.4}{23.0} : \frac{11.3}{12.0} : \frac{45.3}{16.0}$$

$$= 1.89 : 0.942 : 2.83.$$

This ratio must now be converted to a whole-number ratio. To do this the smallest number is divided by itself, making it equal to 1.00. The other numbers must also be divided by the same number to keep the ratio constant.

$$= \frac{1.89}{0.942} : \frac{0.942}{0.942} : \frac{2.83}{0.942}$$

$$= 2.01 : 1.00 : 3.00.$$

The ratio of numbers of atoms should be a whole-number ratio, but the calculation will generally show a small discrepancy from an exact whole-number ratio. This is due to slight experimental errors in the determination of the percentage composition, and to the use of atomic masses which have been quoted only to the first decimal place.

Hence the ratio of the number of mole of atoms is 2:1:3. The empirical formula of the substance is Na_2CO_3 and its name is sodium carbonate.

Many salts are hydrated. When they are crystallized, they contain water in their crystalline structure and this water is present in some definite proportion. The water of crystallization in the salt can be included in the calculation by extending the definition of the mole to include molecules as well as atoms.

Definition of a Mole of a Molecular Substance

A mole of a substance is the amount of substance of specified formula, containing the same number of molecules as there are atoms in 12 g of ^{12}C.

The formula of water is H_2O. The required relative atomic masses are H $= 1.0$ and O $= 16.0$ on the scale $^{12}C = 12$. Hence, the relative molecular mass of $H_2O = 18.0$.

Since one molecule of H_2O has a mass of 18.0 on the scale $^{12}C = 12$, the total mass of N molecules is 18.0 g. Thus:

One mole of H_2O has a mass $= 18.0$ g.

This value can be used to calculate the number of mole of water in a sample of the compound and so the water of crystallization can be included in the empirical formula.

● *In a particular mass of a molecular substance:*

$$number\ of\ mole = \frac{mass\ of\ substance}{mass\ of\ a\ mole}$$

NUMERICAL ILLUSTRATION

It is found by analysis that a hydrated salt has the following composition by mass: 56.2% barium, 29.1% chlorine, 14.7% water.

Calculate the empirical formula of the hydrated salt.

The ratio of the number of mole of the elements and of water is:

$$= \frac{56.2}{137.3} : \frac{29.1}{35.5} : \frac{14.7}{18.0}$$

$$= 0.409 : 0.820 : 0.817$$

$$= \frac{0.409}{0.409} : \frac{0.820}{0.409} : \frac{0.817}{0.409}$$

$$= 1.00 : 2.00 : 2.00$$

$$= 1 : 2 : 2.$$

Hence the empirical formula is $BaCl_2.2H_2O$ and the compound is hydrated barium chloride (or barium chloride dihydrate).

EVIDENCE FOR THE IONIC STRUCTURE OF SALTS

The passage of an electric current through a molten salt (e.g. molten sodium chloride) provides useful information about the particles present in such compounds.

Crystals of common salt can be melted by heating them very strongly in a crucible using an efficient Méker burner. (The melting point of NaCl is 801 °C.)

The molten salt is a clear liquid which can be shown to be a good conductor of electricity in the following way. Two dry cells are connected in series with a small 3-volt globe and wired to a pair of electrodes of either copper or platinum. If these electrodes are dipped into the molten sodium chloride the globe glows, showing that electricity is passing through the molten salt (see fig. 2.3). In addition, cautious inspection of the liquid shows a bubbling while the globe glows. Apparently a chemical reaction occurs.

If the electrodes are removed from the liquid and the salt-encrusted electrodes are touched together, the globe does not glow. From this it can be concluded that solid sodium chloride does not conduct electricity.

Chemical Reactions during Electrolysis

It is difficult to identify the products in the experiment described above because they are formed in such small amounts. However, it is well known that the products are sodium metal and chlorine gas because these elements are produced industrially by the electrolysis of molten sodium chloride.

During the electrolysis, sodium metal forms at the cathode (the negative electrode), and chlorine is produced at the anode (the positive electrode). The cathode is negative because it has an excess of electrons coming from the battery and the anode is positive because it has a deficiency of electrons (see fig. 2.4).

The above evidence can be explained by proposing that molten sodium chloride contains charged particles (ions). The positive ions must be the sodium ions, because the positive ions would migrate to the oppositely charged electrode, i.e. to the cathode, and sodium is formed at the cathode. The negatively charged ions must be chloride ions.

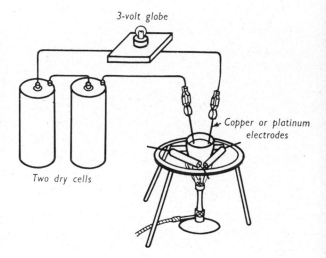

Fig. 2.3. Conductivity of fused sodium chloride.

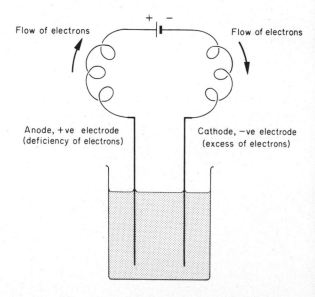

Fig. 2.4. Electrolysis of molten sodium chloride.

The actual charges on these ions can be determined by measuring the amount of electricity which passes through the electrolysis cell in a given time. The quantity of electric charge in coulomb is given by the product of the current (amp) by the time (sec). Since the empirical formula of sodium chloride is NaCl, the sodium and chlorine will be produced in the ratio of 23.0 g of sodium to 35.5 g of chlorine. It has been found that 96 490 C of electricity is required to produce these mole amounts of these elements.

Now, one mole of Na contains about 6.023×10^{23} atoms. Therefore the amount of electrical charge consumed by the formation of one atom of Na can be calculated:

$$\frac{96\ 490}{6.023 \times 10^{23}} = 1.602 \times 10^{-19}\ \text{C}$$

But this is the electric charge of one electron (see chapter 1). It can therefore be concluded that a sodium atom is formed by adding one electron to a sodium ion. The sodium ion is said to be singly positively charged and its symbol is Na^+. During the electrolysis, the singly positively charged sodium ions migrate towards the negatively charged cathode where they each receive an electron and become neutral atoms.

$$Na^+ + e^- \rightarrow Na$$

Similarly, a mole of Cl also contains 6.023×10^{23} atoms. Because 96 490 C of electricity forms this number of atoms, each chloride ion also has the same charge as one electron. They differ from sodium ions in that their charge is negative. During electrolysis, the singly negatively charged chloride ions migrate towards the positively charged anode where they each give up an electron and become neutral atoms.

$$Cl^- \rightarrow Cl + e^-$$

However, chlorine gas is not composed of separate atoms, it is composed of diatomic molecules, Cl_2, i.e., the gas is composed of molecules, each of which consists of two atoms. The above equation is doubled to show this:

$$2Cl^- \rightarrow Cl_2 + 2e^-$$

Multiplying the equation by two is necessary to properly represent the structure of the chlorine gas but it does not alter the fact that each chloride ion gives up one electron.

From experiments on the electrolysis of molten sodium chloride it can be seen that the empirical formula, NaCl, should be written as $Na^+\ Cl^-$ to show that the compound is composed of singly charged ions. It is not composed of atoms. The reason why $Na^+\ Cl^-$ is difficult to melt is that the strong electrostatic attraction between the oppositely charged ions holds them firmly in the rigid lattice of the crystal.

ELECTROLYSIS OF OTHER FUSED SALTS

The experiment above is difficult to perform because of the high temperature required. The following ionic solids are more convenient to experiment with than sodium chloride:

sodium hydroxide	m.p. = 318 °C
potassium hydroxide	m.p. = 360 °C
potassium chloride	m.p. = 715 °C

Of these, potassium hydroxide is probably the most suitable, as it has a low melting point for an ionic solid and does not tend to splutter as much as sodium hydroxide on melting.

● *Caution: exercise great care when working with molten alkalis.*

Alternatively, a mixture of about equal parts of sodium chloride and calcium chloride can be used. This mixture melts at about 500 °C, and functions very satisfactorily in the above experiment.

The electrolysis of other fused salts follows the same general pattern as for fused sodium chloride although different electrode reactions occur. The electrons coming from the negative terminal of the battery flow down the cathode and are consumed in the chemical reaction occurring at the cathode. At the same time an equal number of electrons is produced by the chemical reaction at the anode and flow to the positive terminal of the battery, thus completing the circuit. The migration of negative ions towards the anode and positive ions towards the cathode results in a net movement of charge through the liquid. Thus the migration of ions in the liquid constitutes a flow of electricity through the liquid.

ELECTROLYSIS OF LIQUIDS

Water, H_2O, and concentrated sulphuric acid, H_2SO_4, are liquids at room temperature. Their electrical conductivities can be tested using the same apparatus as for molten salts described above (fig. 2.3) but it is not necessary to heat the liquids. When the electrodes are dipped into these liquids, the globe does not light up. These liquids are either non-conductors or very poor conductors of electricity. Hence they contain no ions or very few ions.

Ionic Compounds

Most compounds which contain a metal *and* a non-metal are composed of *positive ions of the metal* and *negative ions of the non-metal*. Once the charges on one set of ions have been found, they can be used to obtain the charges on other ions. The charge on an ion is called the *electrovalency* of the ion (*electro-* refers to the electrical charge and *-valency* refers to attraction holding the particles together).

Electrovalencies

The empirical formula of sodium chloride is determined by experiments which measure the masses of the two elements in a sample of the compound. The experimentally determined empirical formula is $NaCl$. An empirical formula does not show whether the compound is composed of atoms or ions.

Experiments on the electrolysis of sodium chloride show that the molten compound is composed of ions and that the ions are singly charged. Sodium chloride is Na^+Cl^-.

The empirical formula of sodium oxide is Na_2O (i.e. a mole ratio of $2 : 1$). This compound is also ionic and since there are two sodium ions for each oxide ion, the compound must be composed of Na^+ ions and O^{2-} ions.

The empirical formula of magnesium oxide is MgO (i.e. a mole ratio of $1 : 1$). This compound is also ionic and since the oxide ion is O^{2-}, the magnesium ion must be Mg^{2+}. We could now *predict* the empirical formula of magnesium chloride since we now know the charges on the Mg^{2+} and Cl^- ions. To balance these charges, the empirical formula should be $MgCl_2$. This prediction is correct because this is the empirical formula found by experiment.

From these examples, it can be seen that the charges on the ions can be derived from experimentally determined empirical formulae. The charges are called electrovalencies. It is not necessary to remember a vast number of empirical formulae. Only the electro-valencies of the ions need be remembered.

Some compounds contain a metal and *more than one non-metal*, e.g. Na_2CO_3. Since the sodium ion is Na^+, the group of non-metal atoms must carry a double negative charge, i.e. CO_3^{2-}. This is called the carbonate ion. Similarly, sodium nitrate is found, by experiment, to be $NaNO_3$ and hence the nitrate ion is NO_3^-. A table of electrovalencies is given in fig. 2.5.

Some ions, e.g. copper, have two possible valencies. The distinction used to be made by varying the form of the name (e.g. cuprous, cupric). Modern practice is to incorporate the valency in the name, e.g. copper(I), copper(II), which avoids possible confusion. These should be read 'copper-one' and 'copper-two' respectively.

The table of electrovalencies should be committed to memory as it is essential for writing empirical formulae of ionic substances and hence necessary for any further study of chemistry. Once the symbols and valencies are memorized the empirical formula of any common ionic compound can be written. The only alternative is to learn separately each formula as determined experimentally. In writing the formulae of ionic substances, it must be remembered that these substances do not carry any net electrical charge. Thus if the appropriate ions are written down, the ratio of the numbers of these ions per formula can be decided by inspection (see fig. 2.6).

Salt	Ions	Formula
calcium chloride	Ca^{2+} Cl^-	$CaCl_2$
magnesium sulphate	Mg^{2+} SO_4^{2-}	$MgSO_4$
zinc nitrate	Zn^{2+} NO_3^-	$Zn(NO_3)_2$
ammonium carbonate	NH_4^+ CO_3^{2-}	$(NH_4)_2CO_3$
barium orthophosphate	Ba^{2+} PO_4^{3-}	$Ba_3(PO_4)_2$

Fig. 2.6.

Positive Electrovalencies			Negative Electrovalencies		
+1	+2	+3	−1	−2	−3
sodium Na^+	calcium Ca^{2+}	aluminium Al^{3+}	chloride Cl^-	sulphate SO_4^{2-}	orthophosphate PO_4^{3-}
potassium K^+	barium Ba^{2+}	chromium(III) Cr^{3+}	nitrate NO_3^-	sulphite SO_3^{2-}	nitride N^{3-}
silver Ag^+	magnesium Mg^{2+}	iron(III) Fe^{3+}	nitrite NO_2^-	sulphide S^{2-}	
copper(I) Cu^+	zinc Zn^{2+}		hydroxide OH^-	carbonate CO_3^{2-}	
ammonium NH_4^+	iron(II) Fe^{2+}		hydrogen sulphide HS^-	oxide O^{2-}	
	copper(II) Cu^{2+}		hydrogen sulphate HSO_4^-	silicate SiO_3^{2-}	
	mercury(II) Hg^{2+}		hydrogen sulphite HSO_3^-	zincate $Zn(OH)_4^{2-}$	
	tin(II) Sn^{2+}		hydrogen carbonate HCO_3^-	peroxide O_2^{2-}	
	lead(II) Pb^{2+}		chlorate ClO_3^-	monohydrogen orthophosphate HPO_4^{2-}	
	manganese(II) Mn^{2+}		aluminate $Al(OH)_4^-$		
	chromium(II) Cr^{2+}		hypochlorite OCl^-		
			dihydrogen orthophosphate $H_2PO_4^-$		
			fluoride F^-		
			bromide Br^-		
			iodide I^-		
			acetate CH_3COO^-		

Fig. 2.5. Table of electrovalencies.

Using the electrovalencies set out above, write the empirical formulae of the following compounds:

(a) sodium sulphide
(b) magnesium oxide
(c) copper(II) sulphate
(d) calcium chloride
(e) zinc nitrate
(f) silver sulphate
(g) sodium hydrogen sulphide
(h) magnesium sulphite

(i) potassium orthophosphate
(j) barium hydroxide
(k) iron(II) silicate
(l) calcium orthophosphate
(m) iron(III) chloride
(n) aluminium orthophosphate
(o) chromium(III) sulphate
(p) copper(I) sulphide

(q) ammonium nitrite
(r) sodium hydrogen carbonate
(s) zinc hydrogen sulphate
(t) potassium chlorate
(u) magnesium nitride
(v) mercury(II) nitrate
(w) lead(II) carbonate
(x) barium peroxide

Calculation based on Formulae

Once the empirical formula of a substance is known it is possible to calculate in the reverse direction, i.e., to calculate the percentage composition of a compound.

EXAMPLE:

The empirical formula of magnesium oxide is MgO. The atomic masses are: Mg = 24.3, O = 16.0.

Calculate the percentage magnesium in magnesium oxide.

$$\% \text{ magnesium} = \frac{\text{mass of magnesium}}{\text{mass of magnesium oxide}} \times \frac{100}{1}$$

In one mole of MgO there are 24.3 g of magnesium per (24.3 + 16.0) g of magnesium oxide.

$$\therefore \% \text{ magnesium} = \frac{24.3}{40.3} \times \frac{100}{1}$$
$$= 60.3\%$$

The empirical formula gives the simplest ratio of the numbers of atoms or ions in the formula and this is also the ratio of the numbers of moles of elements in the compound. Hence the relative masses of the elements in the compound can be obtained. In general:

$$\bullet \ \% \ of \ X = \frac{mass \ of \ X \ in \ formula}{total \ mass \ of \ all \ elements} \times \frac{100}{1}$$

NUMERICAL ILLUSTRATIONS

EXAMPLE (i):

% Na in Na_2CO_3

$$= \frac{(2)(23.0)}{(2)(23.0) + (12.0) + (3)(16.0)} \times \frac{100}{1}$$
$$= \frac{46.0 \times 100}{106.0}$$
$$= 43.4\%.$$

EXAMPLE (ii):

% H_2O in $BaCl_2.2H_2O$

$$= \frac{(2)(18.0)}{(137.3) + (2)(35.5) + (2)(18.0)} \times \frac{100}{1}$$
$$= \frac{36.0 \times 100}{244.3}$$
$$= 14.7\%.$$

STUDY QUESTIONS

The following atomic masses should be used where necessary for these problems:

Ag =	107.9	Fe =	55.8	N =	14.0
Al =	27.0	H =	1.0	Na =	23.0
C =	12.0	Hg =	200.6	O =	16.0
Ca =	40.1	K =	39.1	P =	31.0
Cl =	35.5	Mg =	24.3	S =	32.1
Cu =	63.5	Mn =	54.9		

Molecular mass of H_2O = 18.0

1. Calculate the empirical formulae of the substances which have the following compositions by mass:
(a) 40.0% calcium, 12.0% carbon, 48.0% oxygen.
(b) 80.0% copper, 20.0% oxygen.
(c) 30.7% potassium, 25.3% sulphur, 44.0% oxygen.
(d) 16.4% magnesium, 18.9% nitrogen, 64.7% oxygen.
(e) 49.5% manganese, 50.5% oxygen.
(f) 15.8% aluminium, 28.1% sulphur, 56.1% oxygen.
(g) 37.1% sodium, 9.7% carbon, 38.7% oxygen, 14.5% water.
(h) 14.3% sodium, 10.0% sulphur, 19.9% oxygen, 55.8% water.
(i) 18.2% sodium, 8.2% phosphorus, 16.8% oxygen, 56.8% water.
(j) 14.23% iron, 7.14% nitrogen, 2.04% hydrogen, 16.38% sulphur, 32.65% oxygen, 27.55% water.

2. When 0.180 g of carbon is burnt in air, 0.660 g of carbon dioxide is formed. Calculate the empirical formula of carbon dioxide.

3. 1.743 g of mercury(II) oxide is heated carefully in a hard glass test tube. The oxygen is driven off and it is found that 1.614 g of mercury remains. Calculate the empirical formula of mercury(II) oxide.

4. 0.126 g of copper(I) oxide is heated in a stream of hydrogen until only copper remains. The residue of copper weighs 0.112 g. Calculate the empirical formula of copper(I) oxide.

5. 0.237 g of iron is dissolved in hydrochloric acid. The solution is boiled with nitric acid and then ammonia solution is added until the solution is alkaline. The precipitate of iron(III) hydroxide so formed is filtered off and heated to decompose it to iron(III) oxide. The residue of iron(III) oxide weighs 0.339 g. What is the empirical formula of iron(III) oxide?

6. 0.183 g of silver is dissolved in nitric acid. The solution is diluted and then hydrochloric acid solution is added. The precipitate of silver chloride so formed, is filtered off and dried. The precipitate of silver chloride weighs 0.243 g. What is the empirical formula of silver chloride?

7. From the formulae of the following, calculate their percentage composition by mass:
(a) NaCl   (e) NH_4NO_3
(b) $FeCl_3$ (f) $Na_2CO_3.10H_2O$
(c) Al_2O_3 (g) $CuSO_4.5H_2O$
(d) N_2H_4 (h) P_4O_{10}

8. Write down the valency of X in each of the following ionic compounds by stating whether the valency is +1, +2, +3, etc., or −1, −2, −3, etc.:
(a) XO   (e) XSO_4 (i) NaX   (m) ZnX_2
(b) X_2O (f) XNO_3 (j) $(NH_4)_2X$ (n) Na_2X
(c) XCl_3 (g) X_2CO_3 (k) Al_2X_3 (o) K_3X
(d) $X(OH)_2$ (h) $X_3(PO_4)_2$ (l) Mg_3X_2 (p) CaX_2

9. Magnesium metal is produced industrially by the electrolysis of fused magnesium chloride ($MgCl_2$). Write the equations for the reaction at each electrode.

10. State whether each of the following pure liquids would be a good conductor or a very poor conductor of electricity:
(a) liquid H_2O (water);
(b) molten KOH (potassium hydroxide);
(c) liquid H_2SO_4 (hydrogen sulphate);
(d) liquid C_2H_5OH (ethanol).

Answers to Numerical Problems

1. (a) $CaCO_3$ (f) $Al_2(SO_4)_3$
   (b) CuO   (g) $Na_2CO_3.H_2O$
   (c) $K_2S_2O_7$ (h) $Na_2SO_4.10H_2O$
   (d) $Mg(NO_3)_2$ (i) $Na_3PO_4.12H_2O$
   (e) Mn_2O_7 (j) $FeSO_4.(NH_4)_2SO_4.6H_2O$

2. CO_2.

3. HgO.

4. Cu_2O.

5. Fe_2O_3.

6. AgCl.

7. (a) 39.3% sodium, 60.7% chlorine.
   (b) 34.4% iron, 65.6% chlorine.
   (c) 52.9% aluminium, 47.1% oxygen.
   (d) 87.5% nitrogen, 12.5% hydrogen.
   (e) 35.0% nitrogen, 5.0% hydrogen, 60.0% oxygen.
   (f) 16.1% sodium, 4.2% carbon, 16.8% oxygen, 62.9% water.
   (g) 25.4% copper, 12.9% sulphur, 25.6% oxygen, 36.1% water.
   (h) 43.7% phosphorus, 56.3% oxygen.

8. (a) +2.   (e) +2.   (i) −1.   (m) 1.
   (b) +1.   (f) +1.   (j) −2.   (n) −2.
   (c) +3.   (g) +1.   (k) −2.   (o) −3.
   (d) +2.   (h) +2.   (l) −3.   (p) −1.

3: Electronic Configuration of Atoms and the Theory of Valency

So far it has been shown that the electrons occupy the outer region of the atom and that the number of electrons in the neutral atom is equal to the atomic number of the element. The evidence which will be discussed below will show that the arrangement of the electrons around the nucleus of an atom is very closely related to the chemical properties of the element.

Evidence for Electronic Configuration

Evidence which can be related to the arrangement of electrons in an atom can be obtained by heating compounds in the flame of a Bunsen burner.

A few crystals of a salt are picked up on a piece of platinum wire moistened with concentrated hydrochloric acid and introduced into the hot region of a Bunsen burner flame (see fig. 3.1). The colours of the flames obtained with some well known salts are:

sodium chloride	*yellow*
calcium chloride	*brick red*
potassium chloride	*lilac*
barium chloride	*yellowish green*

The colours are characteristic of the metals in the salts listed. For example a yellow flame is obtained with other sodium salts just as for the chloride described above.

Fig. 3.1.
A characteristic flame colour obtained using a salt. One watch glass contains concentrated hydrochloric acid and the other, a few crystals of the salt.

A more accurate analysis of the colour of the light is obtained if the light is passed through a spectroscope (see fig. 3.2 and fig. 3.3).

If sunlight is examined through a spectroscope, a band of coloured light can be observed. This is called a continuous (i.e. unbroken) spectrum (see fig. 3.4, on front endpapers).

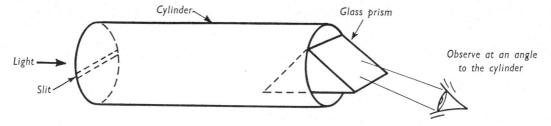

Fig. 3.2. A simple prism spectroscope.

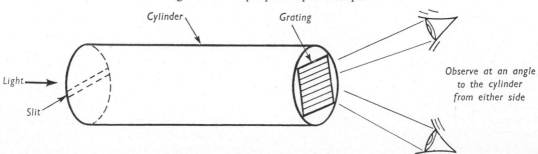

Fig. 3.3. A simple transmission grating spectroscope.

The various colours seen in the spectrum can be *interpreted in two different ways*:

If the energy of the radiated light is considered to be a wave travelling through space, then the different colours can be associated with wave-trains having *different wavelengths*.

The other interpretation is to regard the light as a stream of small pulses of energy called *photons*. The different colours are then associated with the different energies of the photons. It is more helpful to think of light as a stream of photons when interpreting the observed spectra of atoms.

Light observed towards the violet end of the spectrum is due to relatively high energy photons being emitted. Light observed towards the red end of the spectrum is due to relatively low-energy photons being emitted.

If the coloured flames, which are obtained by heating salts in the flame of a Bunsen burner, are each examined through a spectroscope, sets of bright coloured lines will be observed. These lines are the spectra of the characteristic light emitted by each source.

Each coloured line observed in such a spectrum corresponds to photons of a certain energy being radiated. Thus, it follows that the atoms of the respective elements can emit energy *only in certain fixed quantities*. These spectra are represented in the diagram, fig. 3.5 (see front endpapers).

Electronic Energy Levels

Hydrogen emits light if the gas is placed, at low pressure, in a glass tube fitted with electrodes and subjected to a very high voltage electric discharge. Under these conditions, many of the hydrogen molecules are dissociated into atoms in high energy states. These atoms give up their excess energy and it is this energy which is radiated as light. The spectrum of this light contains only a few lines.

Because the hydrogen spectrum consists of lines, it follows that hydrogen atoms can emit photons of *only certain energies*. Hence the energy of the hydrogen atom is restricted to only certain allowed amounts. In chapter 1, a hydrogen atom was shown to consist of a nucleus which carries one positive charge and of an outer region occupied by one electron. Apparently the electron can move around in this region but only in certain ways, each of which is

associated with a particular amount of energy. These permitted energies can be calculated from the positions of the lines in the spectrum and they are represented in the diagram (fig. 3.6). The arrows in the diagram

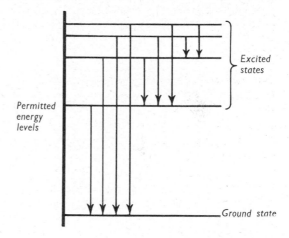

Fig. 3.6. Energy levels in the hydrogen atom.

represent the permitted energy changes of the high energy atoms. Each atom can undergo only one of the changes at any instant but the sample of gas contains very many atoms so all of the possible changes are taking place and all of the corresponding lines can be observed in the spectrum.

A very rough analogy to this pattern of permitted energy levels is provided by a bookcase. If a book is on ground level, it requires expenditure of energy to lift it to one of the shelves and this energy is stored as potential energy. This energy could be recovered by allowing the book to fall back to ground level. Moreover, in placing the book in the bookcase, only certain shelves or energy levels are available. The analogy can be taken further. Usually the distance between the shelves is less when the shelves are higher up in the bookcase. In the same way, the permitted energy levels in an atom become closer as the electron moves outwards from the ground state in the atom.

Investigation of the spectra of other elements shows that the electrons in their atoms are also restricted to certain permitted energy levels. The results of a great deal of experimental work are summarized in the following rules:

1. A number of electrons in an atom can have very nearly the same energy level. Such a set of electrons is called an electron shell. From the nucleus outwards the shells are either numbered, or called the K, L, M, N, O, P, etc., shells.
2. The maximum number of electrons in each shell is limited:
   i.e. 2 electrons for the 1st or K shell.

8	,,	,,	,,	2nd	,,	L	,,
18	,,	,,	,,	3rd	,,	M	,,
32	,,	,,	,,	4th	,,	N	,,

   . . . .

   $2n^2$,,   ,,   ,, n^{th} shell.
3. The outer shell of electrons in an atom never holds more than 8 electrons regardless of the maximum possible number of electrons for the shell indicated in (2) above. For example, the third shell can hold up to 18 electrons but will not hold more than 8 electrons unless there are electrons in the fourth shell; it is then not an outer shell.

Some examples of the numbers of electrons in the energy levels or shells are given in the table (fig. 3.7).

The reason why the energy levels are called shells is illustrated by reference to a lithium atom. A lithium atom has three electrons. The first two electrons have low energies and their energies are very nearly the same. The third electron has a much higher energy. The lowest energy electrons spend most of their time moving in the region close to the nucleus. The higher energy electron spends most of its time in a region further out from the nucleus. It is not possible to say exactly where a particular electron is at any given instant, that is, we cannot determine the actual paths followed by the electrons as they move around the nucleus. Hence we visualize the lithium atom as containing two shells as shown in the diagram fig. 3.8. Because this diagram is difficult to draw, a set of

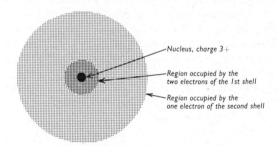

Fig. 3.8. Electron shells in a lithium atom.

Element, symbol and atomic number			Number of electrons in shells			
			1st	2nd	3rd	4th
hydrogen	H	1	1			
helium	He	2	2			
lithium	Li	3	2	1		
beryllium	Be	4	2	2		
boron	B	5	2	3		
carbon	C	6	2	4		
nitrogen	N	7	2	5		
oxygen	O	8	2	6		
fluorine	F	9	2	7		
neon	Ne	10	2	8		
sodium	Na	11	2	8	1	
magnesium	Mg	12	2	8	2	
aluminium	Al	13	2	8	3	
silicon	Si	14	2	8	4	
phosphorus	P	15	2	8	5	
sulphur	S	16	2	8	6	
chlorine	Cl	17	2	8	7	
argon	Ar	18	2	8	8	
potassium	K	19	2	8	8	1
calcium	Ca	20	2	8	8	2
krypton	Kr	36	2	8	18	8

Fig. 3.7.

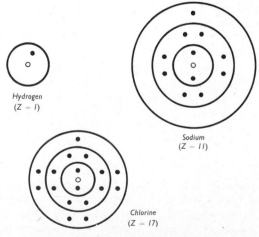

Z is the atomic number of the element.
Fig. 3.9. Shell structure of a few atoms.

concentric circles will be used to represent the energy levels and dots will be placed in these circles to indicate the *number* of electrons in each shell. These dots indicate only the number of electrons, they do not imply that the electrons are located in that particular place in the atom. A small circle will be drawn at the centre to represent the nucleus. Some atoms are represented according to this convention in the diagrams (fig. 3.9).

The Periodic Classification

Examination of the ground state electronic configurations of atoms given in the table (fig. 3.7) reveals that, when the atoms of the elements are arranged in order of increasing atomic number, they are arranged in a sequence in which the electron shells are progressively filled. Furthermore, in the sequence from lithium (Z=3) to neon (Z=10), the number of electrons in the outer shell increases from 1 to 8 and this sequence is repeated for the atoms of sodium (Z=11) to argon (Z=18). Thus there is a periodic repetition of the same outer electronic arrangement for the elements when they are arranged in order of increasing atomic number. This is the basis of the periodic table of the elements.

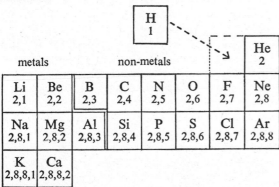

Fig. 3.10. Portion of a periodic table (showing the arrangement of the first 20 elements and their electronic configurations).

A periodic classification of elements consists of a table showing the elements in order of increasing atomic number, with elements of similar electronic arrangements being placed in similar positions in the table. The table (fig. 3.10) shows the first portion of such a table.

After calcium, the next 10 elements (atomic numbers 21 to 30) form a series showing fairly regular gradations from element to element and these elements are called *transition elements*. Reference to the back cover will show an 18 column form of the periodic classification, illustrating one method by which the transition elements can be fitted into the table. It will be observed that the method consists of splitting the table between the second and third groups.

After lanthanum (atomic number = 57) there is a series of 14 very similar elements. This series is called the *lanthanide series* and is shown at the bottom of the table. After actinium (atomic number = 89) there is a second similar series called the *actinide series*.

The elements in the vertical columns labelled Group I, Group II, etc., are called the *main group elements*, and elements in the same main group have similar electron arrangements.

IONIC COMPOUNDS

Compounds of a metal and one or more non-metals are usually composed of ions. The charges on the ions are called their electrovalencies and these have been derived from the experimentally determined empirical formulae. The electrovalencies of some ions taken from fig. 2.5 are:

$$Na^+ \quad Mg^{2+} \quad O^{2-} \quad F^-$$
$$K^+ \quad Ca^{2+} \quad S^{2-} \quad Cl^-$$

Compare these electrovalencies with the electronic arrangements of the neutral atoms in the portion of the periodic table shown in fig. 3.10. It will be seen that the charges on the positive ions correspond to removal of the outer shell of electrons and the charges on the negative ions correspond to increasing the outer shell to eight electrons. Hence the electrovalencies obtained in chapter 2 can be explained by reference to the number of electrons in the outer shells of the atoms. Thus, for sodium chloride, the electronic arrangements of the relevant particles are:

sodium atom	Na	2, 8, 1
sodium ion	Na$^+$	2, 8
chlorine atom	Cl	2, 8, 7
chloride ion	Cl$^-$	2, 8, 8

The bonding in sodium chloride crystals arises from the attraction between the oppositely charged Na^+ and Cl^- ions in the compound. The force of attraction is called an electrostatic force and the bonding is called electrovalency. The charges on the ions in ionic compounds show a close relationship to the electronic arrangements of the neutral atoms of the elements.

There are a number of ways conventionally used to write down the relationship, but it must be emphasized that these are not chemical equations. It is not implied that ions are formed by a *reaction* between *atoms*. The purpose is to show the relationship between the atoms and the ions and the manner in which this affects the relative numbers of positive and negative ions in the compound.

1. SODIUM CHLORIDE

(i) Using symbols and showing the electronic arrangements:

$$Na + Cl \rightarrow Na^+ + Cl^-$$
2,8,1   2,8,7   2,8   2,8,8

(ii) Using symbols and showing only the valency electrons:

$$Na\cdot + .\overset{..}{\underset{..}{Cl}}: \rightarrow Na^+ + :\overset{..}{\underset{..}{Cl}}:^-$$

(iii) Using atomic shell structure diagrams:

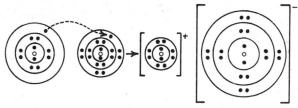

Similarly, other ionic compounds can be represented, but only the first of these three methods will be used for these other compounds.

2. MAGNESIUM CHLORIDE

$$Mg + 2\,Cl \rightarrow Mg^{2+} + 2\,Cl^-$$
2,8,2   2,8,7   2,8   2,8,8

3. SODIUM OXIDE

$$2\,Na + O \rightarrow 2\,Na^+ + O^{2-}$$
2,8,1   2,6   2,8   2,8

4. SODIUM SULPHIDE

$$2\,Na + S \rightarrow 2\,Na^+ + S^{2-}$$
2,8,1   2,6   2,8   2,8,8

5. MAGNESIUM OXIDE

$$Mg + O \rightarrow Mg^{2+} + O^{2-}$$
2,8,2   2,6   2,8   2,8

6. MAGNESIUM SULPHIDE

$$Mg + S \rightarrow Mg^{2+} + S^{2-}$$
2,8,2   2,8,6   2,8   2,8,8

Electrovalency

Electrovalency in compounds usually results when the compound is composed of a metal and a non-metal. Atoms of metals have only a few electrons in their outer shells whereas atoms of non-metals have only a few vacancies for electrons in their outer shells.

MOLECULAR SUBSTANCES

Hydrogen

Mass spectrometers were described in chapter 1 and it was shown how relative atomic masses of elements can be determined. A mass spectrometer can be used to find the relative mass of any particle which can be converted into a positive ion. When hydrogen gas is analysed, particles which have a mass of 2.016 (on the average) are detected. Since the atomic mass of hydrogen is 1.008 it follows that hydrogen is composed of separate particles, each consisting of two atoms of hydrogen joined together. Thus the structure of hydrogen is quite different from the structure of the ionic compounds. Ionic compounds are composed of separate positive and negative ions arranged in a regular pattern throughout a crystal whereas all the particles which comprise hydrogen gas are similar. These particles are electrically neutral, and they are called molecules.

It can be shown that the structure of a molecule of hydrogen can be represented on the basis that the atoms share electrons by remaining close together so that their electron shells overlap. Each atom is one electron short of having a complete electron shell so electron transfer would not lead to electron stability. If the two electrons are shared between the hydrogen atoms then both atoms can be regarded as having complete outer shells, because the two electrons are close to both nuclei. The pair of hydrogen atoms is called a molecule.

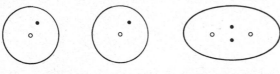

hydrogen atoms   hydrogen molecule

The electrons in the molecule move in a region which includes both nuclei so that each nucleus achieves the effect of having two electrons near it. The shared pair of electrons holds the two nuclei together. The shared pair of electrons is called a single covalent bond.

Hydrogen Chloride

Hydrogen chloride is a gas which consists of particles which have a mass of 36.461 (on the average). The relative atomic masses of hydrogen and chlorine are: H = 1.008 and Cl = 35.453, so that hydrogen chloride must consist of molecules, each of which contains one H-atom and one Cl-atom. The electron arrangements of the atoms of hydrogen and chlorine are shown in

Element	Electronic structure	Number of electrons required to complete the outer shell
hydrogen	1	1 (in the 1st shell)
chlorine	2, 8, 7	1 (in the 3rd shell)

Fig. 3.11.

the table fig. 3.11. These two atoms are bound together by shared electrons. The single electron from the hydrogen atom together with one of the electrons from the chlorine atom comprise a shared pair of electrons, i.e. the atoms are bound together by a single-covalent bond.

Thus in each H-Cl molecule there are two electrons in the 1st energy level of the hydrogen atom and the chlorine atom has eight electrons in its 3rd energy level.

There are various ways of representing electron sharing, but it must be emphasized that the representations are diagrams and not chemical equations. They are only intended to show the relationship between the structure of the molecule and the electronic arrangements of the constituent atoms.

(i) Using symbols and showing the valency electrons:

$$H\cdot + .\ddot{\underset{..}{Cl}}: \rightarrow H:\ddot{\underset{..}{Cl}}:$$

(ii) Using atomic shell structure diagrams.

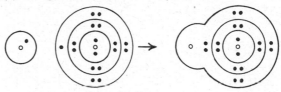

The pair of shared electrons may be regarded as constituting the bond between the atoms and this is conventionally represented by a single stroke, i.e. H-Cl.

Other Molecular Substances

WATER

Water is a compound which is found to consist of molecules, each of which contains two hydrogen atoms and one oxygen atom. Electron sharing between the atoms can be represented:

$$2H\cdot + .\ddot{O}: \rightarrow H:\ddot{O}: \text{ or } H-\ddot{O}: \\ \quad\quad\quad\quad\quad\quad\; H \quad\quad\quad\; H$$

HYDROGEN SULPHIDE AND AMMONIA GAS

The composition of other molecular compounds of hydrogen can be explained by applying the same principle.

(i) hydrogen sulphide gas (two hydrogen atoms per sulphur atom).

$$2H\cdot + .\ddot{S}: \rightarrow H:\ddot{S}: \text{ or } H-\ddot{S}: \\ \quad\quad\quad\quad\quad\quad\; H \quad\quad\quad\;\; H$$

(ii) ammonia gas (three hydrogen atoms per nitrogen atom).

$$3H\cdot + .\ddot{N}\cdot \rightarrow H:\ddot{N}:H \text{ or } \underset{H}{\overset{N}{H\diagup}}-H \\ \quad\quad\quad\quad\quad\quad\;\; H$$

The molecules of each of the compounds described here are made up of groups of atoms which are firmly bound together by pairs of electrons shared between two atoms. Bonding between atoms by electron sharing is called covalency.

It is possible to form more than one covalent bond between two atoms. For example, carbon dioxide is a compound which is found to consist of molecules, each of which contains one carbon atom and two oxygen atoms. The carbon atom obtains a share of four more electrons. Thus two pairs of electrons are shared with each oxygen atom. This can be represented as follows:

Because the bonds between each oxygen atom and the carbon atom consist of two pairs of shared electrons such bonds are called double covalent bonds or, more briefly, double bonds, and they are conventionally represented by a double stroke, i.e. $O=C=O$.

Molecular Elements

Chlorine gas consists of molecules which have a mass of 70.906 (on the average). Thus chlorine molecules contain two atoms. Chlorine molecules can be compared with hydrogen molecules which have been shown to contain two atoms. The structure of these molecules can be explained in similar fashion by assuming that one pair of electrons is shared in each molecule, i.e. the atoms are bound together by a single covalent bond—see fig. 3.12.

	HYDROGEN GAS	CHLORINE GAS
Electronic Formulae	H : H	$:\overset{..}{Cl}:\overset{..}{Cl}:$
Structural Formulae	H—H	Cl—Cl
Molecular Formulae	H_2	Cl_2

Fig. 3.12

Molecules such as H_2 and Cl_2 are said to be *diatomic* or to have an atomicity of 2 (i.e. they contain two atoms per molecule).

Covalency

Covalency in compounds and elements usually results when only non-metal elements are present. Atoms of non-metals have only a few vacancies for electrons in their outer shells. The vacancies of *all* of the atoms in a molecule are simultaneously filled by electron sharing. A shared pair of electrons is called a covalent bond.

Intermediate Types of Bonds

In diatomic molecules of elements such as H_2 or Cl_2, the two atoms in the molecule are identical. The pair of electrons in the covalent bond is shared equally between the two atoms. Conversely, in the crystalline solid $Na^+ Cl^-$ the charges are due to the transfer of an electron from one atom to the other. So far, this electron transfer has been described as though the sodium atom completely lost an electron and the chlorine atom completely captured the electron. This is a very good approximation, but it is not a completely accurate description.

In solid crystalline $Na^+ Cl^-$ the ions are packed close together. The positive charge of the sodium ion exerts an attraction on the electrons around the chloride ion. As a result, the electron spends some of its time in the neighbourhood of the sodium. It is as though the electron has been almost completely, but not quite completely, transferred from the sodium atom to the chlorine atom. Since the electron spends some small proportion of its time between both atoms, this can be described by saying that the bonding in sodium chloride has a very slight degree of covalent bonding. Thus a better description of the bonding is to say that the compound is almost completely ionic, i.e. $Na^+ Cl^-$, but that there is a very slight degree of covalency.

Thus one of the two extremes in types of bonding is the almost-complete electron transfer as in $Na^+ Cl^-$. This is almost completely electrovalent bonding. The other extreme is the equal electron sharing as in H_2 and Cl_2. This is called covalent bonding.

Many molecules have bonds which are intermediate in character between these two extremes. Thus in a molecule of HCl, the bonding is mainly covalent but the electron pair is not equally shared by the two atoms. The chlorine atom has the greater share of the pair of electrons. Because of this inequality of sharing, the bond can be described as having partial ionic character. This can be represented by writing the formula of hydrogen chloride as:

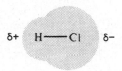

The symbols $\delta+$ and $\delta-$ mean charges which are much smaller than the charge of an electron. The charge on the chlorine is $\delta-$ because it has a slightly greater than equal share of the bonding electrons. The charge on the hydrogen is $\delta+$ because it has a slightly less than equal share of the bonding electrons.

The atoms of the non-metals near the top right-hand corner of the periodic table tend to gain a greater than equal share of the electrons in a covalent bond. Because this tends to create negative charges on these atoms when they are in molecules, the elements are said to be *electronegative*. Thus the bonds in HF and H_2O also have partial ionic character, like the bond in HCl.

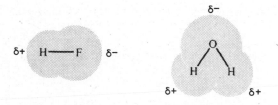

Summary

Compounds of metals with non-metals are almost completely ionic in structure. Molecules consisting of identical non-metal atoms are covalent in structure. Molecules of compounds containing atoms of different non-metal elements are covalent but with some ionic character—the atom of the element closest to the top right-hand corner of the periodic table has a slight negative charge.

EXAMPLES:

NaCl, MgO: largely ionic but with some covalent character.

H_2, Cl_2: pure covalent bonding.

H_2O, HCl: largely covalent, but with some ionic character.

The Shape of Molecules

The shape of a molecule is difficult to describe because the electrons occupy regions, not particular places in the molecule. However, the nuclei of the atoms do occupy well defined positions relative to each other. Hence the shape of a molecule is conventionally defined by a set of lines drawn between the nuclei of the atoms. The shape depends on the number of pairs of electrons around each atom. These electron pairs repel each other because all electrons have charges of the same sign—all electrons are negatively charged.

All of the following atoms have four pairs of electrons around them: C in CH_4, N in NH_3 and O in H_2O. The four pairs of electrons take up positions which are as far apart as possible. This can be represented by imagining the atom at the centre of a tetrahedron as shown in fig. 3.13 with the four pairs of electrons arranged as though they pointed to the corners of the tetrahedron. The lines drawn between atoms in the molecules in fig. 3.13 represent covalent bonds and these lines also define the shapes of these molecules. The shaded areas represent pairs of electrons which are not shared in covalent bonds between two atoms.

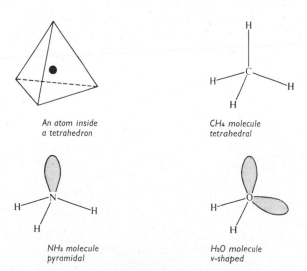

Fig. 3.13. Shapes of some molecules.

Molecular and Empirical Formulae

Hydrogen chloride and sodium chloride are usually represented by the chemical formulae HCl and NaCl respectively, but these formulae have different meanings.

(i) HCl represents a molecule. Hydrogen chloride is composed of vast numbers of separate particles in each of which a hydrogen atom is bound by covalent bonding to a chlorine atom. Thus HCl is a *molecular* formula; *it represents a molecule which is a single particle of matter.* Similarly H_2O, H_2S, NH_3, CO_2, H_2 and Cl_2 are molecular formulae.

(ii) NaCl represents a pair of ions; one Na^+ ion and one Cl^- ion. Sodium chloride is composed of vast numbers of separate sodium ions and separate chloride ions. These ions are not specifically bound to any other particular ion although in the crystal they are arranged in a certain regular order (see next chapter). Thus NaCl is an *empirical* formula; *it represents the ratio of the numbers of the respective ions in the compound.* Similarly $MgCl_2$ (ratio 1 : 2), Na_2O (ratio 2 : 1), MgO (ratio 1 : 1), Na_2S (ratio 2 : 1) and MgS (ratio 1 : 1) are empirical formulae.

Molecular formulae represent molecules and show the number of atoms of each element in a molecule.

Empirical formulae represent the simplest ratio of atoms. They do not give the actual numbers of atoms of each element in any particular sample of the compound. Since ionic compounds are not composed of molecules, the formula of an ionic compound is always an empirical formula.

SUMMARY

The electronic arrangements of atoms can be worked out from experimental evidence. The formulae of substances can be obtained from experimental evidence. It is possible to show that a relationship exists between electronic structures and formulae if it is assumed that electron transfer takes place to form ions or that electron sharing takes place to form molecules. These electron rearrangements are the two extremes which lead to the electrovalency and covalency.

Electron transfer leads to electrovalency. Electrovalent compounds are ionic, and their formation is likely when an atom with few valency electrons (a metal) is combined with an atom with a few vacancies in its valency shell (a non-metal). Thus compounds formed by the combination of a metal with a non-metal are usually ionic.

Electron sharing leads to covalency. Covalent compounds are molecular, and their formation is likely when atoms with few vacancies in their valency shells are combined with each other. Thus compounds formed by the combination of two non-metals are usually molecular. The non-metallic elements themselves are also usually molecular.

STUDY QUESTIONS

1. Complete the following diagrams for the compounds formed by each of the pairs of elements given (show the electron rearrangements that take place):
   Ca + S →
   Ca + Cl →

2. Write down the probable electronic formulae of:
   (a) the gas, methane, CH_4;
   (b) the gas, chlorine monoxide, OCl_2;
   (c) the liquid, carbon disulphide, CS_2;
   (d) the gas, nitrogen, N_2.

3. The following list shows the electronic arrangement of some elements in pairs. Write down the type of valency likely to be present in a compound containing only atoms of the pair given:
   (a) 2,5 ; 1.
   (b) 2,6 ; 2,8,2.
   (c) 2,7 ; 2,6.
   (d) 2,8,6 ; 2,6.
   (e) 2,8,18,18,1 ; 2,8,7.

4. Suggest the probable electrovalency ($+1$, -1, $+2$, -2, etc.) and the probable covalency (1,2, etc.) of each of the elements for which the electronic arrangements are:
   (a) 1.
   (b) 2,8,8,2.
   (c) 2,8,5.
   (d) 2,7.
   (e) 2,8,18,8.

4 : The Structure of Solids

Crystalline Solids

The outstanding feature of crystalline solids is that all crystals of any particular substance show much the same shape. For example, crystals of common salt appear to be cubic in form, although some are distorted because of uneven growth (see fig. 4.1). The regularity of external form suggests a regularity in the arrangement of the particles comprising the solid.

The properties of crystalline solids depend on two features:

(i) the arrangement of the particles in the crystal lattice;

(ii) the nature of the particles which make up the substance, i.e., ions, molecules or atoms.

Fig. 4.1. A crystal of sodium chloride, showing its regular shape. The front edge of the crystal is about one centimetre long.

Determination of Electron Distributions in Crystals

Electrons do not occupy well defined positions in atoms or ions. They move in regions of space about the nuclei. However, the average number of electrons per unit volume can be determined because X-rays are diffracted by electrons. From the way in which the X-rays are diffracted, the distribution of electrons in the crystal can be calculated.

The most widely used experimental method for observing the diffraction of X-rays by crystals is the Weissenberg method. A narrow beam of X-rays is allowed to strike a well-formed crystal as shown in fig. 4.2. The X-rays are scattered by the crystal; some of the scattered rays pass through the slit in the

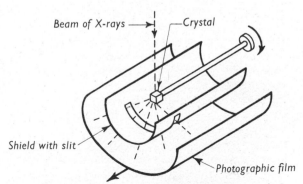

Fig. 4.2. Diffraction of X-rays by a rotating crystal.

shield and strike the photographic film. The crystal is rotated and the photographic film is moved as shown in the diagram. When the film is developed, it shows a regular pattern of spots, the pattern being the same for all crystals of the particular substance. Part of an actual Weissenberg photograph is shown in fig. 4.3 in which it can be seen that different spots have different intensities.

The diffraction pattern arises because the crystalline solid consists of a regular arrangement of atoms or ions. The pattern of spots and their relative intensities are related to the distribution of electrons in the crystal. Although the mathematical analysis of the Weissenberg pattern is complex, it is possible by using computers, to calculate the electron density pattern in the crystal.

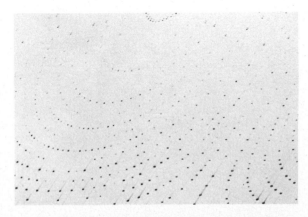

Fig. 4.3. Photograph from a Weissenberg diffraction camera.

The electron density pattern in one of the faces of a sodium chloride crystal is shown in the diagram (fig. 4.4). The lines in this diagram are electron density contour lines. It can be seen that the electrons occupy spherical regions and are closely crowded towards the centre of each region.

From the mathematical analysis of the diffraction pattern, the computer gives the result that there are 9.7 $\pm$ 0.3 electrons in the smaller regions and 17.7 $\pm$ 0.3 electrons in the larger regions. Hence, within the limits of the experimental method, it can be seen that sodium chloride consists of a regular pattern of Na^+ ions (10 electrons) and Cl^- ions (18 electrons). This result is consistent with the electrical conductivity of molten sodium chloride described in chapter 2. Thus it has been experimentally shown that NaCl, whether solid or molten, is composed of Na^+ and Cl^- ions. In the solid these have the regular pattern determined by X-ray diffraction. In the molten state, these ions can move and this makes molten sodium chloride an electrolytic conductor.

The determination of electron distribution in crystals is not confined to ionic substances. Fig. 4.5 represents a projection (in slightly different form) of

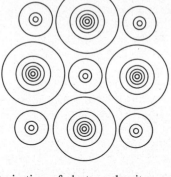

Fig. 4.4. Projection of electron density contours in a crystal of sodium chloride.

the electron density contours in a crystal of glutamic acid hydrochloride. Fig. 4.6 represents a matching projection using symbols. Note that the different atoms in the crystal can be distinguished by their different electron density contours. The positions of the hydrogen atoms do not show up clearly on the electron density pattern. Their estimated positions are shown by crosses.

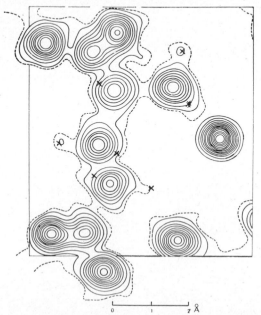

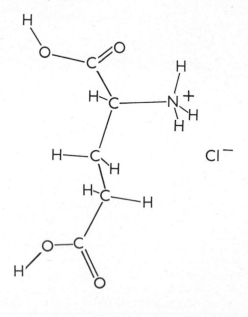

Fig. 4.5 (left): A projection of electron density contours in a crystal of glutamic acid hydrochloride.
Fig. 4.6 (right): A matching projection using symbols to show the relative positions of the centres of the atoms.

IONIC SOLIDS

Crystals of sodium chloride are hard and brittle. They have a high melting point (801 °C) and are very poor conductors of electricity. The structure determined by X-ray diffraction is represented in the diagram, fig. 4.7.

The dotted lines in the diagram help to show the arrangement of the ions in a cubic lattice and must not be taken to represent chemical bonds. The number of ions in a tiny crystal of salt is enormous and the diagram should be imagined as extending in all directions. It should be noted that each ion is surrounded by six ions of opposite charge. This is most easily seen for the sodium ion at the centre of the diagram.

Fig. 4.8. A model representing the structure of sodium chloride.
The large spheres represent chloride ions and the small spheres represent sodium ions.

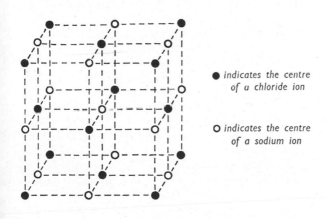

● indicates the centre of a chloride ion

○ indicates the centre of a sodium ion

Fig. 4.7. Structure of sodium chloride.

Fig. 4.7 shows only the position of the centre of each ion. A better picture of the crystal is obtained by representing the ions as spheres (see fig. 4.8).

These structures for sodium chloride are consistent with the observed hardness of the crystal in that they show each ion held in the crystal lattice by the strong attraction of the six oppositely charged ions which surround it. The force between the charged particles is called an electrostatic force.

The brittleness of sodium chloride is also consistent with the diagrams of its structure. Thus, if a deforming force acts on a crystal of sodium chloride and the ions are made to slide past one another, strong

repulsive forces come into play and these forces rupture the crystal (see fig. 4.9).

In order to show that the diagrams are consistent with the observed high melting point of sodium chloride it is necessary to discuss the process of melting or fusion.

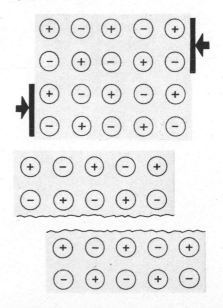

Fig. 4.9 Small portion of an ionic solid before and after the application of a shearing force.

Fusion of Solids

The structures above suggest that the ions are stationary in the crystal but this is not so. They are in a state of constant vibration. If the temperature is increased, the ions vibrate more violently, because the additional heat energy absorbed increases the kinetic energy of vibration of the ions. If the temperature becomes high enough the vibrations will become so violent that the crystal lattice will be destroyed. Thus, even though ions still attract oppositely charged ions, they are free enough to slide past one another, i.e., the solid melts or fuses. When the forces between the particles comprising a solid are strong, considerable energy will be needed to disrupt the crystal lattice and melt the solid. *Thus, the stronger the force holding the particles of a substance in the crystal lattice, the higher the melting point.*

It has been shown that the forces in the sodium chloride crystal are strong electrostatic forces and hence the diagrams are consistent with the high melting point.

Other Ionic Solids

The majority of inorganic compounds are ionic solids. Typical examples are magnesium oxide ($Mg^{2+}O^{2-}$), potassium chloride (K^+Cl^-) and sodium hydroxide (Na^+OH^-). All ionic solids do not necessarily have the same lattice arrangement as that shown for sodium chloride, the structure being dependent on the relative sizes of the ions as well as the relative proportions of the ions.

MOLECULAR SOLIDS

The crystals of many organic substances differ strikingly from salts such as sodium chloride. Naphthalene is a good example of such a compound. Unlike sodium chloride, the crystals are soft and have a low melting point (about 80 °C), although like sodium chloride, naphthalene does not conduct electricity in the solid state. These properties are shared by some inorganic compounds such as hydrogen chloride, oxygen and carbon dioxide in the solid state. These latter substances are gases under ordinary conditions.

It has been shown previously that hydrogen chloride gas consists of molecules in which the atoms are held together by a covalent bond. Oxygen gas and carbon

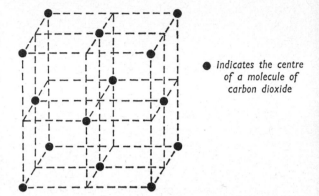

● indicates the centre of a molecule of carbon dioxide

Fig. 4.10. Structure of solid carbon dioxide.

dioxide gas are similar in that the atoms within each molecule are held together by covalent bonds. The water molecule is also similar in this respect although water is a liquid under ordinary conditions.

All of these substances can be obtained in the solid state if the temperature is sufficiently low, and this suggests that there is some force tending to hold the molecules together. The fact that these solids are very easily melted suggests that the forces are weak. These intermolecular forces (i.e. the forces between separate molecules) are called van der Waal's forces, after the Dutch physicist who suggested their existence. Weak intermolecular forces are also called dispersion forces.

Fig. 4.10 shows the arrangement of carbon dioxide molecules in the crystal lattice of solid carbon dioxide (commonly called dry ice). The diagram shows a

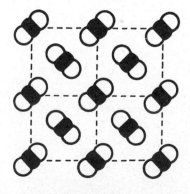

Fig. 4.11. Structure of solid carbon dioxide.

cubic arrangement with one particle (molecule) at each corner and at the centre of each face. The dotted lines in the diagram help to show the arrangement of the molecules and do not represent valency bonds. Fig. 4.11 represents one layer of molecules in a crystal of dry ice.

GIANT MOLECULES

Some inorganic substances are notable for their extreme hardness and very high melting points. Diamond (a pure form of carbon) is a good example of this group. It is extremely hard and can be heated to 3500 °C without melting. Above this temperature it does not melt but goes into the vapour state, i.e., it sublimes. It is a non-conductor of electricity.

Other members of this group include quartz (silicon dioxide) and carborundum (silicon carbide). Carborundum is nearly as hard as diamond and is widely used as an abrasive.

Diamond was among the first substances whose structures were determined by X-ray diffraction. Each carbon atom is surrounded by four other carbon atoms. The arrangement is as if each carbon atom were at the centre of a regular tetrahedron and the four adjacent atoms were at the vertices of the tetrahedron (see fig. 4.12). Each carbon atom has four valency electrons and forms a covalent bond with each of its four nearest neighbours. These bonds are represented by the metal springs in the model.

Fig. 4.12. Model representing the tetrahedral arrangement of covalently bound carbon atoms.

The linking of carbon atoms extends through the whole lattice, giving a three dimensional array in which each carbon atom is firmly bound in the crystal lattice. The crystal thus consists of a giant molecule. Fig. 4.13 shows the arrangement of a few atoms in a crystal of diamond. Because of the small number of atoms shown, only a few of the atoms appear to exert a covalency of four. However, the structures can

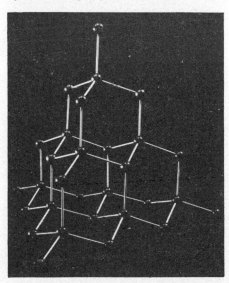

Fig. 4.13. Models representing the structure of diamond.

(left) The spheres represent the centres of carbon atoms.

(right) The carbon atoms are represented as spheres in contact.

be imagined as extending in all directions so that all the carbon atoms do exert a valency of four, except those atoms in the surface of the crystal.

The rigid structure shown in the models is consistent with the observed hardness of diamond and with the observed high temperature of sublimation. Because the force holding the atoms in the crystal lattice is relatively large, the atoms must be made to vibrate with great violence before they can break out of the lattice. Thus the diagram is consistent with the fact that diamond changes directly to a gas rather than to a liquid, i.e., it sublimes.

The structures of quartz (SiO_2) and carborundum (SiC) are similar to diamond.

METALLIC SOLIDS

The physical properties of metals include the following:

1. Metals are usually good conductors of heat and electricity.

2. Most metals are malleable (can be hammered into sheets) and ductile (can be drawn out into wire). However, their hardness ranges from soft to very hard, e.g., sodium is so soft that it can be cut with a knife, whereas chromium is very hard.

3. The densities of metals range from low to very high, although the densities are usually fairly high. This is illustrated in the table (see fig. 4.14). The table indicates that the density of a metal is related to its relative atomic mass.

4. Metals are lustrous, i.e. have a shiny surface, although most metals tarnish and must be scraped to show the lustre.

Metal	Density (g cm^{-3})	Relative Atomic Mass	Density Compared with the Average for a Metal
sodium	0.97	23.00	Very low.
magnesium	1.7	24.31	Low.
iron	7.9	55.85	About average.
lead	11.4	207.19	High.
platinum	21.5	195.09	Very high.

Fig. 4.14.

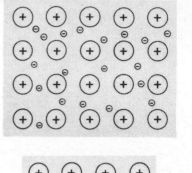

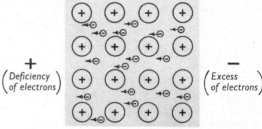

$\underset{\left(\begin{array}{c}\textit{Deficiency}\\\textit{of electrons}\end{array}\right)}{+}$ $\qquad$ $\underset{\left(\begin{array}{c}\textit{Excess}\\\textit{of electrons}\end{array}\right)}{-}$

Fig. 4.15. The structure of sodium metal. If an electrical potential difference is applied as shown in the lower diagram the electrons tend to drift towards the region of higher potential (positive).

5. The melting points of metals range from low to high, e.g., mercury melts at $-39\,°C$ whereas tungsten melts at $3370\,°C$.

Properties and Structure

In discussing the structure of metallic solids, sodium is a suitable example because the electronic arrangement of the sodium atom has been discussed previously. It is generally believed that *metals consist of a lattice of positive ions through which moves a cloud of electrons.* These electrons are the valency electrons of the metal and for sodium they are the outermost electron from each atom. The positive ions tend to repel one another because they carry electrical charges of the same sign. However, they are held together by the cloud of negatively charged electrons. All the valency electrons move freely throughout the whole of the crystal lattice and so bind the whole set of particles into a single unit. It should be noted that a particular valency electron does not belong to any particular ion, but to the whole crystal (see fig. 4.15).

The diagram for metallic structure is consistent with the properties of metals listed above:

1. Conductivity

An electric current in a metal can be shown to consist of a flow of electrons, but proof of this will not be given here. If an electrical potential is applied to the ends of a metallic wire, a general movement of the electrons along the wire is superimposed on their random motion within the lattice. This results in a number of electrons being forced into one end of the wire while a similar number of electrons are forced out at the other end. Thus the electric current passes along the wire.

Heat energy can also be carried by the mobile electrons.

2. Malleability and Ductility

The positive ions in a metal are not held by rigid bonds but are capable of sliding past one another if the metal is deformed. This rearrangement of the ions does not alter the general attractive force in any significant way and hence the metal does not shatter (see fig. 4.16).

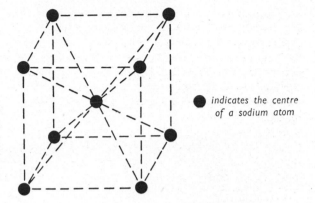

indicates the centre of a sodium atom

Fig. 4.17. Diagram of the sodium lattice.

3. Density

The density of a metal depends on two factors: the lattice arrangement of the ions and the mass of each ion. The lattice for sodium metal is shown in fig. 4.17. Metals generally have lattice arrangements which are similar in many ways to the structure of sodium. Because of this, the densities of metals depend largely on the mass of the ions (i.e. the relative atomic mass), and since most metals have fairly high relative atomic masses, the densities of metals are usually high.

4. Metallic Lustre

Metallic lustre can be explained as being due to the effect of light causing the mobile electrons to be excited.

5. The melting points of metals

The very wide range of melting points in metals suggests that the bonding force can vary widely from metal to metal.

A comparison of the diagrams of sodium chloride, dry ice and diamond with the diagram of sodium metal suggests why the first three solids are non-conductors of electricity. In diamond and dry ice the electrons are firmly bound by sharing between atoms, while in sodium chloride the electrons are firmly bound in their respective ions. Only sodium metal has mobile electrons, and such metals are called metallic conductors or electronic conductors.

Lattice Defects

The diagrams of crystal lattices shown earlier in this chapter are ideal or perfect lattices. Most real crystals

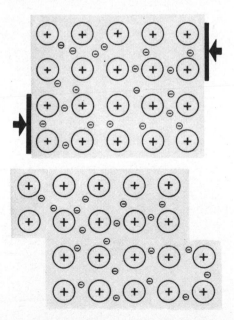

Fig. 4.16. Small portion of a metal before and after the application of a shearing force.

contain defects of one kind or another, e.g., some of the lattice points may be unoccupied. Because of lattice defects the Law of Definite Proportions does not apply to some compounds. In manganese(IV) oxide, for example, some of the oxygen positions are unoccupied while in iron(II) sulphide some of the iron positions are vacant.

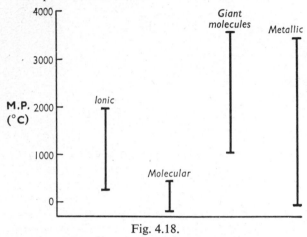

Fig. 4.18.

Types of Crystals and their Melting Points

In this chapter the melting points of a few substances have been given, and the graph (fig. 4.18) is presented to give an idea of the range of melting points of the four types of solids discussed.

Crystals in Solids

As a molten substance is cooled, many crystals begin to form. Eventually, these crystals grow into each other and form a mass of interlocking crystals in the solid (see fig. 4.19).

Fig. 4.19. The crystalline structure of metal can be clearly seen in this fractured brass ingot. Although the crystal can withstand substantial deformation, fracture ultimately occurs.

STUDY QUESTIONS

1. Draw a diagram to represent a small portion of a crystal of carborundum (SiC), which has a similar crystal structure to diamond but each alternate atom is a silicon atom.

2. How can you account for the observed fact that water has a relatively low melting point, but magnesium oxide has a very high melting point?

3. The theory of structure of metals suggests that metals consist of a regular arrangement of positive ions and a cloud of mobile electrons. List those properties of metals which:
   (i) support the theory;
   (ii) do not support the theory.

4. Draw a small portion of a crystal of magnesium oxide, which has the same lattice arrangement as sodium chloride.

5. Explain why sodium chloride is a brittle solid whereas sodium metal is malleable and ductile.

6. Explain why crystals of sodium chloride are non-conductors of electricity, but molten sodium chloride is a good conductor of electricity.

7. What relationship exists between the melting point of naphthalene and:
   (i) the bonding within each molecule of naphthalene (intra-molecular forces);
   (ii) the bonding between adjacent molecules of naphthalene (inter-molecular forces).

8. Carbon atoms in diamond exert a covalency of four. Explain how this arises and how it leads to an explanation of the extreme hardness of diamond.

5 : The Structure of Liquids

The process of melting has been discussed already. It has been seen that when a solid melts, the particles which comprise the solid have sufficient energy to move about but remain close together (see fig. 5.1).

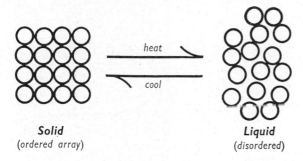

Solid
(ordered array)

Liquid
(disordered)

Fig. 5.1. Structure of solids and liquids

The particles of a liquid continually collide with each other and are in random motion, i.e., they move about in all directions. The speeds of the molecules vary very greatly.

This picture of ceaseless movement of the particles comprising liquids is supported by what is called Brownian Movement, first observed by R. Brown (1827). If a small drop of black printing ink is placed in water and observed under the high power of a microscope, the black particles are seen to be in irregular motion. The smaller the black particles, the more violent is their motion. The motion of the black particles is interpreted as being due to uneven impacts by the particles of the liquid, in their ceaseless motion.

In earlier chapters it has been shown that the particles which comprise most compounds are either ions or molecules. Thus liquids may be conveniently studied in two main classes.

IONIC LIQUIDS

Ionic substances are usually difficult to melt. They exist as liquids only at elevated temperatures. Molten ionic substances are good conductors of electricity and are chemically decomposed by the passage of an electric current. This is called electrolysis and has been discussed in chapter 2.

MOLECULAR LIQUIDS

It has been shown in chapter 2 that water and concentrated sulphuric acid are very poor conductors of electricity and therefore consist essentially of molecules. Hydrogen gas and chlorine gas also consist of molecules and they condense to liquids at low temperatures. Further examples of molecular liquids are ethanol, acetone and diethyl ether.

Evaporation of Liquids

If water or ethanol is placed in a basin and exposed to the atmosphere, the liquid will evaporate entirely. Evaporation of ionic liquids is not so easily observed because ionic substances generally have high melting points and so are not usually met with as liquids. For this reason the discussion that follows will be concerned only with evaporation of molecular liquids.

All the molecules in a liquid experience an attraction from neighbouring molecules. In the case of molecules within the bulk of the liquid, these forces balance each other, but for molecules in the surface of the liquid the forces are directed sideways and downwards (fig. 5.2). Hence for a molecule to escape from

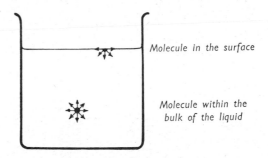

Molecule in the surface

Molecule within the bulk of the liquid

Fig. 5.2 Forces on molecules in a liquid.

the liquid, it must move upwards against an attractive force. Therefore, a molecule must have a certain minimum speed if it is to escape, and only the faster molecules can escape.

If a few drops of ethanol or ether are placed on the back of one's hand, the liquid evaporates rapidly and a cooling sensation is noticed. This is because the faster moving molecules escape and the average speed of the remaining molecules therefore falls. Because the temperature of a liquid depends on the average molecular speed, the temperature falls when the average molecular speed falls.

When water evaporates from an open dish at room temperature, it evaporates so slowly that there is little change in the temperature of the remaining water because heat enters from the surroundings. If, however, the water is evaporated rapidly by placing it in a strong draught, the fall in temperature can be detected.

The quantity of heat absorbed by one kg of a liquid in changing to a vapour at constant temperature is called the *heat of vapourization* for the liquid. The precise value depends on the temperature at which the vapourization takes place, and for water it is about 4060 kJ kg⁻¹ at 100 °C.

Measurement of Gas Pressure

A simple mercury barometer illustrates the fact that air exerts pressure. The atmospheric pressure on the mercury in the bowl supports the column of mercury. The difference between the levels of mercury is about 760 mm (see fig. 5.3). If the pressure of the atmosphere increases, so does the vertical height between the two mercury levels. Similarly, if the atmospheric pressure falls so does the vertical height between the two mercury levels. Hence the height of the mercury column is a convenient measure of air pressure.

A gas pressure which supports a column of 760 mm of mercury is called one atmosphere pressure or a

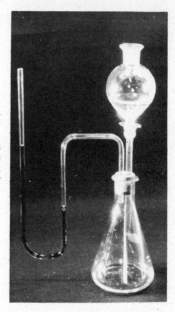

Fig. 5.4. Vapour pressure of alcohol.

The difference in levels of the water in the arms of the U-tube gives a measure of the vapour pressure.

pressure of one standard atmosphere, or often simply standard pressure. Gas pressures are usually stated in mm of mercury pressure or in atmosphere, although other units may be used.

The definition of 1 atm pressure is 101 325 N m⁻².

An approximate conversion factor for changing a pressure measured in mmHg to N m⁻² is 400/3. This simple conversion factor is accurate to better than 1 part in 10⁴. For example, 600 mmHg is:

$$600 \times \frac{400}{3} = 80\,000 \text{ N m}^{-2}$$

This is very close to the correct value of 79 994 N m⁻².

Pressures in this book will usually be given in mmHg but, if desired, they can be easily converted to N m⁻² by using the conversion factor.

GAUGE PRESSURES

Gauge pressures are commonly used in industry but are not much used in chemistry. Gauges often give the pressure in pound force per square inch and the reading is not the true pressure but only the difference from atmospheric pressure. Thus if the air pressure in a motor tyre is 24 lbf in⁻², then the pressure exceeds the atmospheric pressure by this amount, and to obtain the true pressure it would be necessary to add the atmospheric pressure to this figure.

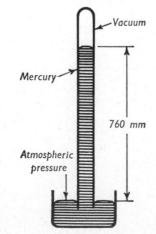

Fig. 5.3.
A mercury barometer.

Vacuum

Mercury

760 mm

Atmospheric pressure

Vapour Pressure

If a sealed flask is connected to a pressure gauge and a small amount of ethanol is placed in the flask, the pressure in the flask will increase. If excess ethanol is added at 20 °C, the pressure rises by about 44 mmHg, and then becomes steady (see fig. 5.4). This increase of pressure is called the saturated vapour pressure of the ethanol at the temperature of the experiment.

Earlier in this chapter evaporation was explained in terms of the escape of molecules from the liquid. In the experiment above, the pressure increase is due to molecules of ethanol which have escaped from the small amount of liquid added. Because the container is closed, the molecules cannot escape from the system and some of the molecules will collide with the surface of the liquid ethanol in the flask and be trapped. After a time the rate of escape from the liquid is equal to the rate of return (i.e. a state of equilibrium has been set up) and so the pressure becomes steady.

● *Provided some liquid is present at equilibrium, the pressure is independent of the amount of liquid.*

Similarly, a pressure increase would be observed if water had been used instead of ethanol but in this case it would be necessary to make sure that the air in the flask was initially dry (i.e. contained no water vapour). The pressure increase due to water vapour in equilibrium with water would be about 17.5 mmHg at 20 °C.

● *The equilibrium pressure for a vapour above a liquid at a given temperature is called the saturated vapour pressure of the liquid at the temperature given.*

The vapour pressure of a liquid increases with an increase of temperature. (See Appendix 3 for water vapour pressures at different temperatures.) The reason for this is that when a liquid is heated, the proportion of fast molecules increases so that molecules escape at a faster rate. Thus when equilibrium is reached there are more molecules of vapour present and so the pressure is higher.

It should be noted that mercury is used in instruments for measuring gas pressure because mercury has a very low vapour pressure and so contributes a negligible amount of pressure to the gas pressure being measured.

Vapour Pressure of Solids

Ice (like water) can be shown to have a vapour pressure. It is small but measurable, and can be measured by direct methods as described for water or alcohol. The equilibrium vapour pressure for ice is:

4.58 mmHg at 0 °C
1.95 mmHg at −10 °C
1.24 mmHg at −15 °C.

The vapour pressures of solids are usually very small indeed and indirect methods are usually employed for their measurement, such as the determination of the loss of weight of the solid in vacuum.

The direct conversion of a solid to the vapour without the production of liquid is known as sublimation.

Boiling

If a liquid in an open vessel is heated, the vapour pressure increases as explained above. Ultimately a temperature is reached at which the pressure of the vapour is sufficient to enable bubbles of the vapour to grow in size within the liquid. This occurs by evaporation of the liquid into the bubbles of vapour. Such growth of vapour bubbles within the liquid will only occur if the vapour pressure is at least equal to atmospheric pressure. The liquid is then said to be boiling (see fig. 5.5).

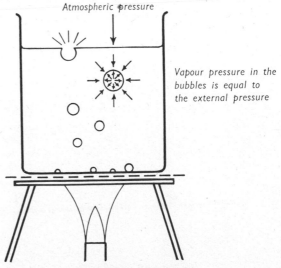

Atmospheric pressure

Vapour pressure in the bubbles is equal to the external pressure

Fig. 5.5. Boiling liquid.

If the atmospheric pressure increases, it is more difficult to boil a given substance, because the bubbles of vapour must grow in size against the increased pressure. Thus a higher temperature is needed to make the vapour pressure equal to the higher external pressure of the atmosphere, i.e., its boiling point rises. If the atmospheric pressure falls then the boiling point falls. Thus:

● *Pure water boils at*
(*i*) *exactly 100 °C if the pressure equals 760 mmHg.*
(*ii*) *above 100 °C if the pressure is greater than 760 mmHg.*
(*iii*) *below 100 °C if the pressure is less than 760 mmHg.*

Different liquids boil at different temperatures, and this is evidently due to the fact that the attractive forces between molecules differ for various substances. Hence different temperatures are required to make the vapour pressures of different substances equal to the atmospheric pressure.

For example, the boiling points of some liquids at a pressure of 760 mmHg are:

chloromethane	—23.8 °C
diethyl ether	34.5 °C
acetone	56.1 °C
water	100.0 °C
octane	125.7 °C
glycol	197.9 °C

The boiling points rise by about 0.04 °C for every mmHg over 760 mmHg pressure and fall by about 0.04 °C for every mmHg below 760 mmHg pressure. This correction is an approximation and should not be used if the pressure is very different from 760 mmHg. Appendix 3 gives the relationship between the vapour pressure and the temperature for water over a wide range of pressure.

STUDY QUESTIONS

1. If a beam of sunlight enters a darkened room and is viewed from one side, tiny flashes of light are seen. These flashes of light arise because the light is scattered by minute specks of dust, too small to be seen directly. The specks of dust (as seen in this indirect fashion), are in random, zig-zag motion.
What explanation can be offered for the motion of the dust particles?

2. Canvas water-bags are sometimes carried on the front of motor cars, especially during long trips in hot weather.
Explain why the water keeps cool in such circumstances.

3. The cooling chamber of a refrigerator becomes cold because a liquid is pumped into it and allowed to evaporate. Explain how this causes a drop in temperature inside the refrigerator.

4. A little water is placed in an empty four gallon kerosene tin and boiled until all the air is expelled. The tin is then sealed and allowed to cool. After a time the tin crumples and is squashed flat. Explain this in terms of the vapour pressure of water inside the tin.

5. Water is boiled in a florence flask until all the air is expelled. The flask is then sealed. The water stops boiling—why? If cold water is now played over the top of the flask, the water inside the flask boils—why?

6. Using the data given in the appendix for the saturated vapour pressures of water at different temperatures, construct a saturated vapour pressure curve for water.

7. In a city which is at an altitude of 1500 m above sea level, the temperature of boiling water is approximately 95 °C.
What explanation can be offered for this?

8. In very cold climates, clothes which have been hung outside to dry, will become stiff because the moisture freezes. However, if the air is dry, the ice will disappear from the clothes without the ice melting.
How can this be explained?

6 : The Properties of Gases and the Gas Laws

In the last chapter the change of state of molecular substances from liquid to gas (or vapour) was explained in terms of the motion of the molecules. The theory put forward suggested that gases consist of molecules which are in very rapid motion, with the molecules widely separated from one another. Ionic substances in the gaseous state will not be considered, because, as mentioned earlier, ionic substances generally have high melting points and very high boiling points. Ionic substances therefore are not commonly met with in the gaseous state.

Fig. 6.1 represents diagrammatically the liquid and gaseous states.

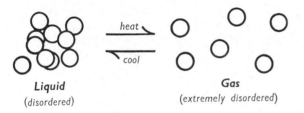

Fig. 6.1. Structure of liquids and gases.

The theory provides an explanation of the observable properties of gases. Such properties include the following:

1. Gases completely fill any containing vessel.

2. Gases may be compressed to much smaller volumes.

3. Gases mix completely with (diffuse into) one another.

These properties can be explained as follows:

1. Molecules are in very rapid motion and will quickly move out to the limits of the containing vessel.

2. Molecules are relatively far apart compared with the spaces between molecules in liquids, and when the gas is compressed, the spaces between the molecules are reduced.

  (Calculations show that the average spacing between molecules of oxygen gas at 0 °C and 760 mmHg is about 13 times the molecular diameter.)

3. If two gases are brought together, the very rapid motion of their molecules, together with their very open structure, will lead to rapid and complete mixing.

All the observed properties of gases considered so far can be explained in terms of the molecular theory and so provide initial support for the validity of the theory.

THE GAS LAWS

It is necessary to know more than these general properties of gases to cope with numerical problems in chemistry. Chemists frequently need to know the quantity of a gas collected. It has been seen already that a gas is compressible, i.e. that its volume is affected by pressure. The volume of a gas is also affected by temperature. Hence the statement of a gas *volume* is not sufficient to specify the *amount* of gas present. The temperature and pressure of the gas must be known.

Furthermore, it is sometimes necessary to compare two volumes of gas where the volumes are measured at different temperatures and pressures. To do this, it is necessary to know the quantitative effect of changes of temperature and pressure on the volume of gas, i.e. the actual volume change brought about by a particular change in temperature and pressure. These quantitative effects are described by Boyle's Law and Charles' Law.

Boyle's Law

It is a well known fact that if the pressure on a gas is increased its volume is decreased. This can be illustrated using a bicycle pump. If the outlet of the pump is firmly closed with a finger and the handle pushed in, the volume of the gas decreases as the pressure on the gas increases.

In 1662 *Robert Boyle* performed the first reliable experiments to determine the quantitative relationship between the volume and pressure of a gas. In a series of carefully conducted experiments he showed that the relationship is independent of the nature of the gas, and the relationship is summarized in his law.

● *At constant temperature, the volume of a given mass of gas is inversely proportional to the pressure.*

This law is illustrated by the diagrams in fig. 6.2.

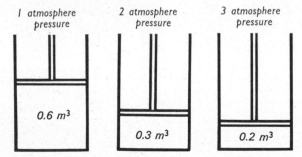

Fig. 6.2. Illustration of Boyle's Law: effect of pressure on a sample of a gas kept at constant temperature.

Boyle's Law may be expressed mathematically in the following way:

For a given mass of a particular gas

$V \propto \dfrac{1}{p}$ provided the temperature is constant.

where $\begin{cases} v \text{ is the volume of the gas} \\ p \text{ is the pressure of the gas} \end{cases}$

$\therefore V = \dfrac{k}{p}$ where k is a constant.

$\therefore pV = k$ (**a constant**).

For a given mass of a particular gas at constant temperature, the value of the constant k depends on the units chosen for measuring p and V.

Consider a sample of a gas kept at constant temperature. If the pressure and volume are measured and called p_1 and V_1 and the volume is then changed to a new volume V_2 at a new pressure p_2 then

$p_1 V_1 = k$ and $p_2 V_2 = k$.

$\therefore p_1 V_1 = p_2 V_2$.

It is suitable to express p_1 in any unit, e.g. millimetre of mercury, atmosphere or pound force per square inch provided p_2 is expressed in the same unit. Likewise V_1 may be expressed in any unit (cm³, dm³, ft³) provided V_2 is expressed in the same unit.

Charles' Law

Almost all substances (including gases) expand on heating. About 1787 *Jacques Charles* performed the first reliable experiments to determine the quantitative relationship between the volume and the temperature of a gas. This relationship is independent of the nature of the gas and is called Charles' Law.

● *At constant pressure, all gases increase in volume by 1/273 of their volume at 0 °C for every one Celsius degree rise in temperature.*

This can be illustrated by considering a volume of one dm³ of gas at a temperature of 0 °C. If the pressure is kept constant, then:

at   1 °C, $V = 1 +$ 1/273 or 274/273 dm³
at   10 °C, $V = 1 +$ 10/273 or 283/273 dm³
at 100 °C, $V = 1 +$ 100/273 or 373/273 dm³
at   −1 °C, $V = 1 −$ 1/273 or 272/273 dm³
at −10 °C, $V = 1 −$ 10/273 or 263/273 dm³
at −100 °C, $V = 1 −$ 100/273 or 173/273 dm³

Thus at −273 °C the gas, if it could be maintained as a gas, would have no volume. Experimentally it is found that all gases liquefy and solidify before this temperature is reached. Nevertheless the reasoning suggests that it is impossible to cool a gas below −273 °C, and that −273 °C is the lowest of all possible temperatures.

Absolute Zero of Temperature

More precisely, the figure is −273.150 °C and this temperature is called *absolute zero of temperature*. The fact that there is an absolute zero of temperature is supported experimentally in another way, viz. that all attempts to reach absolute zero of temperature experimentally have failed, although the temperature has been closely approached.

It is reasonable (and helpful) to change temperatures from the Celsius scale to a new scale in which absolute zero of temperature is called zero (and not −273 as on the Celsius scale). This new scale is called the Absolute scale, and temperatures on the Celsius scale are converted to the absolute scale by adding 273.

Thus: 0 °C = 273 K
              10 °C = 283 K
            −273 °C =   0 K.

In general:

T Kelvin $= t$ Celsius $+ 273$.

The symbol K, used to represent the unit of absolute temperature, is used in honour of Lord Kelvin, who first defined the absolute scale of temperature. The capital letter T is used to imply temperatures on the absolute scale.

If the illustration of Charles' Law (above) is altered so that temperatures are stated on the absolute scale, the table will read:

For one dm³ of gas at 273 K (if the pressure is constant), then:

at 274 K, $V = 274/273$ dm³
at 283 K, $V = 283/273$ dm³
at 373 K, $V = 373/273$ dm³
at 272 K, $V = 272/273$ dm³
at 263 K, $V = 263/273$ dm³
at 173 K, $V = 173/273$ dm³

It can be seen that the volume is proportional to the absolute temperature. Hence Charles' Law may be restated:

● *The volume of a given mass of gas is proportional to its absolute temperature, provided the pressure remains constant.*

This can be expressed in mathematical form as follows:
For a particular sample of any gas at constant pressure:
$V \propto T$

where $\begin{cases} V \text{ is the volume of the gas} \\ T \text{ is its absolute temperature} \end{cases}$

$\therefore V = kT.$

$\therefore \dfrac{V}{T} = k$ (a constant).

For a given mass of a particular gas at constant pressure, the value of the constant k depends only on the unit used to measure the volume.

Consider a sample of gas kept at constant pressure. Suppose the volume of the gas is V_1 and its temperature is T_1. If the volume of the gas is changed to V_2 by changing its temperature to T_2 then

$\dfrac{V_1}{T_1} = k$ and $\dfrac{V_2}{T_2} = k.$

$\therefore \dfrac{V_1}{T_1} = \dfrac{V_2}{T_2}$

Combination of Charles' Law and Boyle's Law

Charles' Law: $V \propto T$ for a given mass of a particular gas at constant pressure.

Boyle's Law: $V \propto \dfrac{1}{p}$ for a given mass of a particular gas at constant temperature.

These two laws can be combined and the combination yields the expression:

$V \propto \dfrac{T}{p}$ for a given mass of a particular gas.

$\therefore V = k\dfrac{T}{p}$ where k is a constant.

$\therefore \dfrac{pV}{T} = k.$

● *This equation is called The Gas Equation.*

For a given mass of a particular gas, the numerical value of the constant depends only on the units used to measure the pressure and the volume.

Another form of the Gas Equation is more commonly used. Consider a sample of a gas and suppose that it has a volume of V_1 at a temperature T_1 and a pressure p_1. Suppose that the pressure is changed to p_2 and the temperature is changed to T_2, causing the volume to change to V_2, then:

$\dfrac{p_1 V_1}{T_1} = k$ and $\dfrac{p_2 V_2}{T_2} = k.$

$\therefore \dfrac{p_1 V_1}{T_1} = \dfrac{p_2 V_2}{T_2}$

This equation can be used to compare gas volumes.

The value of the constant k (above) is the same for one mole of any gas. The value of this constant is approximately:

8.314 J K⁻¹ mol⁻¹

if the unit of pressure is N m⁻² and the unit of volume is m³ mol⁻¹. T must be expressed in K.

Standard Temperature and Pressure

Volumes of gases are usually compared at the one temperature and pressure, viz. 273 K and 760 mmHg. These conditions are referred to as standard conditions or *standard temperature and pressure* (s.t.p.).

NUMERICAL ILLUSTRATION

If a gas occupies 56.2 cm³ at 20 °C and 740 mmHg pressure, what would be its volume at 100 °C and 760 mmHg pressure?

$$\frac{p_1 V_1}{T_1} = \frac{p_2 V_2}{T_2}$$

Let p_1, V_1 and T_1 refer to the conditions of the experiment and let p_2, V_2 and T_2 refer to the required conditions.

$p_1 = 740$ mmHg $p_2 = 760$ mmHg
$V_1 = 56.2$ cm³ $V_2 =$ the unknown
$T_1 = 293$ K $T_2 = 373$ K

N.B. Consistent sets of units must be used for each pressure and each volume. Also, the temperatures must be expressed on the absolute scale.

$$\therefore \frac{740 \times 56.2}{293} = \frac{760 \times V_2}{373}$$

$$\therefore V_2 = \frac{740 \times 56.2 \times 373}{293 \times 760}$$

$$= 69.65.$$

∴ the volume of the gas at 100 °C and 760 mmHg would be 69.7 cm³.

(*Note*: The result is expressed to the same order of accuracy as the data.)

Another useful form of the gas equation is obtained by re-arranging the terms. Thus:

$$V_2 = \frac{p_1 V_1 T_2}{T_1 p_2}$$

$$\text{or } V_2 = V_1 \times \frac{p_1}{p_2} \times \frac{T_2}{T_1}.$$

This form of the equation shows clearly the principles underlying the use of the formula. The new volume V_2 is obtained by correcting the old volume V_1. In the above example, the pressure is to be increased from 740 mmHg to 760 mmHg. This will reduce the volume.

$$\therefore V_2 = 56.2 \times \frac{740}{760}$$

Also, the temperature is to be increased from 293 K to 373 K, and this will increase the volume.

$$\therefore V_2 = 56.2 \times \frac{740}{760} \times \frac{373}{293}$$

etc., as before.

Dalton's Law of Partial Pressures

This law refers to mixtures of gases. The law states:

● *In a mixture of gases, the total pressure of the gases is equal to the sum of the partial pressures of each gas in the mixture, provided the gases do not react chemically.*

The partial pressures are the pressures which each gas would exert if it alone occupied the vessel. Thus each gas acts as though it were the only gas in the vessel. This law has been verified experimentally.

Application of Dalton's Law

It has been seen in the previous chapter that if water is contained in a closed vessel, there is a pressure due to the water vapour. Hence if a gas is collected over water, by bubbling the gas through water, then the gas pressure measured, is in fact the sum of the partial pressures of the gas and water vapour. Water vapour pressures at various temperatures are quoted in Appendix 3. If it is assumed that the gas is saturated with water vapour, then the pressure of the gas alone can be calculated by subtracting the water vapour pressure at the appropriate temperature from the total pressure.

Measurement of Gas Pressures

Fig. 6.3 shows a gas in a tube standing over water. If the levels of water are the same inside and outside the tube containing the gas, then the total pressure of gas and water vapour inside the tube is equal to atmospheric pressure, because these pressures are balancing one another. Therefore the pressure inside the tube is known if the atmospheric pressure is determined by reading a barometer. The pressure of

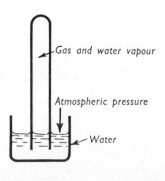

Fig. 6.3. Measurement of gas pressure.

the gas alone can therefore be obtained by subtracting the water vapour pressure from the barometric pressure as described under Dalton's Law.

When a gas is actually collected over water, the water levels inside and outside the tube will almost certainly not be equal. Before any measurements are made, the tube containing the gas should be placed in a vessel of water sufficiently deep to allow the tube to be moved up or down, until the levels are equal.

NUMERICAL ILLUSTRATION

A volume of 43.0 cm³ of hydrogen is collected over water at 15 °C and 762.8 mmHg. Calculate the volume which the hydrogen would occupy if dry and at s.t.p.

(Water vapour pressure at 15 °C = 12.8 mmHg)
$$\text{Pressure of hydrogen} = 762.8 - 12.8$$
$$= 750 \text{ mmHg.}$$

Let p_1, V_1 and T_1 refer to the experimental conditions and let p_0, V_0 and T_0 refer to standard conditions.

$$\frac{p_1 V_1}{T_1} = \frac{p_0 V_0}{T_0}$$

$p_1 = 750$ mmHg $p_0 = 760$ mmHg
$V_1 = 43.0$ cm³ $V_0 =$ the unknown
$T_1 = 288$ K $T_0 = 273$ K

$$\therefore \frac{750 \times 43.0}{288} = \frac{760 \times V_0}{273}$$

$$\therefore V_0 = \frac{750 \times 43.0 \times 273}{288 \times 760}$$

$$= 40.24.$$

∴ the volume occupied by the dry hydrogen at s.t.p. = 40.2 cm³.

THE KINETIC THEORY OF GASES

Many features of this theory have been discussed already in various parts of the text. A gas is believed to consist of a vast number of molecules which are relatively far apart. Each molecule in a gas moves in a straight line until it collides with another molecule or with the walls of the container. The path of a molecule is therefore an irregular sequence of straight lines.

In addition, the speed of a molecule is likely to change with each collision. Some collisions will cause the molecule to speed up, whereas some collisions will cause the molecule to slow down. A gas therefore consists of a vast number of molecules in chaotic motion. The pressure of a gas is attributed to the enormous number of collisions between the molecules and the walls of the container, in each moment of time.

A formal statement of the postulates of the Kinetic Theory of Gases can now be given:

Postulates of the Theory

1. Gases consist of molecules in continuous chaotic motion.
2. The molecules move with great speed in straight lines, colliding with one another and with the walls of the container.
3. The molecules are negligible in size compared to the spaces between them.
4. There is no attraction or repulsion between the molecules.
5. The average kinetic energy of the molecules is proportional to the absolute temperature of the gas. (The kinetic energy of a body is the energy it possesses due to its motion.)

Explanation of Gas Laws by the Kinetic Theory

The kinetic theory can explain the gas laws both qualitatively and quantitatively, but only the qualitative explanation will be given here.

Boyle's Law: If the volume of a sample of gas is reduced, at constant temperature, there will be more collisions on each unit area of the walls of the container per unit time. Thus the pressure of the gas increases.

Charles' Law: If the temperature of a gas is increased, the kinetic energy of the molecules will increase. This means that the increase of temperature increases the average speed of the molecules. At constant *volume* this would result in an increased pressure. However, if the *pressure* is to be kept constant, the gas must occupy a greater volume. Thus if the gas is kept at constant pressure, the molecules will occupy a greater volume because of their increased average speed.

Dalton's Law of Partial Pressures: In a mixture of gases which do not react with one another, each gas will exert a partial pressure because of the collisions of its gas molecules with the walls of the container. The total number of collisions with the walls of the container in unit time, will be equal to the sum of the collisions of each separate gas in the same time.

STUDY QUESTIONS

1. Calculate the volume at s.t.p. which would be occupied by:
   (a) 380 cm³ of gas at 27 °C and 600 mmHg;
   (b) 76.0 cm³ of gas at −23 °C and 750 mmHg;
   (c) 500 cm³ of gas at 21 °C and 745 mmHg;
   (d) 36.2 cm³ of gas at 25 °C and 730 mmHg;
   (e) 5.5 dm³ of gas at 17 °C and 2 atm;
   (f) 2.00 dm³ of gas at 10 °C and 75 cmHg.

2. A sample of gas has a volume of 32.0 cm³ at 15 °C and 740 mmHg. What volume would the gas occupy at:
   (a) 30 °C and 740 mmHg;
   (b) 15 °C and 370 mmHg;
   (c) 30 °C and 370 mmHg;
   (d) 21 °C and 760 mmHg?

3. 40.0 cm³ of hydrogen is collected over water at 25 °C and 765 mmHg. Making allowance for water vapour pressure, correct the volume of the hydrogen to s.t.p.
   (Water vapour pressure at 25 °C = 23.8 mmHg.)

4. 34.3 cm³ of hydrogen is collected over water at 27 °C and 761 mmHg. Calculate the volume which the dry gas would occupy at s.t.p.
   (Water vapour pressure at 27 °C = 26.7 mmHg.)

5. The density of a sample of hydrogen is measured at 27 °C and 760 mmHg. Will the density increase, decrease or remain the same, if the temperature and pressure are changed to:
   (a) 54 °C and 760 mmHg;
   (b) 27 °C and 380 mmHg;
   (c) 54 °C and 380 mmHg;
   (d) 54 °C and 1520 mmHg;
   (e) −150 °C and 380 mmHg?

6. 403 cm³ of carbon dioxide at 22 °C and 758 mmHg is found by experiment to have a mass of 0.734 g. What is the density in g cm⁻³ of carbon dioxide:
   (a) at the conditions of the experiment;
   (b) at s.t.p?

7. The density of dry dust-free air at s.t.p. is 1.293 g dm⁻³. Calculate its density at 17.0 °C and 760 mmHg.

8. A sealed flask containing a gas at 20 °C and 750 mmHg is immersed in boiling water. Find the pressure developed in the flask. (Assume that the flask itself does not expand.)

9. A sealed flask contains 250 cm³ of a gas at 22 °C. It is connected to an evacuated flask so that the total volume occupied by the gas is 300 cm³. To what temperature must the gas be heated in order to restore the original pressure?

10. If the temperature of any gas is gradually reduced (and particularly if the pressure on the gas is high), the gas liquefies.
    (a) In terms of molecular motion, describe the change in structure which occurs on cooling a gas down to the point of liquefaction.
    (b) Examine postulate 4 of the Kinetic Theory and in the light of the above observational facts, discuss the accuracy of the postulate particularly at low temperatures and high pressures.
    (c) Discuss why increase of pressure should aid the process of liquefaction of a gas.

11. It is found by experiment that Boyle's Law does not provide an accurate description of the behaviour of a gas such as nitrogen when the pressure on the gas is several hundred atmospheres. Experimentally it is found that with increase of pressure (at constant temperature) the volume of the gas does not decrease by as much as predicted by Boyle's Law.
    (a) What explanation can be offered for these observations?
    (b) Re-examine the postulates of the Kinetic Theory and state the postulates which are inaccurate at very high pressures.

Answers to Numerical Problems

1. (a) 273 cm³.	(d) 31.8 cm³.
(b) 81.9 cm³.	3. 35.7 cm³.
(c) 455 cm³.	4. 30.2 cm³.
(d) 31.9 cm³.	6. (a) 0.001 82 g cm⁻³.
(e) 10.4 dm³.	(b) 0.001 97 g cm⁻³.
(f) 1.90 dm³.	7. 1.217 g dm⁻³.
2. (a) 33.7 cm³.	8. 955 mmHg.
(b) 64.0 cm³.	9. 81 °C.
(c) 67.3 cm³.	

7 :   Separation of Mixtures

The methods used to separate the constituents of a mixture depend on the kind of mixture to be separated. Some of the more common methods will be described.

1. Filtration of Solids from Liquids

When a substance has been precipitated by a chemical reaction, it is often desired to separate it from the remaining solution. For example, if barium chloride solution ($BaCl_2$ solution) is mixed with dilute sulphuric acid (H_2SO_4 solution), a precipitate of barium sulphate ($BaSO_4$) forms:

$$BaCl_2 + H_2SO_4 \rightarrow BaSO_4(s) + 2HCl$$
(solutions)   (ppt.)   (solution)

The symbol '(s)' indicates that $BaSO_4$ forms as a solid.

The precipitate can be filtered off by a filter paper (see fig. 7.1) and washed with distilled water to rinse off all traces of the solution.

PRACTICAL APPLICATIONS

Commercial filtration plants are similar in principle although often quite different in appearance to the laboratory apparatus represented in the diagrams.

(a) *A paper making machine* filters paper pulp (fibres of cellulose) from a suspension in water on to a moving wire screen so that a continuous sheet is formed. The last traces of water are removed by running the paper over heated rollers.

(b) *A spin dryer* is a form of filter where the rate of filtration is speeded up by spinning the wet clothes rapidly in a perforated basket.

(c) *Filter beds*: Many domestic water supplies are purified by being filtered through a layer of sand, called a filter bed.

2. Filtration of Solids from Gases

(a) *The household vacuum cleaner* illustrates one method. Dust laden air is forced into a fabric or paper bag by a fan. The air escapes through the pores of the bag leaving the dust particles inside.

(b) *The air used by a petrol engine* can be cleaned by drawing it through a filter consisting of a pad of steel or glass wool covered with a thin film of heavy oil. The dust particles stick to the oily surfaces.

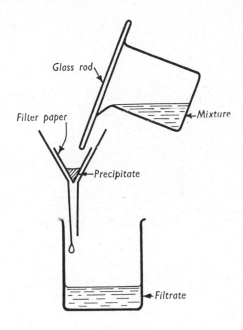

Simple filtration

Fig. 7.1. Methods of filtration

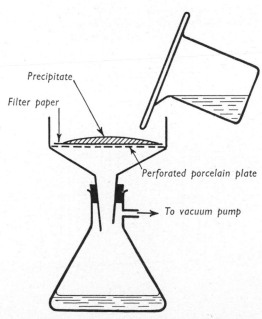

Rapid filtration under suction using a Buchner funnel

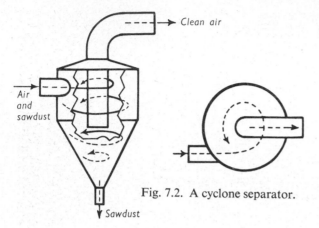

Fig. 7.2. A cyclone separator.

(c) *A cyclone separator* is used to remove sawdust and wood chips from air. The dust laden air is blown into a cylindrical chamber and as it whirls around, the particles are thrown to the outside and fall to the bottom (see fig. 7.2).

(d) *Electrostatic precipitators* are used to remove fine dust from gases. This involves passing the gas between two electrodes which are connected to a very high voltage supply. An electric charge is induced on the dust particles and they are attracted to the oppositely charged electrode (see fig. 7.3).

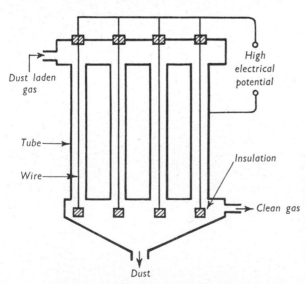

Fig. 7.3. An electrostatic precipitator.

3. Filtration of Liquids from Fibrous or Porous Solids

Many useful oils are extracted from nuts, seeds and kernels. These include coconut, palm kernel, ground-nut (or peanut), cottonseed, soya bean, sunflower and linseed oils. These vegetable oils can be used as foods e.g. in margarine, or in the manufacture of soap, glycerine, paints, varnishes, enamels, polishes, and linoleum and in the preservation of cloth, wood and leather.

Filtration under pressure is the most widely used method of extraction. The seeds or nuts are first ground into a very fine meal. The meal is then placed in fabric bags and subjected to a very high pressure by an hydraulic ram. The oils are expressed and run through the fabric and are collected. The residual cake contains all the fibrous matter of the kernel. This may be used as a cattle food. The oils (from 1 to 10%) still retained by the fibrous matter enhance its food value.

4. Distillation

Distillation can be used to separate a volatile substance from non-volatile substances. For example, pure water is often preferred in making solutions of chemicals and is commonly prepared from tap water by distillation. Water (b.p. = 100 °C) is boiled off during distillation, while impurities such as sodium chloride (b.p. = 1750 °C) remain behind. The vapour is cooled so that it condenses back to the liquid. Water purified in this way is called distilled water (see fig. 7.4).

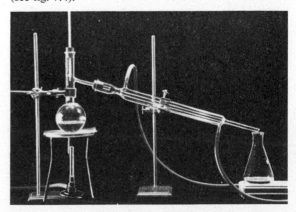

Fig. 7.4. Distillation.
The mixture is boiled in the flask on the left and the distillate is collected from the condenser.

Other liquids can be purified in a similar way. For example: acetone

$$(CH_3—CO—CH_3; B.P. = 56 \ °C)$$

and ethyl acetate

$$(CH_3—CO—O—CH_2—CH_3; B.P. = 77 \ °C)$$

can be distilled out of the mixtures in which they are prepared. These liquids are very useful solvents.

5. Fractional Distillation

This is a further extension of the principle of distillation and can be used to separate two or more volatile liquids which have different boiling points. For example, a mixture of

benzene $(C_6H_6$; b.p. $= 80 \ °C)$ and

toluene $(C_6H_5 —CH_3$; b.p. $= 110 \ °C)$

can be separated by boiling the mixture and allowing the vapours to pass up a *fractionating column*. The toluene vapour (higher b.p.—i.e. less volatile) tends to condense and run back down the column. The benzene vapour (lower b.p.—i.e. more volatile) can be collected from the top of the column (see fig. 7.5).

Ethanol (also called ethyl alcohol; $CH_3—CH_2—OH$) can be obtained by the fermentation of sugar and the water present is largely removed by fractional distillation.

Crude petroleum oil is separated by a similar process. However, there are many constituents in petroleum and usually no effort is made to obtain each one separately. Instead, a number of fractions are distilled off—the main ones being:

  (i) b.p. up to 200 °C—Gas and gasoline (petrol).
  (ii) b.p. up to 271 °C—Kerosene.
  (iii) b.p. up to 350 °C—Gas oil and diesel oil.
  (iv) b.p. above 350 °C—Lubricating oil and paraffin wax.
  (v) b.p. above 350 °C—Fuel oil.
  (vi) residue—Bitumen (asphalt).

Air can be separated into oxygen and nitrogen by fractional distillation of liquid air. The liquid nitrogen (b.p. $= -195°C$) is more volatile than liquid oxygen (b.p. $= -183°C$) and therefore nitrogen gas is collected from the top of the column while oxygen is collected from the bottom. The oxygen used in the manufacture of steel is produced in this way.

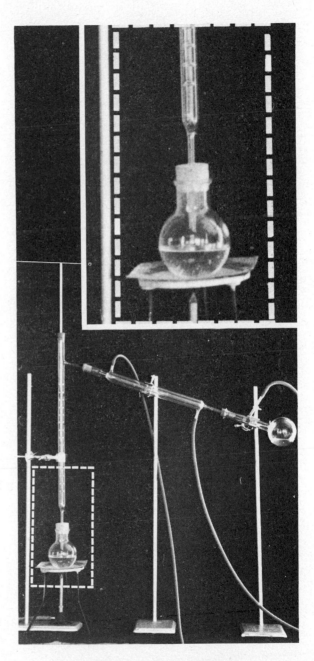

Fig. 7.5. Fractional distillation.

If a mixture of benzene and toluene is boiled in the flask, the mixture becomes richer in toluene as benzene is collected from the condenser. The receiver is closed and a vent leads to the sink to avoid fire risk.

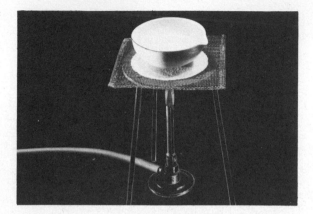

Fig. 7.6. Evaporation.
The solution is boiled in an evaporating basin.

6. Evaporation

This is similar in principle to distillation but the required substance is the solid residue. For example, if it is required to separate a soluble salt from a solution of it in water, the solution may be evaporated to dryness. In such cases it is usually not required to collect the water so the solution is boiled in an open shallow dish called an evaporating basin (see fig. 7.6).

Fig. 7.7. Drying a substance in a desiccator.
Suitable drying agents are calcium chloride, calcium oxide, phosphorus pentoxide or concentrated sulphuric acid.

The burner should be removed when the volume of the solution has become very small, or it may spit and splutter. The salt is then allowed to dry by leaving it exposed to the air and final traces of moisture can be removed by placing the salt in a desiccator (see fig. 7.7).

Common salt (Na^+Cl^-) is often made from sea water by running it into large shallow pans and allowing evaporation to proceed in the sun and wind.

7. Sublimation

Sublimation is the conversion of a solid directly into a vapour and its subsequent condensation directly to a solid without the formation of a liquid phase, and this process can be used to purify some substances.

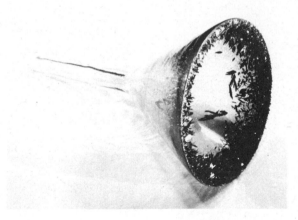

Fig. 7.8. Sublimation of iodine.
Needle-like crystals of iodine form on the inside of the funnel.

For example, iodine can be sublimed by heating it gently in an evaporating basin covered with a funnel (see fig. 7.8). The non-volatile impurities remain at the bottom and most volatile impurities do not subsequently condense with the sublimed iodine which forms as dark lustrous crystals on the sides of the funnel.

Although liquid (i.e. fused) iodine can be obtained, the vapour pressure of the solid is sufficiently high for the solid to vapourize completely and rapidly at temperatures below the melting point.

8. Other Methods

It can be seen that the method used to separate a mixture depends on the properties of the constituents. A few other examples may be mentioned.

Petrol and water do not mix; the petrol floats on the water. Thus the petrol can be drawn off from the upper layer. Petrol storage tanks are always kept full to minimise the fire risk due to inflammable vapours. The lower part of the tank contains salt water. When more petrol is added, water is run out from the bottom of the tank. When petrol is to be removed, water is pumped in at the bottom. The few drops of water

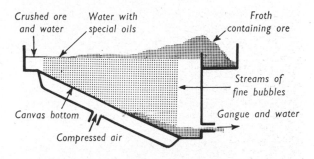

Fig. 7.9. Selective flotation.

Separation of Liquids.
The flask contains kerosene and water. The lighter kerosene floats to the top, and the water can be drawn off from the bottom.

as a froth while the others sink to the bottom. This process is called *selective flotation* (see fig. 7.9).

Sluicing: When substances of very high density are mixed with others of low density they can often be separated by *sluicing*. Alluvial gold can be separated from quartz and sand by swirling with water. The less dense quartz and sand are swept away and the gold

which may become suspended in the petrol can be removed by running it through a fine metal gauze. The water does not wet the petrol soaked gauze and forms globules (because of its surface tension) which do not pass through the gauze.

Mineral deposits such as zinc and lead ores contain many other materials such as sand (silica) and pyrites. The ores are crushed to a very fine powder and placed in water containing special oils (kerosene, creosote, xanthates, etc.). When air is blown through the mixture, the oils cause the air bubbles to become attached to certain minerals but not to others. The particles of ore attached to air bubbles collect on top

Fig. 7.10. This contemporary engraving shows sluicing being used to separate gold from alluvium on the Ballarat field in the 1850s.

remains. A similar principle is used in *winnowing* wheat although it is not so much a difference in density as in shape which makes the separation possible. The wheat grains and husks are allowed to fall through a draught of air. The small rounded grains drop down but the husks are swept away.

Magnetic separation can be used to remove pieces of iron from other non-magnetic materials. Odd pieces of iron (old bolts, nails, etc.) often find their way into minerals which are excavated from the ground. If the mineral has to be crushed these can cause serious damage to valuable equipment. One method used to remove these is to run the material over a conveyor belt. The end roller contains a powerful electromagnet so that as the mineral runs from the end of the belt, the pieces of iron are held as the belt passes around the pulley and are thrown back from the stream of mineral (see fig. 7.11).

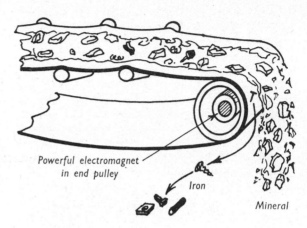

Fig. 7.11. Magnetic separation.

STUDY QUESTIONS

1. Describe how the suspended substance could be recovered in each of the following:
    (a) a suspension of dust in air;
    (b) the smoke from a smoke stack.

2. How could the following mixtures be treated so as to obtain each constituent separately:
    (a) iron dust and copper filings;
    (b) kerosene and water;
    (c) sugar and water;
    (d) salt and sand;
    (e) copper filings and powdered sulphur?

3. Briefly describe three ways in which the process of filtration can be speeded up.

4. How would you prepare a sample of pure silver chloride, starting with a solution of silver nitrate?

5. Explain what is meant by the term *sublimation*, and explain how impurities are removed from iodine by sublimation.

6. Describe how the process of fractional distillation is used to separate liquids.

8 : Solutions and Crystalline Salts

There are many substances which do not react with water but which are soluble in water. For example, sucrose which has the molecular formulae $C_{12}H_{22}O_{11}$, is a sugar which is obtained from the juice of sugar cane or sugar beet. It is a white crystalline solid which seems to disappear when placed in water. If the water is evaporated away, the white crystals reappear. The process in which a solid dissolves in a liquid is called *dissolution* and the process in which the solid is formed is called *crystallization*. The homogeneous mixture obtained by dissolution of a solid in a liquid is called a *solution*. The dissolved substance is called the *solute* and the substance in which it is dissolved is called the *solvent*.

The process of dissolution can be described in the following way: the particles of the solid are removed

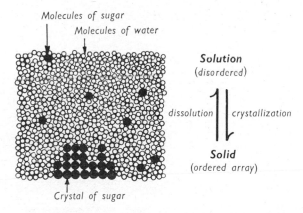

Molecules of sugar
Molecules of water

Solution
(disordered)

dissolution crystallization

Solid
(ordered array)

Crystal of sugar

Fig. 8.1. Structure of solids and solutions.

from the close packed ordered pattern of the crystal and move around with the particles of the solvent. A solution is an intimate combination of the particles of solute and solvent (see fig. 8.1).

● *The solute and solvent cannot be separated by filtration.*

COLLOIDS

Sometimes, when substances are precipitated in chemical reactions, the precipitates remain suspended in the liquid. Thus if iron(III) chloride solution is added drop by drop to boiling water, a red suspension is formed. This consists of minute particles of precipitated iron(III) hydroxide which will not settle. These particles are so small that they will pass through a filter paper and in fact the suspension looks like a red solution. This suspension of very small particles is called a colloid. However, the fine particles can be coagulated by the addition of a small amount of a solution of a salt such as sodium chloride or alum. The particles of iron(III) hydroxide clump together and become so large that all of the iron(III) hydroxide settles to the bottom of the vessel and can be easily filtered.

Thus one property of colloids which can often be used to identify them is their property of coagulating or flocculating on the addition of an electrolyte. A colloid cannot be separated into its components by filtration because the suspended particles are too small, but after the addition of an electrolyte the suspended matter can be removed by filtration.

In a solution, the solute particles are individual ions or molecules which are of the order of 10^{-10} m in diameter. In a colloid, the suspended particles are groups of thousands of ions or molecules of the order of 10^{-7} to 10^{-9} m in diameter.

SOLUTIONS

The term *solution* is not confined to solids in liquids, but includes systems of solids, liquids or gases in liquids.

It is difficult to make any generalization concerning the solubility of substances in liquids but one useful working rule is as follows. *Covalent substances* are usually soluble in *covalent liquids* such as ether, ethanol, acetone, carbon tetrachloride, etc. *Ionic substances* are not usually soluble in such liquids but are usually soluble in water, which allows the ions to dissociate, (see chapter 11).

Thus I_2 (molecular) is readily soluble in CCl_4 (molecular) but not very soluble in water, whereas Na^+Cl^- (ionic) is soluble in water but insoluble in CCl_4.

In cases where this rule does not apply it is usually found that chemical reaction occurs. For example, HCl (molecular) is soluble in CCl_4, as expected, but is also very soluble in water. When HCl dissolves in H_2O a reaction occurs which is represented by the equation:

$$HCl + H_2O \rightarrow H_3O^+ + Cl^-$$

SOLUTIONS OF SOLIDS IN LIQUIDS

The solubility of a solid in a liquid depends on three main factors:
 (i) the nature of the solid (solute);
 (ii) the nature of the liquid (solvent);
(iii) the temperature.

Saturation

There is a limit to the amount of a particular solid which can be dissolved in a given amount of a liquid. For example, if sodium chloride is placed in water some of it will dissolve. Dissolution continues until a definite amount has dissolved and any additional salt present will remain undissolved, provided that the temperature remains constant and that the amount of liquid remains unchanged. The solid and the solution are then in equilibrium and the solution is said to be saturated. Definition:

● *A saturated solution is a solution which is in equilibrium with undissolved solute at a given temperature.*

The solubility of a solid in a liquid may be specified in several ways. One way the solubility can be specified is by the number of units of mass of solute which saturates 100 units of mass of solvent.

● *For example, if a saturated solution contained 5 g of solute in 20 g of solvent, then the solubility would be 25.*

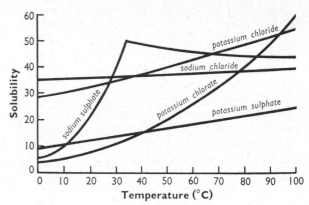

Fig. 8.2. Solubility curves.

Solubility Curve

A solubility curve is a graphical representation of the variation of solubility with temperature. A few solubility curves are shown in the diagram (see fig. 8.2). The solubility of a solid in a liquid usually *increases* in a regular way with *increase* of temperature, but there are some cases where it decreases (e.g. calcium hydroxide) or changes direction suddenly (e.g. sodium sulphate).

It is sometimes more helpful to say that for most substances the solubility *decreases* in a regular way with *decrease* of temperature. If a hot concentrated

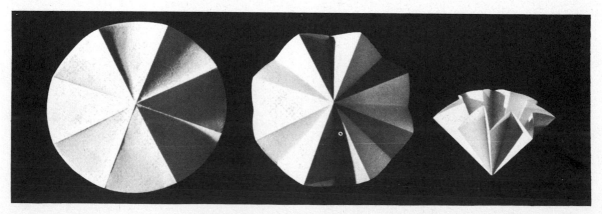

Folding a fluted filter paper for filtration of a hot solution showing from left to right:
Step 1. Fold into eight segments. Step 2. Fold each segment in half. Step 3. The fluted paper ready for use.
The right hand picture shows rapid filtration using a fluted filter paper and a short-stemmed funnel.

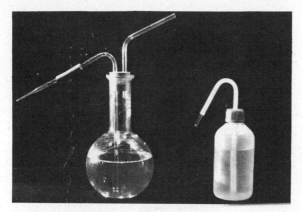

Fig. 8.3. Wash bottles.

solution is cooled, crystals of the solute will appear at some stage of the cooling. The solute is said to crystallize from the solution.

Recrystallization

This can be used to purify a soluble substance if its solubility changes considerably with temperature.

RECRYSTALLIZATION OF POTASSIUM CHLORATE

The solubility of potassium chlorate ($K^+ClO_3^-$) is about 60 at 100 °C but only about 5 at room temperature. Suppose a sample of potassium chlorate is contaminated by fairly small amounts of impurities—

some of which are soluble and some insoluble in water. The sample is placed in hot water and filtered while hot. This removes the insoluble impurities. The filtrate is a clear solution but when it is cooled, most of the potassium chlorate crystallizes out of solution. These crystals are filtered off and washed with a little cold water from a wash bottle (see fig. 8.3). The soluble impurities pass through with the filtrate and the pure salt is retained on the filter paper. Some of the original potassium chlorate is lost by this process but most of it is recovered and in a much purer state.

Supersaturation

It is possible to form solutions which contain more solute than is necessary to saturate the solution. Such solutions are said to be supersaturated.

If crystals of hydrated sodium thiosulphate (the formula is $Na_2S_2O_3,5H_2O$, i.e., water is incorporated in the crystals) are heated carefully in a clean flask without the addition of water, a liquid is obtained which may not crystallize on cooling. It is advisable to put a plug of clean cotton-wool in the mouth of the flask to prevent any dust particles from entering the flask. If crystallization does not occur on cooling, the solution obtained is a supersaturated solution. If a small crystal of the salt is dropped into the flask the whole solution crystallizes very rapidly (see fig. 8.4).

Technique for the Filtration of a Hot Solution

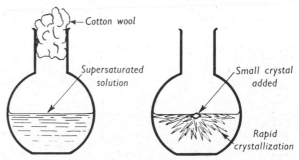

Fig. 8.4. Crystallization of a supersaturated solution.

Sometimes crystallization can also be initiated by dust particles or by scratching the inside of the flask with a glass rod. Heat is evolved during the crystallization, and the flask becomes quite warm.

● *A supersaturated solution is not in equilibrium with undissolved solute.*

Crystals of hydrated salts.
Top left: Copper(II) sulphate; top right: alum; bottom left; nickel sulphate; bottom right: chrome alum.

Hydrated Salts

When salts are crystallized from aqueous solution, water is often incorporated in the structure of the crystal. For example, when a solution of copper(II) sulphate ($Cu^{2+}SO_4^{2-}$) is evaporated to dryness (as described in Chapter 7) blue crystals are obtained. The empirical formula of the blue crystalline substance is $CuSO_4,5H_2O$. This means that for each ion pair, $Cu^{2+}SO_4^{2-}$, there are 5 molecules of water in the crystal structure.

If the blue crystals are heated gently in a hard glass test tube (at 200 °C) they crumble to a white powder and a colourless liquid (water) condenses on the cool part of the tube. If the white powder is allowed to cool and is then treated with a few drops of water, it turns blue.

The blue crystals ($CuSO_4,5H_2O$) are called hydrated copper(II) sulphate and the white powder ($CuSO_4$) is called anhydrous copper(II) sulphate. Other examples of hydrated salts are:

Na_2CO_3,H_2O
  sodium carbonate monohydrate
$Na_2CO_3,10H_2O$
  sodium carbonate decahydrate
  (washing soda)
$FeSO_4,7H_2O$
  iron(II) sulphate heptahydrate
$MgSO_4,7H_2O$
  magnesium sulphate heptahydrate
  (Epsom salt)

$Na_2SO_4,10H_2O$
  sodium sulphate decahydrate
  (Glauber's salt)
$CaSO_4,2H_2O$
  calcium sulphate dihydrate
  (gypsum)

Many crystalline salts do not form hydrates. They crystallize in the *anhydrous* form, e.g.:

$NaCl$ sodium chloride (common salt)
$KClO_3$ potassium chlorate.

Anhydrous substances do not contain water of crystallization, and they may or may not be crystalline. An example of a non-crystalline anhydrous substance is magnesium carbonate.

Hydrated Ions

In the structure of blue crystalline copper (II) sulphate, $CuSO_4,5H_2O$, four of the H_2O molecules are located close to the copper(II) ion. This suggests that they are bound by valency bonds, and the copper(II) ion is really $Cu,4H_2O^{2+}$ or $Cu(H_2O)_4^{2+}$. The fifth H_2O molecule in hydrated copper(II) sulphate appears to be associated with both the copper(II) ion and the sulphate ion. The crystal can thus be represented more accurately as $Cu(H_2O)_4^{2+}SO_4^{2-},H_2O$.

Many other metallic ions are hydrated both in solution and in their crystalline salts. The hydration of some cations may be represented as follows:

$Zn(H_2O)_4^{2+}$ zinc $Ca(H_2O)_6^{2+}$ calcium
$Fe(H_2O)_6^{2+}$ iron(II) $Na(H_2O)_n^{+}$ sodium
$Fe(H_2O)_6^{3+}$ iron(III) $K(H_2O)_n^{+}$ potassium

It appears that the number of closely bound H_2O molecules is variable and in some cases (e.g. Na^+ and K^+) cannot be determined. For this reason the ions are often represented without the H_2O molecules.

In solution, anions as well as cations are hydrated, i.e. all the ions in an aqueous solution are surrounded by spheres of closely bound water molecules. Although these molecules of water of hydration are often omitted when writing chemical equations, they play an important role in determining the solubility of salts in water.

For example, the equlibrium between a solution of potassium chloride and solid potassium chloride can be represented by the simple equation:

$$K^+ + Cl^- \rightleftharpoons KCl(s)$$

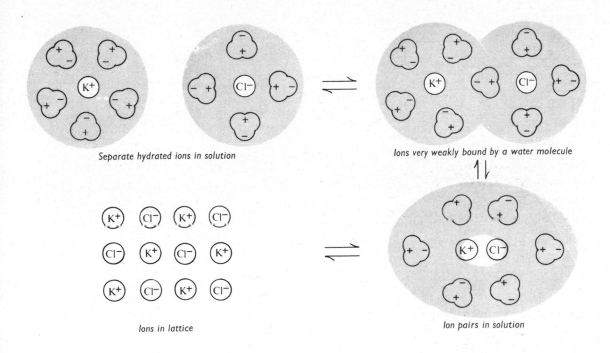

Fig. 8.5. The steps in the equilibrium between a solution and a solid.

The actual change from ions in solution to ions in a crystalline lattice of a solid, involves a number of steps in which bound water molecules are eliminated from the closely bound spheres around each ion (see fig. 8.5).

Heat of Solution

Crystalline salts usually dissolve in water with the absorption of heat. Examples of such salts include the hydrated salts copper(II) sulphate pentahydrate, sodium sulphate decahydrate, sodium carbonate decahydrate and the anhydrous salts sodium nitrate and ammonium nitrate. Ammonium nitrate dissolves with the absorption of so much heat that the solution becomes quite cold and dew forms on the outside of the vessel.

Some salts such as anhydrous copper(II) sulphate, anhydrous sodium carbonate, anhydrous sodium

A model of a hydrated cation.

sulphate and sodium carbonate monohydrate dissolve in water with the evolution of heat.

A few salts such as sodium chloride dissolve in water without much heat change.

These observations concerning heats of solution can be explained by proposing that the dissolution of a salt involves two processes.

(1) *Ions are torn out of the crystal lattice*, and because this requires the separation of charged particles, energy will be required to bring it about.

(2) *The ions become hydrated* and this process releases energy.

The relative amounts of energy involved in these two processes will determine whether the overall process of dissolution will evolve or absorb heat. Thus, if the energy of hydration is greater than the energy needed to destroy the crystal lattice, the dissolution of the salt is accompanied by evolution of heat. However, for many salts and particularly for hydrated salts, the energy of hydration is less than the energy required to destroy the crystal lattice, and the dissolution of the salt is accompanied by absorption of heat.

For sodium chloride, the two processes involve about the same amount of energy and there is little heat change on dissolution.

Deliquescence

Some substances absorb water from the atmosphere to such an extent that they dissolve in the water absorbed, and form an *aqueous solution*.

Examples of such substances are:

$CaCl_2,6H_2O$ calcium chloride hexahydrate

$FeCl_3,6H_2O$ iron(III) chloride hexahydrate

$Cu(NO_3)_2,3H_2O$ copper(II) nitrate trihydrate

$Zn(NO_3)_2,6H_2O$ zinc nitrate hexahydrate

$NaOH$ sodium hydroxide (caustic soda).

Such substances must be stored in tightly-stoppered bottles or in desiccators.

Hygroscopic Substances

Some substances absorb some water from the atmosphere but not so much that a solution is formed. For example:

CuO copper(II) oxide

Na_2CO_3 sodium carbonate (anhydrous).

Anhydrous sodium carbonate slowly forms the monohydrate (Na_2CO_3,H_2O) on standing in air.

Efflorescence

Many hydrated salts lose water to the atmosphere. For example:

$Na_2SO_4,10H_2O$ sodium sulphate decahydrate

$Na_2CO_3,10H_2O$ sodium carbonate decahydrate.

On standing in air, crystals of sodium carbonate decahydrate change from clear glassy crystals to a white powder consisting of the monohydrate (Na_2CO_3,H_2O).

SOLUBILITY OF GASES IN LIQUIDS

The solubility of a gas in a liquid depends on:
(i) the nature of the gas;
(ii) the nature of the liquid;
(iii) the temperature;
(iv) the pressure of the gas above the solution.

The solubilities of a few gases in water are given in the table (fig. 8.6). The figures represent the number

Gas	Temperature		
	0 °C	20 °C	40 °C
hydrogen	0.22	0.018	0.016
nitrogen	0.24	0.016	0.012
oxygen	0.49	0.31	0.23
carbon dioxide	1.71	0.88	0.53
hydrogen chloride	506	442	386
ammonia	1300	710	—

Fig. 8.6. Absorption coefficients of gases in water.

of volumes of gas expressed at s.t.p. dissolved by unit volume of water under a pressure of one atmosphere *of the gas*. These figures are called *absorption coefficients*. The gases which are very soluble in water are also the gases which react with water.

Most gases can be completely driven out of solution by boiling. This is not only due to the decrease in solubility at the higher temperature, but also depends on other effects. For example, ammonia escapes slowly from a solution exposed to the atmosphere because ammonia molecules escape from the surface and diffuse away into the air. Thus the partial pressure of the ammonia above the solution is always low and equilibrium is never established between the solution

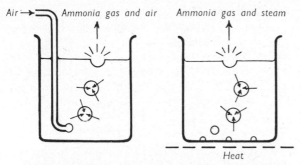

Fig. 8.7. Removal of ammonia gas from ammonia solution. Ammonia gas diffuses into the bubbles.

and the ammonia gas above it. If air is bubbled through the solution, the process is speeded up because of the greater surface of solution exposed to the air and also because the stream of air carries the gas away faster than it could diffuse.

Boiling has much the same effect. Ammonia diffuses into the bubbles of water vapour which escape from the liquid (see fig. 8.7).

Azeotropic Mixtures

Solutions which cannot be separated into their components by boiling, evaporation or distillation are called azeotropic mixtures. This name is derived from Greek and means "not affected by boiling".

If a dilute solution of hydrochloric acid is gently boiled, most of the vapour driven off consists of water, and the remaining solution becomes more concentrated. However, if concentrated hydrochloric acid is boiled gently, mostly hydrogen chloride gas is driven off and the solution becomes more dilute. No matter what the original concentration of the acid may have been, boiling will eventually leave a mixture which contains about 20% hydrogen chloride by mass. If boiling is continued, a mixture of this composition distils over unchanged and so this mixture is called an azeotropic mixture.

Ethanol and water is another common example of an azeotropic mixture. When it is distilled, a solution containing about 96% ethanol is obtained. This is sold under the name "rectified spirit". The remaining water can be removed by chemical drying agents such as calcium oxide, leaving almost dry ethanol. This is sold under the name "absolute alcohol".

STUDY QUESTIONS

1. Explain what is meant by the statement that the solubility of potassium nitrate in water at 48 °C is 80.

2. 6 g of potassium chlorate is weighed into a clean dry test tube. 20 cm³ of water is added and the mixture is heated until all of the salt has dissolved. As it cools slowly with constant stirring, crystals are observed to begin forming at 66 °C. What is the solubility of potassium chlorate in water at this temperature?

3. Plot a solubility curve for potassium nitrate by drawing the curve of best fit for the following data:

Temperature (°C)	18	33	44	57	60	67
Solubility	30	50	70	90	110	130

Use the graph to determine:
(a) the solubility at 30 °C;
(b) the temperature at which 26 g of potassium nitrate would just dissolve in 50 g of water;
(c) the temperature at which the saturated solution contains 25% potassium nitrate by mass;
(d) whether any of the sets of figures given in the data seem to be unreliable, and if so, what the figures ought to be.

4. Explain why it is difficult to purify a sample of sodium chloride by a simple process of recrystallization from water.

5. Devise an experiment to show that the solubility of calcium hydroxide in water decreases with increase of temperature. (The solubility is 0.176 at 10 °C and 0.077 at 100 °C.)

6. Devise an experiment to show that:
(a) crystalline sodium carbonate contains water in its structure;
(b) black copper(II) oxide is hygroscopic;
(c) crystalline potassium chloride is anhydrous.

7. When hydrated calcium chloride is exposed to air it forms and *aqueous* solution.
Devise an experiment to establish the truth of this statement.

8. 3.460 g of a saturated sodium chloride solution at 15 °C is evaporated to dryness. 0.914 g of solid sodium chloride is obtained. What is the solubility of sodium chloride in water at 15 °C?

Answers to Numerical Problems

2. 30.
3. (a) 45. (b) 34 °C. (c) 21 °C.
   (d) Solubility of 90 at 53 °C and not 57 °C as given.
4. 35.9.

9 : Molecular Formulae

For ionic substances, the only formula which can be written is an empirical formula, e.g. Na^+Cl^-. For covalent or molecular substances, however, both an empirical formula and a molecular formula can be written.

● *For a molecular substance, the empirical formula gives the simplest ratio of the number of atoms of each element in the substance.*

This can be calculated in exactly the same way as that described in chapter 2.

● *The molecular formula gives the actual number of atoms of each element in a molecule of the substance.*

e.g. The empirical formula of benzene is CH (i.e. one carbon atom per one hydrogen atom) and its molecular formula is C_6H_6 (i.e. six atoms of carbon and six atoms of hydrogen in each molecule of benzene).

Definition of Relative Molecular Mass:
● *The relative molecular mass of an element or a compound is the weighted mean of the masses of the molecules of the substance on the scale in which an atom of ^{12}C is taken as 12 exactly.*

The relative masses of the molecules of many covalent substances are obtained directly from the mass spectrometer. For example, when hydrogen gas is examined in a mass spectrometer, it is found that the average mass of the molecules is 2.016 on the scale $^{12}C = 12$.

Hydrogen gas must therefore be diatomic—the molecular formula being H_2 (A = 1.008 and M = 2.016). Similarly, oxygen, nitrogen and chlorine gases are diatomic—O_2, N_2 and Cl_2.

When carbon monoxide and carbon dioxide are examined the molecular masses are found to be 28.0 and 44.0 respectively. Since the relative atomic masses of the elements are: C = 12.0 and O = 16.0, the molecular formulae of these gases must be CO and CO_2 respectively. Similarly, it is found that the molecular formula of hydrogen chloride is HCl, water is H_2O and ammonia is NH_3.

Because the mass spectrometer is an expensive machine requiring skilled operation, it is desirable to know of other simpler methods for determining the molecular masses of molecular substances. The necessary theory of such a method will now be developed and then methods of determination will be outlined. It will then be shown that a knowledge of empirical formulae and molecular masses leads to molecular formulae. It will also be shown that approximate molecular masses are sufficient for this purpose.

Definition of a Mole of a Molecular Substance

In chapter 2, the definition of a mole of an element and the definition of a mole of a molecular substance were given. The latter definition is repeated here.

● *The mole is the amount of substance of specified formula, containing the same number of molecules as there are atoms in 12 g of ^{12}C.*

This definition makes it possible to obtain the mass of a mole by simply adding the unit 'g' to the relative molecular mass. For example, the molecular mass of CO is 28.0 on the ^{12}C scale. Thus:

One atom of C weighs   12.0 units
One molecule of CO weighs 28.0 units

If there are N atoms of C and N molecules of CO, where N is 6.02×10^{23}, then the total masses will still be in the same ratio as the masses of the single particles. Since the mass of C is now 12.0 g, the mass of CO must be 28.0 g.

● *Thus: 1 mole of CO has a mass = 28 g.*

Similarly, the mole amount of any element or compound can be obtained from the relative molecular mass.

1 mole of H_2 has a mass	= 2.0 g
1 mole of O_2 has a mass	= 32.0 g
1 mole of N_2 has a mass	= 28.0 g
1 mole of H_2O has a mass	= 18.0 g
1 mole of NH_3 has a mass	= 17.0 g
1 mole of HCl has a mass	= 36.5 g

It is useful to calculate the volume occupied by the mole amounts of some gases at s.t.p. This can be done if the density is known (from physical measurements) and if the molecular mass is known (from mass spectrometric measurements).

EXAMPLE: for hydrogen—

density at s.t.p. $= 0.089\,88$ g dm^{-3}

$M = 2.016$

Now: density $= \dfrac{\text{mass}}{\text{volume}}$

$\therefore$ volume $= \dfrac{\text{mass}}{\text{density}}$

Therefore the volume at s.t.p. occupied by one mole (2.016 g) of hydrogen:

$= \dfrac{2.016}{0.089\,88} = 22.43$ dm^3 at s.t.p.

The results obtained with some other gases are listed in the table (see fig. 9.1).

Gas	Density at s.t.p. (g dm^{-3})	M	Volume occupied by one mole at s.t.p. (dm^3)
hydrogen	0.08988	2.016	22.43
oxygen	1.429	32.00	22.39
nitrogen	1.251	28.01	22.39
argon	1.784	39.94	22.39
carbon dioxide	1.977	44.01	22.26
hydrogen chloride	1.639	36.46	22.25

Fig. 9.1.

It can be seen that one mole of any gas occupies approximately the same volume at s.t.p. as one mole of any other gas also at s.t.p. This volume is given the name *molar volume of a gas*:

● The **molar** *volume of any gas is the volume occupied by one mole of the gas at a stated temperature and pressure.*

The value 22.4 dm^3 or 22 400 cm^3 at s.t.p. is the value generally used but the numerical value would be different at other temperatures and pressures (e.g. 30.6 dm^3 at 100 °C and one atmosphere pressure). However, it is usual to compare gases at s.t.p. so the molar volume is usually used only for this temperature and pressure.

The value 22.4 dm^3 at s.t.p. is an approximation but it is usually correct to within 2 to 3%.

Hence a useful rule for calculating relative molecular masses can be stated:

● *One mole of any gas occupies approximately 22.4 dm^3 at s.t.p.*

The basis for this rule is summed up in the diagram (fig. 9.2). The value of N is given by the Avogadro constant, which is:

● $N = (6.022\,52 \pm 0.000\,28) \times 10^{23}$ mol^{-1}

Whenever this number of molecules is present, the masses of the gases are simply their relative molecular masses expressed as masses in gram and their volumes are always approximately 22.4 dm^3 at s.t.p.

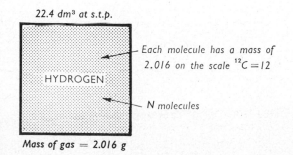

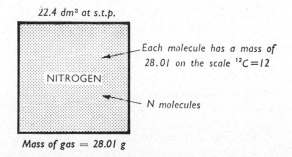

Fig. 9.2. Molar volumes at s.t.p. of hydrogen and nitrogen.

Measurements needed for the Calculation of Molecular Mass

To calculate the molecular mass of a gas it is necessary to find the mass of 22.4 dm³ of the gas at s.t.p. To find this it is necessary to find the mass and volume at s.t.p. of a sample of the gas. However, it would be difficult to make measurements at exactly 0 °C and 760 mmHg. So in practice, four measurements are made on a sample of the gas:

    (i) mass (m, in gram);
    (ii) volume (v, in cm³);
    (iii) temperature (T, in Kelvin);
    (iv) pressure (p, in mmHg).

The calculation may be made in two steps:
(a) Calculate the volume of the sample of gas at s.t.p. (v_0, in cm³).
(b) Calculate the mass in gram of 22.4 dm³ of the gas at s.t.p.
(See numerical illustration 1.)

The densities of many gases at s.t.p. are listed in tables of Physical and Chemical Constants. Several books of such tables are published and so the relative molecular mass of a gas could be calculated from these figures. In this way, the careful experimental work of others can be used to obtain a reliable result. The densities listed in the table (fig. 9.1) could be used.

If the density is given in g dm⁻³ at s.t.p., then the mass of 1 dm³ of the gas at s.t.p. is known. Hence the molecular mass is given by the rule:

    M = density × 22.4
        (where the density is in g dm⁻³)
or M = density × 22 400
        (where the density is in g cm⁻³).
(See numerical illustration 2.)

It should be realized that the results calculated by such methods as described here are of limited accuracy because the figure 22.4 dm³ or 22 400 cm³ is the average for many gases. Therefore the numerical value of the result should not be expressed to a very high degree of accuracy.

Determination of Molecular Formulae from Approximate Values of Relative Molecular Masses

A numerical illustration will be used to show how an approximate value of the relative molecular mass can be used to derive the molecular formula of a compound. (See numerical illustration 3.)

NUMERICAL

1

It is found by experiment that 0.333 g of oxygen gas has a volume of 250 cm³ at 17 °C and 753 mmHg. Calculate the relative molecular mass of the gas.
(i) Volume at s.t.p.

$$v_0 = \frac{(250)\ (273)\ (753)}{(290)\ (760)}$$

$$= 233 \text{ cm}^3.$$

(ii) Relative Molecular Mass
  0.333 g of oxygen gas occupies 233 cm³ at s.t.p.
  M g of oxygen gas occupies 22 400 cm³ at s.t.p.

$$\therefore \text{ M} = \frac{(0.333)\ (22\ 400)}{(233)}$$

$$= 32.0.$$

Hence the relative molecular mass of the gas is 32.0.

2

Reference tables give the density of the noble gas krypton as 3.743 g dm⁻³ at s.t.p.
  $\therefore$ M of krypton = (3.743) (22.4)
                    = 83.8.

3

The following figures refer to ethyl acetate (a covalent compound containing carbon, hydrogen and oxygen).

The composition by mass of ethyl acetate as determined by analysis is:
54.5% carbon, 9.2% hydrogen, 36.3% oxygen.

The relative molecular mass of ethyl acetate as determined experimentally is approximately 88.

Atomic Masses (from tables):
C = 12.01, H = 1.008, O = 16.00.

Calculate the empirical formula, the molecular formula and the accurate molecular mass of ethyl acetate.

Determination of Molecular Formulae from Accurate Values of Relative Molecular Masses

Mass spectrometers can be used to obtain very accurate values of relative molecular masses. In these circumstances, it is possible to determine a molecular

ILLUSTRATIONS

3 (*continued*)

(i) *Empirical Formula*:

The ratio of the number of mole of atoms of carbon : hydrogen : oxygen

$$= \frac{54.5}{12.01} : \frac{9.2}{1.008} : \frac{36.3}{16.00}$$
$$= 4.54 : 9.13 : 2.27$$
$$= \frac{4.54}{2.27} : \frac{9.13}{2.27} : \frac{2.27}{2.27}$$
$$= 2.00 : 4.02 : 1.00$$
$$= 2 : 4 : 1.$$

The empirical formula of ethyl acetate is C_2H_4O.

(ii) *Molecular Formula*:

This must be some multiple of the empirical formula, i.e. $(C_2H_4O)_n$, where n is a whole number. (n must be a whole number because it is really a number of atoms.)
$\therefore$ M = $(2 \times 12.01 + 4 \times 1.008 + 16.00)n = 44.05n$.
But the molecular mass is approximately 88

$$\therefore 44.05n \approx 88$$
$$\therefore n \approx \frac{88}{44.05}$$
$$\therefore n = 2 \text{ (n must be a whole number).}$$

The molecular formula is therefore $C_4H_8O_2$.

An accurate value of the molecular mass can be calculated knowing the molecular formula:
$\because$ n = 2 and M = 44.05n.
$\therefore$ M = 88.10.

Alternatively the molecular mass can be calculated by adding up the atomic masses:
$\therefore$ M = $(4 \times 12.01 + 8 \times 1.008 + 2 \times 16.00)$
= 88.10.

formula without the need to determine the empirical formula first. Thus, knowing the accurate molecular mass and accurate atomic masses of the elements, the number of atoms of the various elements present in the molecule can be inferred.

Covalencies in Molecular Formulae

From the molecular formulae of the common covalent compounds of the non-metals it is possible to draw up a table of covalencies (see fig. 9.3). The table can then be used as an aid in writing the empirical formulae of covalent compounds.

Element and symbol		Covalency
hydrogen	H	1 e.g. HCl etc. (see below)
chlorine	Cl	1 e.g. HCl
oxygen	O	2 e.g. H_2O
carbon	C	4 e.g. CH_4, CO_2
sulphur	S	2, 4 or 6 e.g. H_2S, SO_2 and SO_3
phosphorus	P	3 or 5 e.g. PCl_3 and PCl_5 also P_2O_3 and P_2O_5
silicon	Si	4 e.g. SiO_2
nitrogen	N	3 e.g. NH_3

Some of these elements also exhibit other covalencies but these are less common and have been omitted from this list.

Fig. 9.3. List of Common Covalencies.

Experimental Determination of the Relative Molecular Mass of a Gas

A flask of known volume is evacuated and weighed, filled with the gas at a measured temperature and pressure and then reweighed. Thus the mass, volume, temperature and pressure of the sample of gas have been determined.

This method of determining molecular masses of gases is called *Regnault's Method* or the *Globe Method* and it can give very accurate results if fairly elaborate precautions are taken.

A simple determination which may be carried out, but which does not yield very accurate results will now be described.

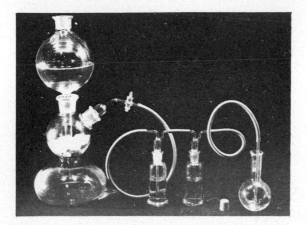

Fig. 9.4. Filling a flask with dry carbon dioxide. The gas is generated in the Kipp's apparatus (left) by reaction between marble chips and hydrochloric acid. The gas is bubbled through water to dissolve acid spray and then dried by concentrated sulphuric acid. These liquids are contained in the Dreschel bottles (centre). The gas is collected in the flask making sure that the delivery tube is kept near the bottom.

(i) Weigh a stoppered flask of about 200 to 400 cm³ capacity. Because the flask contains air, the mass determined is really the mass of the stoppered flask plus the air it contains.

(ii) Fill the flask with dry carbon dioxide from a Kipp's apparatus (see fig. 9.4).

(iii) Weigh the stoppered flask full of carbon dioxide.

(iv) Repeat steps (ii) and (iii) until constant mass is recorded. This ensures that all the air has been displaced from the flask.

(v) Hold a thermometer in the carbon dioxide in the flask and record the temperature.

(vi) Record the atmospheric pressure from a barometer. The carbon dioxide in the flask will be at this pressure.

(vii) Fill the flask with water to the level of the bottom of the stopper. Pour this water into a graduated cylinder and record the volume. The volume of air in step (i) and of carbon dioxide in step (iv) in the flask will be equal to this.

NUMERICAL ILLUSTRATION

Mass of stoppered flask plus air = 69.887 g
Mass of stoppered flask plus CO_2 = 70.137 g
Volume of gas in flask = 404 cm³
Temperature = 19 °C
Pressure = 770 mmHg

$$\therefore \text{ Volume at s.t.p. of gas in flask} = \frac{(404)\,(273)\,(770)}{(292)\,(760)}$$
$$= 383 \text{ cm}^3$$

The density of air at s.t.p. = 0.001 293 g cm⁻³ (this figure is taken from tables of gas densities).

$\therefore$ Mass of air in flask = 383 × 0.001 293
= 0.495 g

$\therefore$ Mass of stoppered flask = 69.392 g
$\therefore$ Mass of CO_2 = 0.745 g

$\begin{cases} 0.745 \text{ gram of } CO_2 \text{ occupies } 383 \text{ cm}^3 \text{ at s.t.p.} \\ M \text{ gram of } CO_2 \text{ occupies } 22\,400 \text{ cm}^3 \text{ at s.t.p.} \end{cases}$

$\therefore M = \dfrac{(0.745)\,(22\,400)}{383} = 43.6$ (approximately).

$\therefore$ The relative molecular mass of carbon dioxide is about 43.6.

If the expression for M is worked out using four figure logs, the result is calculated as 43.57 but the reliability of the results, e.g., the volume of gas in the flask, clearly do not justify such accuracy in the relative molecular mass.

STUDY QUESTIONS

The following relative atomic masses should be used where necessary for these problems:

C = 12.0	H = 1.0	O = 16.0
Cl = 35.5	N = 14.0	S = 32.1

1. Mass spectrometric analysis gives the molecular masses of the following substances:
(a) two oxides of nitrogen have M's of 30.0 and 46.0;
(b) two hydrides of carbon have M's of 16.0 and 30.0;
(c) two oxides of sulphur have M's of 64.1 and 80.1.
Suggest possible molecular formulae for these compounds.

2. Write down the valency of X in each of the following covalent compounds:
(a) XCl   (c) XO   (e) H_2X (g) HX
(b) XO_2 (d) XH_3 (f) Cl_2X (h) CX_2

3. Calculate the number of mole of each substance, in each of the masses given:
(a) 36.0 g of water (H_2O);
(b) 4.00 g of carbon dioxide (CO_2);
(c) 51.0 g of ammonia (NH_3).

4. What is the mass in gram of:
(a) 0.500 mole of carbon monoxide (CO);
(b) 2.00 mole of sulphur dioxide (SO_2);
(c) 0.100 mole of hydrogen chloride (HCl)?

5. What is the volume in dm^3 at s.t.p. of:
(a) 2.00 mole of carbon dioxide (CO_2);
(b) 22.0 g of carbon dioxide (CO_2);
(c) 73.0 g of hydrogen chloride (HCl);
(d) 1.00 g of carbon monoxide (CO)?

6. What is the mass in gramme of each of the following volumes of gases, all measured at s.t.p.:
(a) 11.2 dm^3 of carbon monoxide (CO);
(b) 2.80 dm^3 of sulphur dioxide (SO_2);
(c) 500 cm^3 of carbon dioxide (CO_2);
(d) 1.00 cm^3 of hydrogen (H_2)?

7. What is the volume at 27 °C and 380 mmHg of:
(a) 2.00 mole of sulphur dioxide (SO_2);
(b) 10.0 gram of nitrogen monoxide (NO)?

8. What is the mass in gramme of:
(a) 1.00 dm^3 of carbon dioxide measured at 17 °C and 770 mmHg;
(b) 500 cm^3 of methane (CH_4) measured at 15 °C and 765 mmHg?

9. Calculate the molecular masses of the following gases from their molecular formulae, and then using the densities given, calculate the molar volume at s.t.p. for each gas:
(a) chlorine (Cl_2); 3.214 g dm^{-3}.
(b) carbon monoxide (CO); 1.250 g dm^{-3}.
(c) nitrogen monoxide (NO); 1.340 g dm^{-3}.

10. The densities of the following gases at s.t.p. are:
(a) dinitrogen monoxide; 1.977 g dm^{-3}.
(b) ethylene; 0.001 26 g cm^{-3}.
(c) methane; 0.000 717 g cm^{-3}.
(d) butane; 2.67 g dm^{-3}.
Calculate the molecular mass of each gas.

11. 1.000 g of a gas occupies 523 cm^3 at 20 °C and 764 mmHg. Calculate the molecular mass of the gas.

12. A flask has a capacity of 200 cm^3 at room temperature and pressure. When it is completely evacuated it weighs 84.845 g. When it is filled with nitrogen at room temperature (17 °C) and pressure (770 mmHg) it weighs 85.084 g. Calculate the molecular mass of nitrogen.

13. 0.222 g of ether ($C_4H_{10}O$) was vapourized and the vapour occupied 97.4 cm^3 at 100 °C and 728 mmHg. Calculate the molecular mass of ether from these figures. Also calculate the molecular mass from the molecular formula.

14. 0.124 g of a volatile liquid was vapourized and the vapour occupied 71.8 cm^3 at 80 °C and 753 mmHg. Calculate the molecular mass of the vapour of the volatile liquid.

15. 0.128 g of ethanol (C_2H_6O) was vapourized and the vapour occupied 89.6 cm^3 at 127 °C and 767 mmHg. Calculate the molecular mass of the vapour of ethanol from these figures. Also calculate the molecular mass from the molecular formula.

16. Aluminium chloride has the following composition by mass: 20.2% aluminium, 79.8% chlorine. When 0.571 g of the chloride was vapourized at 200 °C, the volume of the vapour was 83.2 cm^3 at 750 mmHg. For aluminium chloride, calculate:
(a) the empirical formula;
(b) the molecular formula;
(c) an accurate value for the molecular mass.

17. Methyl formate has the following composition by mass: 40.0% carbon, 6.7% hydrogen, 53.3% oxygen. 0.148 g of the substance was vapourized and the vapour occupied 74.6 cm^3 at 95 °C and 756 mmHg. What is the molecular formula of methyl formate?

Answers to Numerical Problems

1. (a) NO and NO_2.
   (b) CH_4 and C_2H_6.
   (c) SO_2 and SO_3.
2. (a) 1.
   (b) 4.
   (c) 2.
   (d) 3.
   (e) 2.
   (f) 2.
   (g) 1.
   (h) 2.
3. (a) 2.00.
   (b) 0.091.
   (c) 3.00.
4. (a) 14.0 g.
   (b) 128 g.
   (c) 3.65 g.
5. (a) 44.8 dm^3.
   (b) 11.2 dm^3.
   (c) 44.8 dm^3.
   (d) 0.80 dm^3.
6. (a) 14.0 g.
   (b) 8.01 g.
   (c) 0.982 g.

(d) 0.000 089 g.
7. (a) 98.5 dm^3.
   (b) 16.4 dm^3.
8. (a) 1.87 g.
   (b) 0.341 g.
9. (a) M = 71.0
      molar volume 22.1 dm^3.
   (b) M = 28.0
      molar volume 22.4 dm^3.
   (c) M — 30.0
      molar volume 22.4 dm^3.
10. (a) 44.3.
    (b) 28.2.
    (c) 16.1.
    (d) 59.8.
11. 45.7.
12. 28.1.
13. 73, 74.0.
14. 51.
15. 47, 46.0.
16. (a) $AlCl_3$.
    (b) Al_2Cl_6.
    (c) 267.0.
17. $C_2H_4O_2$.

10 : Chemical Equations and Calculations

A chemical equation is the most convenient way of representing a chemical reaction. Equations became widely used when the atomic theory of matter became generally accepted because they are the best way of representing changes which take place between the small discrete particles which compose all matter. Some equations have been used so far in this book because an explanation without them would have been cumbersome and involved.

WRITING AND BALANCING EQUATIONS

Before an equation can be written, the reaction must be studied. It is most important to realize that an equation is written to summarize observed data and that the equation must conform to the changes which are observed to take place. It is not a correct approach to predict what should happen by writing an equation.

For example, an equation can be written showing $NaCl$ and H_2O reacting to form $NaOH$ and HCl, but no such reaction occurs. The oceans have contained $NaCl$ dissolved in H_2O for millions of years and they have not yet turned to caustic soda solution and released clouds of hydrogen chloride gas. In fact the reverse reaction occurs:

$$HCl + NaOH \rightarrow NaCl + H_2O$$

Example

The reaction between sodium chloride and silver nitrate.

1. OBSERVATION

Both of the reactants are white crystalline solids which are soluble in water giving colourless solutions. If a solution of sodium chloride is mixed with a solution of silver nitrate, a white curdy precipitate forms and a colourless solution remains.

2. EXPLANATION

What is the white precipitate? What (if anything) is contained in the remaining colourless solution? The answers to these questions must be decided by experiment before the equation can be written. It can be shown that the precipitate is silver chloride and that the solution contains sodium nitrate.

3. WRITING THE EQUATION

The reaction can be represented verbally:

sodium chloride + silver nitrate
→ silver chloride + sodium nitrate.

The formulae of each of these substances can be written by using the valencies of the ions concerned (see fig. 10.1).

The equation may be rewritten by replacing the name of each substance by its formula:

$$NaCl + AgNO_3 \rightarrow AgCl + NaNO_3$$

The equation must now be checked to see it if is balanced, i.e., equal numbers of each ion, atom or group of atoms must appear on each side of the equation. In this equation, *one* of each of the ions is shown on the left hand side. *One* of each of these same ions appears on the right hand side and the equation is therefore balanced as it stands. The steps taken in the final balancing of an equation are illustrated in greater detail with the further examples.

Further Examples

● *Example (a): Barium chloride solution ($BaCl_2$) and sulphuric acid (H_2SO_4) react to form a precipitate, which can be shown to be barium sulphate ($BaSO_4$), and a solution, which can be shown to contain hydrochloric acid (HCl).*

(*i*) *Verbal Equation:*
barium chloride + hydrogen sulphate
→ barium sulphate + hydrogen chloride.

(*ii*) *Unbalanced Equation:*
$$BaCl_2 + H_2SO_4 \rightarrow BaSO_4 + HCl$$
The equation is unbalanced because it shows two Cl's and two H's on the left hand side and only one of each on the right hand side.

(*iii*) *Balanced Equation:*
$$BaCl_2 + H_2SO_4 \rightarrow BaSO_4 + 2HCl$$
The equation has been balanced by placing the number 2 in front of the formula HCl. It would be incorrect to alter the formula, say to H_2Cl_2, because hydrochloric acid is produced and H_2Cl_2 is not its formula. To alter the formula to something else would mean that the equation no longer represents the reaction.

● *Example (b): Caustic soda ($NaOH$) and sulphuric acid (H_2SO_4) can be made to react to form sodium sulphate (Na_2SO_4) and water (H_2O).*

(*i*) *Verbal Equation:*
sodium hydroxide + hydrogen sulphate
→ sodium sulphate + water.

(*ii*) *Unbalanced Equation:*
$$NaOH + H_2SO_4 \rightarrow Na_2SO_4 + H_2O$$

(*iii*) *Balanced Equation:*
$$2NaOH + H_2SO_4 \rightarrow Na_2SO_4 + 2H_2O$$

● *Example (c): Zinc (Zn) displaces hydrogen gas (H₂) from dilute sulphuric acid (H₂SO₄) forming zinc sulphate (ZnSO₄) solution.*

(*i*) *Verbal Equation:*
zinc + hydrogen sulphate
$$\rightarrow \text{zinc sulphate} + \text{hydrogen.}$$

(*ii*) *Unbalanced Equation:*
$$Zn + H_2SO_4 \rightarrow ZnSO_4 + H_2$$

(*iii*) *Balanced Equation:*
This is the same as before, because the equation is found to be balanced as it stands. However, this must always be checked *after* the formulae have been written in.
$$Zn + H_2SO_4 \rightarrow ZnSO_4 + H_2.$$
It should be noted here that the formula of hydrogen gas is H_2 and not H. Whenever a molecular substance is included in an equation the molecular formula should be written. Because substances which are gases under ordinary conditions are composed of molecules, this situation will be most often met with in reactions involving gases.

● *Example (d): Sodium carbonate (Na₂CO₃) reacts with hydrochloric acid (HCl) to form sodium chloride (NaCl), water (H₂O) and carbon dioxide gas (CO₂).*

(*i*) *Verbal Equation:*
sodium carbonate + hydrogen chloride
$$\rightarrow \text{sodium chloride} + \text{water} + \text{carbon dioxide.}$$

(*ii*) *Unbalanced Equation:*
$$Na_2CO_3 + HCl \rightarrow NaCl + H_2O + CO_2$$

(*iii*) *Balanced Equation:*
$$Na_2CO_3 + 2HCl \rightarrow 2NaCl + H_2O + CO_2$$

● *Example (e): Hydrogen gas (H₂) reacts with chlorine gas (Cl₂) to form hydrogen chloride gas (HCl).*

(*i*) *Verbal Equation:*
hydrogen + chlorine → hydrogen chloride.

(*ii*) *Unbalanced Equation:*
$$H_2 + Cl_2 \rightarrow HCl$$

(*iii*) *Balanced Equation:*
$$H_2 + Cl_2 \rightarrow 2HCl$$

APPLICATIONS OF EQUATIONS
1. Mass Relationships

The relative masses of each of the reactants and products of a chemical reaction can be calculated by using the atomic masses of each of the elements to find the various mole amounts. This will be illustrated for *the reaction of sodium chloride with silver nitrate* (see fig. 10.2).

These masses can be introduced into the equation as follows:

NaCl	+	AgNO₃	→	AgCl	+	NaNO₃
1 mole		1 mole		1 mole		1 mole
58.5 g		169.9 g		143.4 g		85.0 g

For the reaction between caustic soda and sulphuric acid the relative masses are:

2NaOH	+	H₂SO₄	→	Na₂SO₄	+	2H₂O
2 mole		1 mole		1 mole		2 mole
2 × 40.0 g		98.1 g		142.1 g		2 × 18.0 g

Salt	Ions		Formula		Atomic Masses	Mass of One Mole
sodium chloride	Na⁺	Cl⁻	NaCl	. . . .	(23.0) + (35.5)	58.5 g
silver nitrate	Ag⁺	NO₃⁻	AgNO₃	. . . .	(107.9) + (14.0) + (3 × 16.0)	169.9 g
silver chloride	Ag⁺	Cl⁻	AgCl	. . . .	(107.9 + (35.5)	143.4 g
sodium nitrate	Na⁺	NO₃⁻	NaNO₃	. . . .	(23.0) + (14.0) + (3 × 16.0)	85.0 g

Fig. 10.1. Empirical formulae.

Fig 10.2. Mole amounts.

2. Mass—Volume Relationships

One mole of any gas occupies 22.4 dm³ at s.t.p. Thus, where a gas is involved in a reaction, either as a reactant or a product it is possible to give the volume of gas as well as its mass. *For the reaction between zinc and sulphuric acid the relative masses and volume are*:

$$Zn \quad + \quad H_2SO_4 \quad \rightarrow \quad ZnSO_4 \quad + \quad H_2$$

Zn	H₂SO₄	ZnSO₄	H₂
1 mole	1 mole	1 mole	1 mole
65.4 g	98.1 g	161.5 g	2.0 g
			$\left\{ \begin{array}{l} 22.4\ dm^3 \\ at\ s.t.p. \end{array} \right.$

For the reaction between sodium carbonate and hydrochloric acid:

$$Na_2CO_3 \quad + \quad 2HCl \quad \rightarrow \quad 2NaCl \quad + \quad H_2O \quad + \quad CO_2$$

Na₂CO₃	2HCl	2NaCl	H₂O	CO₂
1 mole	2 mole	2 mole	1 mole	1 mole
106.0 g	2 × 36.5 g	2 × 58.5 g	18.0 g	44.0 g $\left\{ \begin{array}{l} 22.4\ dm^3 \\ at\ s.t.p. \end{array} \right.$

3. Volume Relationships

Where a reaction involves only gases, it is often more useful to calculate the relative volumes, rather than use the molar volume at s.t.p. This will be illustrated for *the reaction between hydrogen gas and chlorine gas*:

$$H_2 \quad + \quad Cl_2 \quad \rightarrow \quad 2HCl$$

H₂	Cl₂	2HCl
1 mole	1 mole	2 mole
1 molar volume	1 molar volume	2 molar volumes

Because the molar volume at s.t.p. is approximately 22.4 dm³ for all gases, the relationship can be expressed:

$$1 \text{ volume of } H_2 + 1 \text{ volume of } Cl_2$$
$$\rightarrow 2 \text{ volumes of HCl}$$

The volume relationship applies only if each volume is expressed at the same temperature and pressure.

The combustion of carbon monoxide in oxygen to form carbon dioxide:

$$2CO \quad + \quad O_2 \quad \rightarrow \quad 2CO_2$$

2CO	O₂	2CO₂
2 mole	1 mole	2 mole
2 volumes	1 volume	2 volumes

Reactions involving the combustion of measured quantities of gases are usually carried out in a graduated tube inverted over mercury. A pair of metallic wires are sealed into the top of the tube and are used to create an electric spark to ignite the gas mixture. Such a tube is called a eudiometer (see fig. 10.3).

Calculations from Mass Relationships

● *Example (a): Calculate the mass of* AgCl *precipitated, when excess silver nitrate solution is added to a solution containing 1.00 g of sodium chloride.*

$$NaCl \quad + \quad AgNO_3 \quad \rightarrow \quad AgCl \quad + \quad NaNO_3$$

NaCl	AgNO₃	AgCl	NaNO₃
1 mole	1 mole	1 mole	1 mole

It is not necessary to calculate the mole amounts of *both* reactants and *both* products as the question involves only two substances.

$$\begin{array}{l} 1 \text{ mole of NaCl} \rightarrow 1 \text{ mole of AgCl} \\ \left\{ \begin{array}{l} 58.5 \text{ g of NaCl} \rightarrow 143.4 \text{ g of AgCl} \\ 1.00 \text{ g of NaCl} \rightarrow w \text{ g of AgCl} \end{array} \right. \end{array}$$

$$\therefore w = \frac{1.00 \times 143.4}{58.5} = 2.451$$

The mass of precipitated silver chloride would be 2.45 g.

● *Example (b): Calculate the mass of barium sulphate precipitated when a solution containing 1.20 g of barium chloride is treated with excess sulphuric acid.*

$$BaCl_2 \quad + \quad H_2SO_4 \quad \rightarrow \quad BaSO_4 \quad + \quad 2HCl$$

BaCl₂	H₂SO₄	BaSO₄	2HCl
1 mole		1 mole	
$\left\{ \begin{array}{l} 208.3 \text{ g} \\ 1.20 \text{ g} \end{array} \right.$		233.4 g w g	

$$\therefore w = \frac{1.20 \times 233.4}{208.3} = 1.345$$

The mass of barium sulphate would be 1.35 g.

● *Example (c): What mass of each of the reactants would be necessary to prepare 1.00 g of sodium sulphate from caustic soda and sulphuric acid?*

$$2NaOH \quad + \quad H_2SO_4 \quad \rightarrow \quad Na_2SO_4 \quad + \quad 2H_2O$$

2NaOH	H₂SO₄	Na₂SO₄	2H₂O
2 mole		1 mole	
$\left\{ \begin{array}{l} 2 \times 40.0 \text{ g} \\ w \text{ g} \end{array} \right.$		142.1 g 1.00 g	

$$w = \frac{1.00 \times 2 \times 40.0}{142.1} = 0.5630$$

The mass of caustic soda would be 0.563 g.

$$2NaOH \quad + \quad H_2SO_4 \quad \rightarrow \quad Na_2SO_4 \quad + \quad 2H_2O$$

2NaOH	H₂SO₄	Na₂SO₄	2H₂O
	1 mole	1 mole	
	$\left\{ \begin{array}{l} 98.1 \text{ g} \\ w \text{ g} \end{array} \right.$	142.1 g 1.00 g	

$$w = \frac{1.00 \times 98.1}{142.1} = 0.6904$$

The mass of sulphuric acid would be 0.690 g.

Note that this question is really two problems. Each quantity must be calculated separately.

NUMERICAL ILLUSTRATIONS

Calculations from Mass-Volume Relationships

● *Example: Calculate the volume of carbon dioxide formed, (i) at s.t.p., (ii) at 20 °C and 770 mmHg, when 1.00 g of sodium carbonate is treated with excess hydrochloric acid.*

$$Na_2CO_3 \;+\; 2HCl \;\rightarrow\; 2NaCl \;+\; H_2O \;+\; CO_2$$

1 mole   1 mole

$\begin{cases} 106.0\ g & 22\,400\ cm^3\ \text{at s.t.p.} \\ 1.00\ g & v\ cm^3\ \text{at s.t.p.} \end{cases}$

$$v = \frac{1.00 \times 22\,400}{106.0} = 211.3.$$

The volume at s.t.p. would be 211 cm³.

The volume at 20 °C and 770 mmHg can then be calculated:

$$= \frac{211 \times 293 \times 760}{273 \times 770} = 224.$$

The volume at 20 °C and 770 mmHg would be 224 cm³.

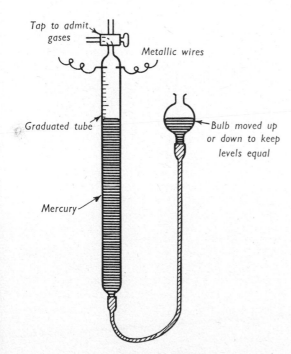

Tap to admit gases

Metallic wires

Graduated tube

Bulb moved up or down to keep levels equal

Mercury

Fig. 10.3. A eudiometer.

Calculations from Volume Relationships

● *Example (a): Calculate the volume of oxygen necessary to react with 20 cm³ of carbon monoxide. All volumes are measured at the same temperature and pressure.*

$$2CO \;+\; O_2 \;\rightarrow\; 2CO_2$$

2 mole   1 mole

$\begin{cases} 2\ \text{volumes} & 1\ \text{volume} \\ 20\ cm^3 & v\ cm^3 \end{cases}$

$$v = \frac{1 \times 20}{2} = 10$$

The volume of oxygen would be 10 cm³ at the temperature and pressure at which the volume of the carbon monoxide was measured.

● *Example (b): Calculate the volume of gas remaining after 10 cm³ of carbon monoxide is mixed with 40 cm³ of oxygen and ignited. All volumes are measured at room temperature and pressure.*

To solve this problem it must be appreciated that there is more oxygen present than is needed for complete combustion of the carbon monoxide. The steps in the calculation are to find:
   (i) the volume of carbon dioxide formed;
   (ii) the volume of oxygen remaining unused.
The final volume is the sum of these two volumes.

$$2CO \;+\; O_2 \;\rightarrow\; 2CO_2$$

2 mole   2 mole

$\begin{cases} 2\ \text{volumes} & 2\ \text{volumes} \\ 10\ cm^3 & v\ cm^3 \end{cases}$

$$\therefore v = \frac{2 \times 10}{2} = 10$$

The volume of carbon dioxide would be 10 cm³ at room temperature and pressure.

$$2CO \;+\; O_2 \;\rightarrow\; 2CO_2$$

2 mole   1 mole

$\begin{cases} 2\ \text{volumes} & 1\ \text{volume} \\ 10\ cm^3 & v\ cm^3 \end{cases}$

$$\therefore v = \frac{1 \times 10}{2} = 5$$

The volume of oxygen used would be 5 cm³ at room temperature and pressure. Therefore 35 cm³ of oxygen would remain unused.

At the end of the experiment, the gas mixture would consist of the carbon dioxide formed (10 cm³) and the oxygen left over (35 cm³). The total volume of gas would be 45 cm³ at room temperature and pressure.

The Information given by Equations

The equations written in this chapter are sometimes called stoichiometric equations. Such equations are written after a reaction has been studied and give the following information:

(a) The substances present before the reaction (reactants) and the substances formed during the reaction (products).

(b) The relative amounts of each reactant and product. This can be expressed as the relative numbers of mole. The relative volumes (of gases only) or the relative masses can be found by using the molar volume of gases or by using mole amounts.

Information not given by Equations

(a) The physical state of the reactants and products. Thus, an equation does not show whether the substances are solid, liquid, gas or whether in solution, and it does not show whether the substances are ionic or molecular. In some circumstances, particularly where heats of reaction are being considered, it is necessary to specify the physical state of the reactants and products. The following symbols are then used:

    g gaseous
    l liquid
    s solid
    aq aqueous solution.

Example: $2H_2(g) + O_2(g) \rightarrow 2H_2O(l)$

This convention is also useful to show that one of the products is a solid precipitate or a gas, e.g.
$$NaCl + AgNO_3 \rightarrow AgCl(s) + NaNO_3$$
$$Na_2CO_3 + 2HCl \rightarrow 2NaCl + H_2O + CO_2(g)$$

(b) The conditions necessary for the reaction. The equation does not show the temperature, pressure of gases, concentration of solutions or whether a catalyst is used.

(c) The rate of the reaction. For example, magnesium and iron both dissolve in dilute sulphuric acid:
$$Mg + H_2SO_4 \rightarrow MgSO_4 + H_2(g)$$
$$Fe + H_2SO_4 \rightarrow FeSO_4 + H_2(g)$$
However, the reaction is much faster with magnesium than with iron (see fig. 10.4). The equations give no information concerning this.

Fig. 10.4. Metals dissolving in dilute sulphuric acid showing rapid effervescence with magnesium (left) and slow effervescence with iron (right).

(d) The extent of the reaction. Not all reactions are complete. For example, if dilute hydrochloric acid is added to a solution of silver nitrate and also to a solution of lead nitrate, insoluble chlorides are precipitated:
$$AgNO_3 + HCl \rightarrow AgCl(s) + HNO_3$$
$$Pb(NO_3)_2 + 2HCl \rightarrow PbCl_2(s) + 2HNO_3$$

However, the reaction is much more nearly complete for silver chloride than for lead chloride. The equations give no information concerning the extent or completeness of precipitation. Even a reversible sign, $\rightleftharpoons$, instead of an arrow does not tell the extent of a reaction although it does emphasize that the reaction is not complete.

(e) The mechanism of the reaction. Reaction mechanisms are often quite complicated. For example, hydrogen gas and chlorine gas do not react when mixed in complete darkness. However, when light falls on the mixture reaction takes place.

● *Caution: an explosion will result if the mixed gases are exposed to direct sunlight.*

It seems that the light causes Cl_2 molecules to split into separate Cl atoms, which then react with H_2 molecules.

$$\text{light}$$
$$Cl_2 \rightarrow 2Cl \qquad \text{(i)}$$
$$H_2 + Cl \rightarrow HCl + H \qquad \text{(ii)}$$
$$Cl_2 + H \rightarrow HCl + Cl \qquad \text{(iii)}$$

Then steps (ii) and (iii) are repeated until the chain of reactions is stopped. Some of the reactions which may stop the chain reaction are:

$$H + Cl \rightarrow HCl$$
$$\text{or:} H + H \rightarrow H_2$$
$$\text{or:} Cl + Cl \rightarrow Cl_2.$$

Thus the stoichiometric equation,
$$H_2 + Cl_2 \rightarrow 2HCl,$$
shows only the reactants and products. In fact it appears that H_2 molecules do not combine directly with Cl_2 molecules.

Summing Up

(a) WRITING AN EQUATION:

(i) Before an equation can be written, the reactants and products must be known.

(ii) The equation is first set out verbally using chemical names although with practice this may be done mentally instead of being written out in full.

(iii) The formula of each of the substances is then written down.

(iv) Finally, steps must be taken to balance the equation.

(b) MEANING OF AN EQUATION:

(i) An abbreviated version of the verbal equation.

(ii) The stoichiometric relationships between the reactants and products.

STUDY QUESTIONS

The following relative atomic masses should be used where necessary for these problems:

Ag = 107.9.	Cu = 63.5.	N = 14.0.
Ba = 137.3.	Fe = 55.8.	Na = 23.0.
C = 12.0.	H = 1.0.	O = 16.0.
Ca = 40.1.	Hg = 200.6.	P = 31.0.
Cl = 35.5.	Mg = 24.3.	S = 32.1.

1. Write balanced stoichiometric equations for the following reactions. (If you are not familiar with the reactions you should take steps to determine the products before writing the equations.)
(a) zinc is dissolved in dilute sulphuric acid;
(b) magnesium is dissolved in dilute hydrochloric acid;
(c) carbon dioxide is dissolved in water;
(d) marble chips are dissolved in dilute hydrochloric acid;
(e) copper(II) oxide is dissolved in hot dilute sulphuric acid;
(f) carbon dioxide is absorbed by lime water;
(g) mercury(II) oxide is heated;
(h) magnesium is burnt in oxygen;
(i) phosphorus is burnt in excess oxygen.

2. Calculate the mass of sodium chloride needed to react completely with 2.000 g of silver nitrate in solution.

3. Calculate the mass of sodium sulphate formed by the neutralization of 10.00 g of caustic soda (in solution), with dilute sulphuric acid.

4. 5.00 g of copper(II) oxide is dissolved in excess dilute sulphuric acid. What mass of copper(II) sulphate ($CuSO_4$) does the solution contain? What mass of copper(II) sulphate pentahydrate ($CuSO_4.5H_2O$) could be obtained from the solution?

5. What mass of magnesium is required to produce 2.00 dm^3 at s.t.p. of hydrogen, by reaction with dilute sulphuric acid?

6. 5.00 g of mercury(II) oxide is decomposed by heat. The oxygen is collected at 20 °C and 770 mmHg pressure. What volume does the oxygen occupy?

7. 100 kg of limestone is heated in a kiln forming calcium oxide and carbon dioxide. Calculate the mass of calcium oxide formed.

8. 1.20 g of carbon is burnt in excess oxygen forming carbon dioxide. Calculate:
(a) the mass of oxygen which combines with the carbon;
(b) the volume at s.t.p. of carbon dioxide formed.

9. 1.00 dm^3 at s.t.p. of carbon dioxide is absorbed in calcium hydroxide solution, forming a precipitate of calcium carbonate. Calculate the mass of precipitate formed.

10. The calcium carbonate formed in Q. 9 is heated strongly, decomposing it to calcium oxide and carbon dioxide. Calculate the volume at s.t.p. of carbon dioxide formed. How does this compare with the volume of carbon dioxide in Q. 9? Explain this comparison.

11. 0.558 g of iron is dissolved in excess hydrochloric acid forming a solution of iron(II) chloride and liberating hydrogen. Calculate the volume of hydrogen formed at 20 °C and 765 mmHg pressure.

12. Sodium chloride may be formed by allowing caustic soda to react with hydrochloric acid. Calculate the mass of caustic soda and the mass of hydrogen chloride needed to prepare 1.000 g of sodium chloride.

13. When sodium chloride is treated with concentrated sulphuric acid, sodium hydrogen sulphate is formed and hydrogen chloride gas is liberated. Calculate the mass of sodium hydrogen sulphate and the volume of hydrogen chloride at s.t.p. formed by reacting 0.500 g of sodium chloride with excess sulphuric acid.

14. The reaction used to prepare phosphorus is represented by the stoichiometric equation:
$Ca_3(PO_4)_2 + 3SiO_2 + 5C \rightarrow 3CaSiO_3 + 5CO + 2P$
Calculate the mass of calcium orthophosphate needed to produce one kg of phosphorus.

15. A solution containing 8.00 g of barium chloride is mixed with a solution containing 6.00 g of sodium sulphate. Barium sulphate and sodium chloride are formed. Caculate the mass of precipitate formed.

16. 5.00 g of sodium chloride is dissolved in water and added to a solution containing 8.00 g of silver nitrate. What mass of precipitate would be produced? What substances remain in the solution? Calculate the mass of these substances.

17. A solution containing 0.375 g of iron(II) chloride is boiled with a mixture of hydrochloric acid and nitric acid:
$6FeCl_2 + 2HNO_3 + 6HCl \rightarrow 6FeCl_3 + 2NO(g) + 4H_2O$
The solution is treated with caustic soda solution:
$FeCl_3 + 3NaOH \rightarrow Fe(OH)_3(s) + 3NaCl$
The precipitate is filtered off, washed and heated strongly:
$2Fe(OH)_3 \rightarrow Fe_2O_3 + 3H_2O(g)$
Calculate the mass of iron(III) oxide formed.

18. Nitrogen monoxide (NO) combines with oxygen (O_2) to form nitrogen dioxide (NO_2). Calculate the volume of nitrogen dioxide formed from 5 cm³ of nitrogen monoxide with excess oxygen. All volumes are measured at the same temperature and pressure.

19. Methane (CH_4) burns in oxygen (O_2) forming carbon dioxide (CO_2) and water (H_2O). 20 cm³ of methane is mixed with 70 cm³ of oxygen at room temperature and pressure. The mixture is ignited. Calculate the volume (at the same temperature and pressure) of:
(a) carbon dioxide formed;
(b) oxygen left over.

20. Methane (CH_4) reacts with excess chlorine (Cl_2) forming carbon tetrachloride (CCl_4) and hydrogen chloride (HCl). 10 cm³ of methane is mixed with 100 cm³ of chlorine and allowed to react. Calculate the volume of the HCl gas formed. (All volumes are at the same temperature and pressure.)

21. 20 cm³ of carbon monoxide is mixed with 20 cm³ of oxygen, ignited and allowed to cool back to the original temperature and pressure. Calculate:
(a) the total volume initially;
(b) the volume of carbon dioxide formed;
(c) the volume of oxygen used;
(d) the total final volume;
(e) the overall decrease in volume.

22. 15 cm³ of ethylene (C_2H_4) is mixed with excess oxygen (O_2). The mixture is ignited forming carbon dioxide and water, and is then allowed to cool back to the original temperature and pressure. Calculate the volume of oxygen used.

23. 10 cm³ of propane gas (C_3H_8) is mixed with 110 cm³ of oxygen. The mixture is ignited forming carbon dioxide and water, and is then allowed to cool back to the original temperature and pressure. Calculate:
(a) the volume of carbon dioxide formed;
(b) the volume of oxygen left over.

24. 5 cm³ at s.t.p. of butane gas (C_4H_{10}) is burnt in excess oxygen. Calculate the volume of carbon dioxide and steam (both expressed at s.t.p.) formed.

25. Octane (C_8H_{18}) is a constituent of motor petrol. Calculate the volume of oxygen (expressed at s.t.p.) needed per kilogram of octane burnt.
The octane is burnt in an engine which discharges its exhaust at 150 °C and 1.20 atmosphere pressure. Calculate the volumes of carbon dioxide and of steam discharged from the exhaust per kilogram of octane burnt.

Answers to Numerical Problems

2. 0.689 g.
3. 17.8 g.
4. 10.0 g, 15.7 g.
5. 2.17 g.
6. 274 cm³.
7. 56 kg.
8. (a) 3.20 g. (b) 2.24 dm³.
9. 4.47 g.
10. 1.00 dm³.
11. 239 cm³.
12. 0.684 g, 0.624 g.
13. 1.03 g, 192 cm³.
14. 5 kg.
15. 8.96 g.
16. 6.75 g. In solution: 2.25 g NaCl, 4.00 g NaNO₃.
17. 0.236 g.
18. 5 cm³.
19. (a) 20 cm³. (b) 30 cm³.
20. 40 cm³.
21. (a) 40 cm³. (b) 20 cm³. (c) 10 cm³. (d) 30 cm³. (e) 10 cm³.
22. 45 cm³.
23. (a) 30 cm³. (b) 60 cm³.
24. 20 cm³, 25 cm³.
25. (a) 2 460 dm³. (b) 4 310 dm³.

11 : Conductivity of Electrolytes

Electrical conductors may be divided into two classes:

(a) ELECTRONIC CONDUCTORS (mainly metals). These conduct electricity but are not decomposed by it. Such conductors have been considered already in chapter 4.

(b) ELECTROLYTIC CONDUCTORS or ELECTROLYTES. These are substances which conduct electricity and are decomposed by it.

Some examples of electrolytes have been considered already, e.g., fused salts such as sodium chloride and fused alkalis such as sodium hydroxide. It has been seen also, that pure liquid hydrogen chloride and concentrated sulphuric acid (100%) are either non-conductors or very poor electrical conductors.

There are two major groups of electrolytes:

(i) *fused salts and alkalis*;

(ii) *aqueous solutions of acids, alkalis and salts.*

The fact that these aqueous solutions are electrolytes can be demonstrated as described in the experiment below.

Demonstration of Conductivity of Electrolytes

Two flat copper electrodes are insulated from one another, by clamping them on either side of a piece of cork as shown in fig. 11.1. These electrodes are then wired into a circuit as shown in fig. 11.2. A suitable source of power is a 6-12 volt portable transformer rectifier. This will provide direct current at either

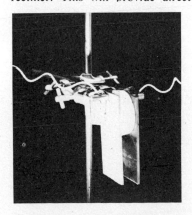

Fig. 11.1.
Copper electrodes.

Two flat copper plates are held in a clamp and separated by a flat piece of cork. Pieces of cork insulate the copper plates from the metal clamp.

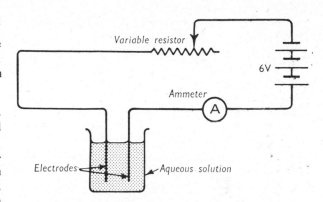

Fig. 11.2. Conductivity of electrolytes.

6 volt or 12 volt. The variable resistor is used to adjust the current in case a full scale deflection of the ammeter occurs. An ammeter giving a full scale deflection with a current of 5 ampere is suitable.

Beakers containing aqueous solutions of various acids, alkalis and salts are to be introduced around the electrodes. If the solution is an electrolyte, a current will flow in the circuit and this will be shown by the ammeter. It is necessary to determine first whether any current is conducted by the water itself. Hence a beaker of distilled water is introduced around the electrodes. If the ammeter needle moves at all, the movement is the barest perceptible movement. Having shown that the distilled water contributes very little to the conductivity, a variety of aqueous solutions may now be tested. Some liquids which are not solutions, e.g., ethanol, may also be tested.

● *Three points of technique should be observed.*

1. The electrodes must be washed with distilled water between each test. This can be done by raising and lowering a beaker of water around the electrodes, until no deflection of the ammeter is seen.

2. The area of the electrodes immersed must be about the same for all liquids. It is very easy to show experimentally that the area of electrodes immersed makes a significant difference.

3. It is easy to show that the concentration of the solution used affects the deflection of the ammeter needle, but even a fairly large variation in concentration does not have a very large effect.

● *Two things are to be observed during each test with an aqueous solution:*

(i) that chemical reactions occur at the electrodes, very often in the form of a gas being evolved;

(ii) that the amount of deflection of the ammeter needle varies.

The first observation provides evidence that the substance being tested is an electrolyte, although solutions which are very poor conductors may have to be tested for several minutes to observe much product.

The second observation enables the solutions to be divided fairly sharply into good and poor conductors.

The very small deflection of the ammeter needle obtained when distilled water is tested should be interpreted with care. Elaborate experiments have been performed on very carefully purified water and these indicate that water is an electrolyte but an extremely poor one, much poorer than is indicated by tests such as those described above. The difference can be attributed to traces of dissolved substances such as carbonic acid derived from the carbon dioxide in the air. It can be shown that if great care is exercised in purifying the water and protecting it from exposure to air, the current recorded is extremely small. On the basis of such experiments, water is classified as a poor electrolyte. Similar experiments have shown that pure hydrogen sulphate is also a poor conductor of electricity (see fig. 11.3).

Explanation of Conductivity

It has been seen in earlier chapters, that the conductivity of a liquid can be explained in terms of migration of ions. In fact, the conductivity of some liquids, e.g., molten sodium chloride, and the lack of conductivity of other liquids, e.g., liquid HCl, are important pieces of evidence supporting the theory of valency.

It is reasonable to explain the conductivity of *aqueous* solutions as being similarly due to the migration of ions. Solutions which are good conductors contain relatively large numbers of ions and poor conductors contain relatively few ions. When a substance dissolves in water, ions may be obtained in the solution in two ways:

(a) DISSOCIATION

This is a process in which ions, already in existence, merely become free to move. Ionic substances like sodium chloride are composed of ions whether in the solid state or in the liquid state. Solid Na^+Cl^- is a very poor conductor of electricity because the ions have very restricted mobility in the crystal lattice. However, fused Na^+Cl^- is a good conductor of electricity because the ions are free to move when the solid is melted.

Dissolution of Na^+Cl^- in water, can be pictured as the removal of ions from the crystal lattice by molecules of water. The hydrated ions then move about with the molecules of water. This is illustrated in

Good Conductors		Poor Conductors		Non-Conductors	
HCl solution		pure water		ethanol	*No ions*
H_2SO_4 solution		concentrated H_2SO_4		acetone	
HNO_3 solution		ammonia solution			
NaOH solution		acetic acid solution			
KOH solution	*Relatively large number of ions—*	carbonic acid solution	*Relatively few ions—*		
NaCl solution		sulphurous acid solution			
Na_2CO_3 solution		citric acid solution			
$CuSO_4$ solution			*Weak electrolytes*		
NH_4Cl solution	*Strong electrolytes*	In general: solutions of			
		(i) weak acids			
In general: solutions of		(ii) weak bases			
(i) strong acids		Also water and			
(ii) alkalis		concentrated H_2SO_4			
(iii) salts					

Fig. 11.3. Summary of results and conclusions from conductivity experiments.

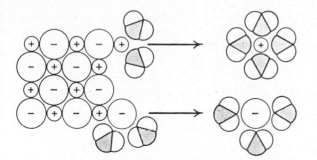

Fig. 11.4. Dissolution of sodium chloride in water.

fig. 11.4. Because sodium chloride is completely in the form of ions in solution, a solution of Na^+Cl^- in water is a good conductor; that is, a *strong electrolyte*.

The dissociation of the ions on dissolution can be represented:

$$Na^+Cl^- + xH_2O \rightarrow Na(H_2O)_n^+ + Cl(H_2O)_m^-$$

where x, n and m are numbers which have not been determined. The equation is therefore commonly written as:

$$Na^+Cl^- \rightarrow Na^+ + Cl^-$$

This equation is the same as the equation used to represent the dissociation of sodium chloride on fusion.

Because sodium chloride is completely ionized in solution, the solution is called a *strong electrolyte*.

(b) IONIZATION

This is a process in which ions are formed. It was shown in chapter 5 that substances like anhydrous hydrogen chloride are non-conductors and do not contain ions. They are covalent or molecular substances.

Some substances, such as water and concentrated sulphuric acid, are very poor conductors. It can be concluded that such substances are essentially covalent, although they do contain a very few ions, and they are called *very weak electrolytes*.

When substances such as hydrogen chloride and concentrated sulphuric acid are dissolved in water, the solutions formed are very good conductors. Hence reactions must occur which produce ions. There is good evidence for believing that in dilute solutions, the reactions can be represented as:

$$HCl + H_2O \rightarrow H_3O^+ + Cl^-$$
$$H_2SO_4 + 2H_2O \rightarrow 2H_3O^+ + SO_4^{2-}$$

The two acids are said to ionize on dissolution in water. This is illustrated in fig. 11.5. Because these acids are completely ionized in solution these solutions are *strong electrolytes*.

Acetic acid is an example of an acid which is a very poor conductor in aqueous solution and this suggests that its solution contains relatively few ions. The ionization of acetic acid is represented:

$$CH_3CO_2H + H_2O \rightleftharpoons H_3O^+ + CH_3COO^-$$

Because such acids are only slightly ionized in solution, the solutions are *weak electrolytes*.

The concepts discussed above form part of the Ionic Theory.

Postulates of the Ionic Theory

1. An electrolyte contains charged particles called ions:
   (a) positive ions are called cations;
   (b) negative ions are called anions.

2. The charge on an ion is proportional to the valency of the ion.

3. A solution in which the solute is completely ionized is a strong electrolyte. A solution in which the solute is only partly ionized is a weak electrolyte.

NOTE: The dissociation or ionization of all the electrolytes examined can be represented as shown in

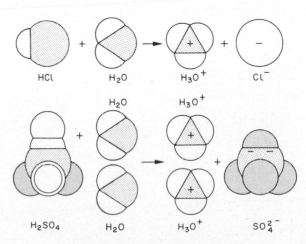

Fig. 11.5. Ionization of hydrogen chloride and hydrogen sulphate in water.

the table, fig. 11.6. Non-electrolytes are not shown in fig. 11.6, because they do not contain ions.

The conductivity of a solution is not sufficient evidence in itself to classify some substances as strong or weak electrolytes. If a substance is only slightly soluble, its solution will be a poor conductor because only a small amount is dissolved, irrespective of whether the dissolved substance is a strong or weak electrolyte. Because all of the substances used in the experiments described above are soluble in water, it follows that the good conductors are strong electrolytes and the poor conductors are weak electrolytes.

Strong Electrolytes	Weak Electrolytes
$HCl + H_2O \rightarrow H_3O^+ + Cl^-$	$2H_2O \rightleftharpoons H_3O^+ + OH^-$
$H_2SO_4 + 2H_2O \rightarrow 2H_3O^+ + SO_4^{2-}$	$NH_3 + H_2O \rightleftharpoons NH_4^+ + OH^-$
$HNO_3 + H_2O \rightarrow H_3O^+ + NO_3^-$	$CH_3CO_2H + H_2O \rightleftharpoons H_3O^+ + CH_3COO^-$
$NaOH \rightarrow Na^+ + OH^-$	$* H_2CO_3 + H_2O \rightleftharpoons H_3O^+ + HCO_3^-$
$KOH \rightarrow K^+ + OH^-$	$* H_2SO_3 + H_2O \rightleftharpoons H_3O^+ + HSO_3^-$
$NaCl \rightarrow Na^+ + Cl^-$	
$Na_2CO_3 \rightarrow 2Na^+ + CO_3^{2-}$	
$CuSO_4 \rightarrow Cu^{2+} + SO_4^{2-}$	* Only the primary (or first) ionizations of these acids are shown because the secondary ionizations are very slight.
$NH_4Cl \rightarrow NH_4^+ + Cl^-$	

Fig. 11.6

STUDY QUESTIONS

1. Classify each of the following substances as non-conductors, electronic conductors or electrolytic conductors.
   (a) sodium metal
   (b) aqueous caustic soda
   (c) molten caustic soda
   (d) copper
   (e) sulphur
   (f) graphite
   (g) solid carbon dioxide
   (h) solid common salt.

2. Classify each of the following electrolytes as strong electrolytes or weak electrolytes.
   (a) water
   (b) ammonia solution
   (c) sodium sulphate solution
   (d) concentrated sulphuric acid
   (e) molten sodium chloride
   (f) dilute sulphuric acid
   (g) copper(II) nitrate solution
   (h) carbonic acid solution.

3. Write equations to illustrate the dissociation or ionization of the following substances, when they are dissolved in water.
   (a) nitric acid
   (b) sodium sulphate
   (c) ammonia
   (d) copper(II) nitrate
   (e) acetic acid
   (f) sulphuric acid
   (g) ammonium sulphate
   (h) magnesium sulphate.

4. The conductivities of aqueous solutions of the following substances are as indicated:
   perchloric acid, $HClO_4$: very good conductor;
   hypochlorous acid, $HOCl$: poor conductor;
   ether, $C_4H_{10}O$: non-conductor.
What inferences can be drawn from these observations? Write equations (where appropriate) to illustrate your answer.

5. Conductivity experiments have been discussed for a number of aqueous solutions. What generalizations can be made from these observations concerning:
   (a) the conductivity of aqueous solutions of ionic substances;
   (b) the conductivity of aqueous solutions of molecular substances;
   (c) the cation formed by acids in aqueous solution;
   (d) the anion formed by alkalis in aqueous solution.

6. If pure dry hydrogen chloride gas is dissolved in the liquid, toluene, the solution is a non-conductor of electricity. However, if some water is now added to the solution, the liquid so formed becomes a very good conductor of electricity.
   What inferences can be drawn from these observational facts?

7. Calcium hydroxide is not very soluble in water but it is sufficiently soluble to form a very mildly alkaline solution called lime water. This solution is not a very good conductor of electricity. Can calcium hydroxide be classified as a strong or weak electrolyte from this evidence?

12 : Acids and Bases

The terms 'acid' and 'base' are widely used in chemistry, and go back to ancient times. The history of acid—base concepts, throws light on how chemical concepts develop and broaden in scope.

Early History
The only acid known in ancient times was vinegar. When other acids, such as sulphuric acid, were discovered in the twelfth and thirteenth century it was natural that their sour taste should lead to them being called acids.

Towards the end of the eighteenth century, Lavoisier recognized that oxides of non-metals produced acidic solutions with water and he proposed the first theory of acids, in which oxygen was regarded as an essential constituent of acids. In fact, the word oxygen means acid-producer.

---

1. sour taste;
2. turn blue litmus red;
3. react with alkalis to form a salt and water, e.g.—
   $$H_2SO_4 + 2NaOH \rightarrow Na_2SO_4 + 2H_2O$$
4. react with *some* oxides (basic oxides) to form a salt and water, e.g.—
   $$H_2SO_4 + CuO \rightarrow CuSO_4 + H_2O$$
5. react with *most* metals to form a salt and hydrogen, e.g.—
   $$H_2SO_4 + Zn \rightarrow ZnSO_4 + H_2$$
6. react with carbonates to form a salt, water and carbon dioxide, e.g.—
   $$H_2SO_4 + Na_2CO_3 \rightarrow Na_2SO_4 + H_2O + CO_2$$

Fig. 12.1. Some properties of acids.

---

1. bitter taste;
2. turn red litmus blue;
3. have a soapy feel;
4. react with *some* oxides (acidic oxides) to form a salt and water, e.g.—
   $$2NaOH + CO_2 \rightarrow Na_2CO_3 + H_2O$$
5. react with acids to form a salt and water, e.g.—
   $$2NaOH + H_2SO_4 \rightarrow Na_2SO_4 + 2H_2O$$

Fig. 12.2. Some properties of bases.

In the early nineteenth century Davy showed that the acid, hydrogen chloride, is a compound of hydrogen and chlorine only. It does not contain oxygen. This led to the view that hydrogen is the element always present in acids.

Acid—Base Theories
For most of the nineteenth century, an acid was regarded as a compound which contained hydrogen that can be replaced, wholly or partly, by a metal, forming a salt. The table, fig. 12.1, lists some properties of acids and illustrates the notion of replaceable hydrogen in an acid.

Bases were regarded as compounds containing an OH group, so that all bases could be written as MOH. The table, fig. 12.2, lists some properties of soluble bases e.g. sodium hydroxide. It was recognized that many bases are insoluble in water, e.g. iron(III) hydroxide.

This theory had value in systematizing properties of acids and bases. However it was found that some compounds which contain the OH group are not bases, e.g., ethanol, C_2H_5OH is not a base. Moreover, the theory did not explain why some acids are strong acids e.g., sulphuric acid, and other acids are weak acids, e.g. acetic acid.

The next step in acid-base theory was taken in 1887 by Arrhenius who recognized that all electrolytes in solution contain ions. This led to the definition of an acid as a substance that ionizes in water to give hydrogen ions. For example,

$$HCl \rightarrow H^+ + Cl^-$$
acid

A base was defined as a substance that ionizes in water to give hydroxide ions. For example,

$$NaOH \rightarrow Na^+ + OH^-$$
base

The sour taste and other properties of acids thus became associated with the hydrogen ion, and reactions of acids could be represented as reactions of the hydrogen ion, e.g. the reaction of zinc metal with an acid:

$$Zn + 2H^+ \rightarrow Zn^{2+} + H_2$$

In a similar way, the properties of bases were attributed to the properties of the hydroxide ion, e.g. the reaction of sodium hydroxide with carbon dioxide gas was represented:

$$CO_2 + 2OH^- \rightarrow CO_3^{2-} + H_2O$$

The Arrhenius theory provided a useful development of acid-base theory in that it emphasized the similarity in reactions of different acids. It also provided a means of comparing acid strengths. Thus, a weak acid is one that is only partly ionized in water and therefore provides relatively few hydrogen ions, whereas a strong acid is completely ionized in water and provides many hydrogen ions.

$$HCl \rightarrow H^+ + Cl^- \text{ (strong acid)}$$
$$CH_3CO_2H \rightleftharpoons H^+ + CH_3COO^- \text{ (weak acid)}$$

Although this theory proved to be fairly satisfactory, it is obvious that it is restricted to solutions in water. Moreover, there are many substances like ammonia (NH_3), sodium carbonate (Na_2CO_3) and calcium oxide (CaO) which are able to neutralize acids and have other basic properties, yet they were not bases on this theory. Thus, if ammonia, sodium carbonate, trisodium phosphate (Na_3PO_4) and calcium oxide are tested with moist litmus paper it is found that they all turn the litmus blue.

THE PROTONIC THEORY OF ACIDS AND BASES

An acid-base theory suggested by Lowry and independently by Brönsted in 1923 is widely used today. This theory emphasizes the complementary nature of acids and bases; it defines these terms by a reaction between an acid and a base. The theory does not substantially change the earlier concepts of acids but it provides a much broader definition of a base. The Lowry-Brönsted theory defines an acid and a base in terms of donation and acceptance of a proton (H^+).

● *Acid*
An acid is an ion or a molecule which can donate a proton.

● *Base*
A base is an ion or a molecule which can accept a proton.

● *Acid-Base Reaction*
When an acid and a base react with each other, a proton is transferred from the acid to the base. Notice that an acid cannot function as an acid (a proton donor) unless a base (a proton acceptor) is present.

Example

Hydrogen chloride gas reacts with ammonia gas to form ammonium chloride. The HCl donates a proton to the NH_3.

$$\underset{\text{acid}}{HCl} + \underset{\text{base}}{NH_3} \rightarrow NH_4^+ + Cl^-$$

When ammonium chloride is heated, it decomposes into hydrogen chloride and ammonia. The NH_4^+ donates a proton to the Cl^-.

$$\underset{\text{acid}}{NH_4^+} + \underset{\text{base}}{Cl^-} \rightarrow HCl + NH_3$$

Thus HCl and NH_4^+ are acids and NH_3 and Cl^- are bases. The acid-base pair, HCl and Cl^-, have formulae which differ by one proton. They are called a *conjugate pair*. HCl is the *conjugate acid* to the base Cl^- and Cl^- is the *conjugate base* to the acid HCl. Similarly, the acid NH_4^+ and the base NH_3 are a conjugate pair.

The acid-base reaction involving these conjugate pairs can be summed up by writing

$$\underset{\text{acid}_1}{HCl} + \underset{\text{base}_2}{NH_3} \rightleftharpoons \underset{\text{acid}_2}{NH_4^+} + \underset{\text{base}_1}{Cl^-}$$

The subscripts 1 and 2 show which pair of formulae is a conjugate pair. Thus $acid_1$ and $base_1$ are a conjugate pair and $acid_2$ and $base_2$ are the other conjugate pair.

The General Acid-Base Reaction

By defining acids as proton donors and bases as proton acceptors, all acid-base reactions can be represented:

$$acid_1 + base_2 \rightleftharpoons acid_2 + base_1$$

Thus the reaction of an acid and base *must* produce, another acid and another base.

Acid-Base Reactions in Aqueous Solution

Water, H_2O, can accept a proton to form H_3O^+. Thus water can function as a base and its conjugate acid is H_3O^+; e.g. when HCl is dissolved in water, the HCl donates a proton:

$$\underset{\text{acid}_1}{HCl} + \underset{\text{base}_2}{H_2O} \rightarrow \underset{\text{acid}_2}{H_3O^+} + \underset{\text{base}_1}{Cl^-}$$

In addition to its reactions as a base, water can also function as an acid. Water, H_2O, can donate a proton and form OH^-. Thus H_2O is an acid and OH^- is its conjugate base; e.g., when NH_3 is dissolved in water the NH_3 accepts a proton.

$$H_2O + NH_3 \rightleftharpoons NH_4^+ + OH^-$$
acid$_1$ base$_2$ acid$_2$ base$_1$

Since water is both an acid and a base it is not surprising that it reacts with itself, although to only a very slight extent.

$$H_2O + H_2O \rightleftharpoons H_3O^+ + OH^- \text{ (very slight)}$$
acid$_1$ base$_2$ acid$_2$ base$_1$

Strengths of Acids and Bases

Strong acids, like HCl, are virtually completely ionized in aqueous solution.

$$HCl + H_2O \rightarrow H_3O^+ + Cl^- \text{ (complete)}$$
acid$_1$ base$_2$ acid$_2$ base$_1$

Weak acids, like CH_3COOH, are only slightly ionized in aqueous solution.

$$CH_3CO_2H + H_2O \rightleftharpoons H_3O^+ + CH_3COO^- \text{ (slight)}$$
acid$_1$ base$_2$ acid$_2$ base$_1$

The relative strengths of acids and the relative strengths of their conjugate bases are related in a simple way. In the above examples, HCl is a stronger acid than CH_3COOH because it reacts more completely with water. If the reverse reactions are considered, it follows that Cl^- is a weaker base than CH_3COO^- because Cl^- does not react with H_3O^+ but CH_3COO^- reacts to a considerable extent.

This can be summed up by a tabulation of conjugate pairs. The strongest acid has the weakest conjugate base and vice versa (see fig. 12.3).

Conjugate Pairs			
	Acid	Base	
Stronger acids	HCl	Cl^-	
	HNO_2	NO_2^-	
	CH_3CO_2H	CH_3COO^-	Stronger bases
	NH_4^+	NH_3	

Fig. 12.3. Relative strengths of some conjugate acids and bases.

Ionization of Polyprotic Acids

Acids which can yield more than one proton per molecule of acid are said to be polyprotic. Some common examples of diprotic acids are sulphuric acid (H_2SO_4), carbonic acid (H_2CO_3) and sulphurous acid (H_2SO_3). Phosphoric acid (H_3PO_4) is an example of a triprotic acid.

Conductivity measurements on solutions of *polyprotic acids* over a wide range of concentrations indicate that they ionize in stages. For sulphuric acid the measurements indicate that in fairly concentrated solutions, only one proton is produced per H_2SO_4 molecule:

$$H_2SO_4 + H_2O \rightarrow H_3O^+ + HSO_4^-$$

If the solution is diluted considerably, a further ionization occurs:

$$HSO_4^- + H_2O \rightleftharpoons H_3O^+ + SO_4^{2-}$$

This second reaction is incomplete even in quite dilute solutions and so the entities present in dilute sulphuric acid (apart from water molecules) include H_3O^+ ions, HSO_4^- ions and SO_4^{2-} ions. However, the reactions of dilute sulphuric acid are generally those of a solution containing H_3O^+ ions (e.g. neutralization of alkalis) and SO_4^{2-} ions (e.g. precipitation of barium sulphate) and the approximation is often made that both ionizations are complete:

$$H_2SO_4 + 2H_2O \rightarrow 2H_3O^+ + SO_4^{2-}$$

Other polyprotic acids also ionize in stages. Thus carbonic acid ionizes in two stages, but there is a very significant difference in that carbonic acid is a weak acid and the first (or primary) ionization is only very slight, even in dilute aqueous solutions:

$$H_2CO_3 + H_2O \rightleftharpoons H_3O^+ + HCO_3^-$$

Experimental evidence suggests that, as for sulphuric acid, the secondary ionization is very much less than the primary ionization.

$$HCO_3^- + H_2O \rightleftharpoons H_3O^+ + CO_3^{2-}$$

Thus this last reaction is very slight indeed, and it is for this reason that only the primary ionization of carbonic acid was shown in chapter 11.

In a similar way, phosphoric acid ionizes in three stages:

$$H_3PO_4 + H_2O \rightleftharpoons H_3O^+ + H_2PO_4^- \text{ (partial)}$$
$$H_2PO_4^- + H_2O \rightleftharpoons H_3O^+ + HPO_4^{2-} \text{ (slight)}$$
$$HPO_4^{2-} + H_2O \rightleftharpoons H_3O^+ + PO_4^{3-} \text{ (very slight)}$$

Acidic, Basic and Neutral Solutions

Pure water has a very small conductivity, due to self ionization.

$$H_2O + H_2O \rightleftharpoons H_3O^+ + OH^-$$

Because the conductivity of water is constant, the concentrations of the hydrogen ions and the hydroxide ions must be constant. Thus:

$$[H_3O^+][OH^-] = \text{a constant.}$$

The value of this constant is 1×10^{-14} at 298 K.

● *The square brackets around the formulae are conventionally used to represent concentrations in mol dm^{-3}*

The concentrations of the hydrogen ions and the hydroxide ions are equal in pure water and therefore each is equal to 1×10^{-7} mol dm^{-3}. Thus:

$$[H_3O^+] = [OH^-] = 10^{-7}.$$

If an acid is added to water, the concentration of the hydrogen ions increases but the concentration of the hydroxide ions decreases. It has been found that the *product* of the concentrations is still equal to 10^{-14}. Similarly if a base is added to water, the concentration of the hydroxide ions increases and the concentration of the hydrogen ions decreases. However, the *product* of the concentrations equals 10^{-14}, e.g., if sufficient acid is added to make $[H_3O^+] = 10^{-1}$ then $[OH^-] = 10^{-13}$. Alternatively, if sufficient base is added to make $[OH^-] = 10^{-2}$ then $[H_3O^+] = 10^{-12}$.

Definitions:

● *A solution is neutral if* $[H_3O^+] = 10^{-7}$
● *A solution is acidic if* $[H_3O^+] > 10^{-7}$
● *A solution is basic if* $[H_3O^+] < 10^{-7}$

Hydrogen Ion Concentration and pH

It is somewhat inconvenient to express the concentrations of the hydrogen ion by way of the very small numbers shown above. A notation, called the pH notation, has been devised to avoid the difficulty.

● *The pH of a solution is defined:*
$$pH = -log_{10}[H_3O^+]$$

The effect of this is that:
if $[H_3O^+] = 10^{-7}$, then the pH of the solution $= 7$
if $[H_3O^+] = 10^{-1}$, then the pH of the solution $= 1$
if $[H_3O^+] = 10^{-12}$, then the pH of the solution $= 12$.
It should be noted that if the concentration of the hydrogen ion is increased, then the pH of the solution becomes smaller. Thus:

a neutral solution has a pH $= 7$
an acidic solution has a pH < 7
a basic solution has a pH > 7.

Acid-Base Indicators

Acid-base indicators are substances which undergo a definite colour change within a fairly narrow range of pH. Litmus is sometimes used as an indicator, particularly in the form of litmus paper. Although it changes colour when the pH of the solution is approximately 7, it is used for rough work only, for it is not very sensitive to slight changes in pH. However, litmus can be used to detect whether a solution is acidic or alkaline (basic); it is red in acidic solutions and blue in alkaline solutions.

The table, fig 12.4, shows a few common acid-base indicators together with their colours and the pH range at which each colour change takes place.

INDICATOR	APPROX. pH RANGE FOR COLOUR CHANGE	COLOUR AT LOWER pH	COLOUR AT HIGHER pH
methyl orange	3.1 — 4.4	red	yellow
methyl red	4.2 — 6.2	red	yellow
bromo-thymol blue	6.0 — 7.6	yellow	blue
phenolphthalein	8.3 — 10.0	colourless	pink

NOTE. A lower pH value indicates:
(a) a more acidic solution if the pH is less than 7;
(b) a less basic solution if the pH is greater than 7.

Fig. 12.4. Some common indicators.

Amphiprotic Substances

A substance which can function as both an acid and a base is called an amphiprotic substance. Amphiprotic substances can donate protons or accept protons depending on the other reagents present. Water is an example of an amphiprotic substance.

$$H_2O + H_2O \rightleftharpoons H_3O^+ + OH^- \text{ (very slight)}$$

THE HYDROGEN CARBONATE ION

In the first (or primary) ionization of H_2CO_3, the HCO_3^- is a base:

$$\underset{acid_1}{H_2CO_3} + \underset{base_2}{H_2O} \rightleftharpoons \underset{acid_2}{H_3O^+} + \underset{base_1}{HCO_3^-} \text{ (slight)}$$

In the secondary ionization, HCO_3^- functions as an acid:

$$\underset{acid_1}{HCO_3^-} + \underset{base_2}{H_2O} \rightleftharpoons \underset{acid_2}{H_3O^+} + \underset{base_1}{CO_3^{2-}} \text{ (very slight)}$$

Thus HCO_3^- is amphiprotic; it can be a proton donor or a proton acceptor. The salt sodium hydrogen carbonate, $NaHCO_3$, contains the amphiprotic HCO_3^- anion. Therefore, this salt can be reacted with both acids and bases. For example:

(i) with HCl,

$$HCl + NaHCO_3 \rightarrow NaCl + H_2CO_3$$

The equation written with empirical formulae obscures the essential nature of the reaction. HCl solution contains the acid H_3O^+ and the reaction is:

$$\underset{acid_1}{H_3O^+} + \underset{base_2}{HCO_3^-} \rightarrow \underset{acid_2}{H_2CO_3} + \underset{base_1}{H_2O}$$

The Na^+ and Cl^- ions are 'spectator' ions.

(ii) with NaOH,

$$NaOH + NaHCO_3 \rightarrow Na_2CO_3 + H_2O$$

The NaOH solution contains the base OH^-. Thus:

$$\underset{acid_1}{HCO_3^-} + \underset{base_2}{OH^-} \rightarrow \underset{acid_2}{H_2O} + \underset{base_1}{CO_3^{2-}}$$

The Na^+ ions are 'spectator' ions.

The amphiprotic nature of HCO_3^- has been used as a typical example. In a similar way, HSO_3^-, HSO_4^-, $H_2PO_4^-$ and HPO_4^{2-} are amphiprotic.

Hydrolysis of Amphiprotic Substances

Hydrolysis is a reaction with water. If sodium hydrogen carbonate is dissolved in water, two reactions could occur because both HCO_3^- and H_2O are amphiprotic. These two possible reactions are:

$$\text{(a) } \underset{acid}{HCO_3^-} + \underset{base}{H_2O} \rightleftharpoons H_3O^+ + CO_3^{2-}$$

or

$$\text{(b) } \underset{acid}{H_2O} + \underset{base}{HCO_3^-} \rightleftharpoons H_2CO_3 + OH^-$$

Indicators can be used to detect the presence of H_3O^+ ions or OH^- ions in solution. A solution containing an excess of H_3O^+ ions is *acidic*. A solution containing an excess of OH^- ions is *alkaline*. The table, fig. 12.4, gives the colours of some common indicators in acidic, alkaline and neutral solutions.

A solution of $NaHCO_3$ is slightly alkaline to litmus. Therefore, the basic function of HCO_3^- predominates in its reaction with water and the reaction (b) shown above, occurs to a detectable extent. The reaction (a) can be neglected.

A solution of $NaHSO_4$ is acidic to litmus. Therefore the acidic function of HSO_4^- predominates in its reaction with water:

$$HSO_4^- + H_2O \rightleftharpoons H_3O^+ + SO_4^{2-}$$

The reason why some amphiprotic anions give acidic solutions and some give alkaline solutions can be seen from the fig. 12.5 which includes some amphiprotic

	Acids	Conjugate Pairs	Bases
Stronger Acids	H_2SO_4	HSO_4^-	
	HCl	Cl^-	
	HNO_3	NO_3^-	
	H_3O^+	H_2O	
	H_2SO_3	HSO_3^-	
	HSO_4^-	SO_4^{2-}	
	H_3PO_4	$H_2PO_4^-$	
	CH_3CO_2H	CH_3COO^-	
	H_2CO_3	HCO_3^-	
	HSO_3^-	SO_3^{2-}	
	$H_2PO_4^-$	HPO_4^{2-}	
	NH_4^+	NH_3	
	HCO_3^-	CO_3^{2-}	
	HPO_4^-	PO_4^{3-}	
	H_2O	OH^-	Stronger
	NH_3	NH_2^-	Bases

Fig. 12.5. Relative strengths of conjugate acids and bases.

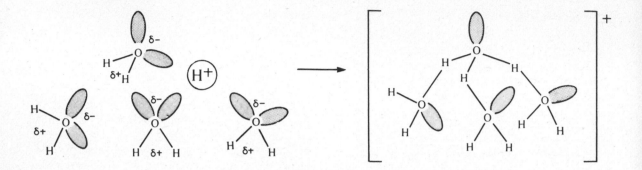

Fig. 12.6. Formation of $H_9O_4{}^+$ ion in solution.

species. Note that HCO_3^- is a very weak acid but a moderately strong base. Also note that HSO_4^- is extremely weak as a base but moderately strong as an acid. Thus the reactions of these anions with water are summarized. Similar conclusions can be drawn for the other amphiprotic anions.

Symbols Used for the Proton

A proton is the nucleus of a hydrogen atom; its symbol is H^+. However, this species cannot exist in aqueous solution because of its positive charge. This positive charge is attracted to the electrons on water molecules and causes several water molecules to become linked together as shown in fig. 12.6. The number of H_2O molecules joined up is in some doubt but the best formula to represent the hydrated proton is $H_9O_4^-$. This is a somewhat cumbersome formula and it is usually simplified when writing chemical equations.

When a proton transfer is to be emphasized the formula H_3O^+ is most convenient. H_3O^+ is called the oxonium ion. However, for simplicity, the formula H^+ is often written and the ion is called the hydrogen ion. Thus the following equations all mean the same thing:

$$HCl(g) + 4H_2O(l) \rightarrow H_9O_4^+ \text{ (aq)} + Cl^-\text{(aq)}$$
$$HCl + H_2O \rightarrow H_3O^+ + Cl^-$$
$$HCl \rightarrow H^+ + Cl^-$$

The choice of representation of the hydrated proton in aqueous solution depends on the purpose which the equation is intended to serve. Generally, the simplest suitable form is chosen.

SUMMARY

The protonic theory of acids and bases defines an acid as a proton donor and a base as a proton acceptor. An acid can only function as an acid if a base is present. The reaction of an acid with a base produces the conjugate acid and the conjugate base.

$$\text{acid}_1 + \text{base}_2 \rightleftharpoons \text{acid}_2 + \text{base}_1$$

If an acid is a strong acid, then its conjugate base is a weak base. A weak acid has a strong conjugate base.

Polyprotic acids ionize in stages and the primary ionization is always much greater than the secondary ionization, etc.

Amphiprotic species can react either as acids or as bases. For example, the HCO_3^- ion can either donate or accept protons:

$$HCO_3^- + OH^- \rightarrow H_2O + CO_3^{2-}$$
$$H_3O^+ + HCO_3^- \rightarrow H_2CO_3 + H_2O$$

An acidic solution is defined as a solution in which the concentration of the hydrogen ion is greater than 10^{-7} mol dm^{-3}. The pH of such a solution is less than seven.

STUDY QUESTIONS

1. Write equations to illustrate the reaction of dilute nitric acid with:
(a) caustic soda solution;
(b) copper(II) oxide;
(c) sodium carbonate;
(d) calcium carbonate.

2. Write equations to illustrate that:
(a) carbonic acid is a weak acid;
(b) sulphurous acid is a weak acid;
(c) nitric acid is a strong acid.

3. Classify each of the following acids as strong or weak acids, from the information given.
N.B. Each acid is quite soluble in water.

Acid	Conductivity of aqueous solution
nitric acid	good conductor
formic acid	poor conductor
trichloracetic acid	good conductor
perchloric acid	good conductor
oxalic acid	poor conductor

4. Write equations to illustrate the reaction of potassium hydroxide solution with:
(a) carbon dioxide;
(b) sulphur dioxide.

5. Write equations to illustrate the reaction between dilute hydrochloric acid and:
(a) magnesium;
(b) magnesium oxide.

6. A solution of hydrochloric acid is mixed with a solution of caustic soda, in suitable proportions for accurate neutralization.
(a) Write the molecular equation for the reaction.
(b) What ions are present before the reaction?
(c) What ions are present after the reaction?
(d) What molecules are formed during the reaction?
(e) What ions are used up during the reaction?
(f) What ions are not used up during the reaction?
(g) Write a balanced equation in which you show only the ions used up and only the molecules formed from these ions.

7. Classify the following entities are Lowry-Brönsted acids, bases or amphiprotic substances:
(a) HCl
(b) NH_3
(c) NH_4^-
(d) Cl^-
(e) HCO_3^-
(f) CH_3COO^-
(g) CO_3^{2-}
(h) OH^-

8. Explain why
(a) PO_4^{3-} is a strong base;
(b) NO_3^- is a very weak base.

9. Write equations to illustrate the formation of the three sodium orthophosphates, by reaction of orthophosphoric acid with different proportions of caustic soda solution.
Predict the effect of each of the three sodium orthophosphates on litmus. Check your prediction by reference to some suitable text.

10. When calcium oxide, $Ca^{2+}O^{2-}$, is placed in water, an alkaline solution containing calcium hydroxide, $Ca^{2+}(OH^-)_2$, is formed. Write an equation for the reaction which occurs and label the Lowry-Brönsted acids and bases.

11. If a few drops of aqueous ammonia are added to water and then phenolphthalein is added, the liquid turns pink. If a fairly concentrated solution of ammonium chloride is then added, the phenolphthalein becomes colourless.
Provide an explanation for these observations.

13 : Electrolysis

● *Definition: Electrolysis is the chemical change brought about by the passage of an electric current through an electrolyte.*

Electrolysis is carried out in a cell containing two electrodes connected to a source of direct current. An electric current from such a source, consists of a flow of electrons. The electrons flow from the negative terminal of the battery as shown in fig. 13.1.

EXAMPLES OF ELECTROLYSIS

1. Molten Sodium Chloride with Unreactive Electrodes

This has been discussed in chapter 2, but will be reviewed briefly.

The electrolysis is performed commercially with a carbon anode. The products are sodium metal at the cathode and chlorine gas at the anode. Any theory or explanation of this process must be consistent with these observations.

● *The explanation or theory, given in chapter 2 is that the sodium ions are positive and migrate to the cathode where they are discharged.*

CATHODE REACTION:

$$Na^+ + e^- \rightarrow Na$$

● *Chloride ions are negative and migrate to the anode. They discharge at the same time as the sodium ions, and in equal number.*

ANODE REACTION:

$$2Cl^- \rightarrow Cl_2 + 2e^-$$

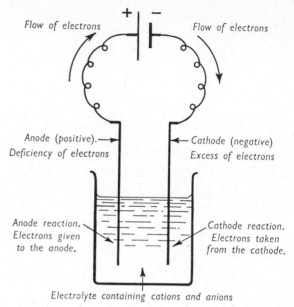

Fig. 13.1. Electrolysis.

Thus the sodium ions (or cations) discharge at the cathode consuming some of the excess of electrons. Simultaneously an equal number of electrons is produced on the anode, by discharge of the chloride ions (or anions).

The migration of cations to the cathode and anions to the anode, constitutes a flow of electric current through the molten salt.

2. Concentrated Aqueous Sodium Chloride with Unreactive Electrodes

Carbon electrodes are suitable for this electrolysis. Platinum electrodes can also be used.

OBSERVATIONS:

When the current flows, it is observed that bubbles of gas appear at each electrode. If the electrodes are made to point upwards and each electrode is covered with a test tube full of the concentrated solution of sodium chloride, the gas at each electrode can be collected and identified (see fig. 13.2). The gas collected at the cathode is odourless and burns when ignited and is evidently hydrogen. The gas collected at the anode is greenish-yellow in colour, has a pungent odour and bleaches moist litmus paper. The gas is evidently chlorine.

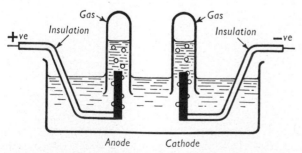

Fig. 13.2. Electrolysis of concentrated sodium chloride solution.

Further information about the electrolysis of a concentrated solution of sodium chloride, can be obtained by placing litmus solution (purple) in the electrolysis cell. As the electrolysis proceeds, the litmus in the vicinity of the cathode goes blue, whereas the litmus in the vicinity of the anode is bleached. Evidently an alkali is produced at the cathode and the bleaching of the litmus at the anode is due to the chlorine.

EXPLANATION:

Before attempting an explanation, it is necessary to consider what particles (or entities) are present in the solution. The sodium chloride is dissociated:

$$NaCl \rightarrow Na^+ + Cl^-$$

Water is also present and is very slightly ionized:

$$2H_2O \rightleftharpoons H_3O^+ + OH^-$$

The requirements of the cathode reaction are that it produces hydrogen, consumes electrons and leads to production of an alkali in the vicinity of the electrode.

During the electrolysis of the solution, sodium ions will migrate to the cathode. However, the sodium ions are apparently not discharged because sodium metal does not appear at the cathode. The discharge of the few hydrogen ions present is possible, despite their low concentration, and would partly explain the products. However, it does not explain immediately the developing alkalinity around the electrode.

All requirements for the cathode reaction can be met in a direct fashion, by considering that water itself takes part in the electrode reaction in the following way:

CATHODE REACTION:

$$2H_2O + 2e^- \rightarrow H_2 + 2OH^-$$

Hence hydrogen gas is produced and a mixture of sodium ions and hydroxide ions accumulates around the cathode. These sodium ions and hydroxide ions comprise caustic soda solution and the alkalinity around the cathode is due to the hydroxide ions produced.

Chloride ions will migrate to the anode. The requirements of the anode reaction are that it produces chlorine gas and electrons. Evidently, the reaction occurring is:

ANODE REACTION:

$$2Cl^- \rightarrow Cl_2 + 2e^-$$

The number of electrons produced on the anode must equal the number of electrons consumed at the cathode, and the two equations for the electrode reactions do show this.

The ions conducting the current through the solution are mainly the sodium ions and chloride ions. It has been explained already that the migration of these ions constitutes the flow of an electric current.

The net result of the electrolysis can be represented by the equation:

$$2Na^+ + 2Cl^- + 2H_2O \rightarrow 2Na^+ + 2OH^- + H_2 + Cl_2$$

The electrolysis of a concentrated (nearly saturated) solution of sodium chloride is carried out industrially to obtain caustic soda solution, hydrogen gas and chlorine gas, all three of which are valuable products. The electrolytic cell is usually provided with a porous partition, to help separate the products.

3. Dilute Aqueous Sodium Chloride with Unreactive Electrodes

Carbon electrodes are again suitable for this electrolysis. Platinum electrodes can also be used.

The electrolysis of a dilute solution of sodium chloride is carried out in the same way as for a concentrated solution. Once again additional experimental evidence can be obtained by placing litmus solution in the electrolyte, before the electrolysis is begun.

OBSERVATIONS:

The observations are that a gas is collected at the cathode and the litmus in the vicinity of the cathode turns blue. At the same time a gas is collected at the anode, and the litmus in the vicinity of the anode turns red. Tests on the gas collected at the cathode show that it is hydrogen. The gas at the anode (in contrast to concentrated aqueous NaCl) is colourless and odourless. Furthermore, it re-ignites a glowing string and is evidently oxygen. The colour changes observed for the litmus indicate that an alkali is produced in the vicinity of the cathode and an acid is produced in the vicinity of the anode.

If more elaborate apparatus is used, the volumes of hydrogen and oxygen can be measured, and it is found that the gases are in the ratio of two volumes of hydrogen to one volume of oxygen (see Hofmann's voltameter, fig. 13.3).

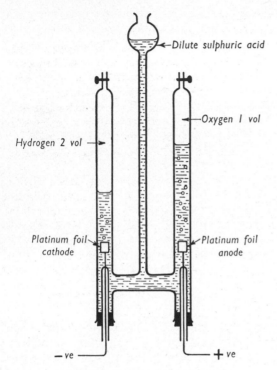

Fig. 13.3. Hofmann's voltameter.

EXPLANATION:

Before suggesting a theory to explain these facts, it is worth considering what particles (or entities) are present. In dilute solution the sodium chloride is completely dissociated:

$$NaCl \rightarrow Na^+ + Cl^-$$

Water is also present and is very slightly ionized:

$$2H_2O \rightleftharpoons H_3O^+ + OH^-$$

Thus we see that the same entities are present as in a concentrated solution of sodium chloride. Furthermore, the same products are formed at the cathode. Thus apparently sodium ions migrate to the cathode and accumulate there. These sodium ions arriving by migration are partnered by hydroxide ions produced when water takes part in the electrode reaction:

CATHODE REACTION:

$$2H_2O + 2e^- \rightarrow H_2 + 2OH^-$$

Hence hydrogen gas is evolved at the cathode and sodium ions and hydroxide ions accumulate around

the cathode, just as with a concentrated solution of NaCl.

Chloride ions, together with the few hydroxide ions, will migrate to the anode. The requirements of the anode reaction are that it yields oxygen gas, produces electrons and accounts for the developing acidity around the anode. It would appear that chloride ions are not discharged, because chlorine gas is not produced. Discharge of hydroxide ions is possible despite their very low concentration. However, this would not lead directly to an explanation of the increasing acidity around the anode. All the requirements of the anode reaction are met in a direct fashion by representing the overall process as:

ANODE REACTION (a):

$$2H_2O \rightarrow O_2 + 4H^+ + 4e^-$$

The hydrogen ions are in fact hydrated but they are written as H^+ in order to simplify the equation. If desired the equation could be written:

ANODE REACTION (b):

$$6H_2O \rightarrow O_2 + 4H_3O^+ + 4e^-$$

The hydrogen ions produced at the anode will partner the chloride ions arriving by migration. These hydrogen ions and chloride ions comprise hydrochloric acid solution, accounting for the developing acidity.

The number of electrons produced by the anode reaction equals the number consumed by the cathode reaction. The equations for the electrode reactions could be made to show this, by multiplying the equation representing the cathode reaction by two. Thus:

CATHODE REACTION:

$$4H_2O + 4e^- \rightarrow 2H_2 + 4OH^-$$

ANODE REACTION:

$$2H_2O \rightarrow O_2 + 4H^+ + 4e^-$$

It can be seen that two molecules of hydrogen are produced for each molecule of oxygen. Hence, two mole of H_2 are produced for every one mole of O_2 and so there are two volumes of hydrogen produced for each volume of oxygen, provided the gases are measured at the same temperature and pressure.

The ions conducting the current in solution are mainly the sodium ions and chloride ions, although these do not take part in the electrode reactions.

The electrolysis of aqueous sodium chloride (dilute and concentrated) draws attention to *the effect which decreasing concentration of an ion, has on the possibility of its discharge*. Thus in concentrated solutions of sodium chloride (greater than 12%), chlorine is produced. For solutions of sodium chloride between 6% and 12% concentration, both chlorine and oxygen are produced, and below 6%, only oxygen is produced at the anode. There are, however, other factors besides concentration which determine the electrode reaction which occurs. These factors include the size and nature of the electrode material, the current used, and the temperature of the solution.

4. Aqueous Sodium Sulphate with Unreactive Electrodes

The electrolysis of a solution of sodium sulphate is similar in all respects to the electrolysis of a dilute solution of sodium chloride.

OBSERVATIONS:

The experimental evidence is that two volumes of hydrogen are produced at the cathode for each volume of oxygen produced at the anode. In addition, the solution becomes alkaline in the vicinity of the cathode and acidic in the vicinity of the anode.

EXPLANATION:

The entities present in the solution are as shown:

$$Na_2SO_4 \rightarrow 2Na^+ + SO_4^{2-}$$
$$2H_2O \rightleftharpoons H_3O^+ + OH^-$$

The cathode reaction must fulfil the same conditions as in the electrolysis of sodium chloride solution (dilute or concentrated). The cathode reaction is evidently the same in each case:

CATHODE REACTION:

$$2H_2O + 2e^- \rightarrow H_2 + 2OH^-$$

The sodium ions migrate to the cathode where they partner the hydroxide ions being produced. Hence the solution around the cathode becomes alkaline.

The anode reaction must fulfil the same requirements as for the electrolysis of a dilute solution of sodium chloride, viz., the production of oxygen and electrons, together with the production of acid in the neighbourhood of the anode. Evidently the anode reaction is the same as for a dilute solution of sodium chloride.

ANODE REACTION:

$$2H_2O \rightarrow O_2 + 4H^+ + 4e^-$$

Therefore, the sulphate ions which migrate to the anode are not discharged, but partner the hydrogen ions which are produced at the anode. Thus the solution in the vicinity of the anode becomes acidic.

It is possible to explain the volume relationship of the hydrogen and oxygen in the same way as for the electrolysis of a dilute solution of sodium chloride.

The ions conducting the current in solution are mainly sodium ions and sulphate ions.

It is found by experiment that electrolysis of sodium sulphate solution gives the same products, irrespective of the concentration of the solution. It is therefore reasonable to infer that (unlike chloride ions) the sulphate ions will not discharge even in a concentrated solution of sodium sulphate.

5. Dilute Sulphuric Acid with Unreactive Electrodes

This electrolysis can be carried out in an electrolytic cell similar to that used in the earlier experiments. Platinum electrodes are suitable. Hoffman's voltameter (fig. 13.3) is often used.

OBSERVATIONS:

The experimental evidence is that two volumes of hydrogen are produced at the cathode for each one volume of oxygen produced at the anode. Furthermore, analysis shows that the sulphuric acid is not consumed during the electrolysis.

EXPLANATION:

Because the solution of sulphuric acid is dilute, the ionization of the acid can be represented:

$$H_2SO_4 + 2H_2O \rightarrow 2H_3O^+ + SO_4^{2-}$$

Water is also present and it is very slightly ionized:

$$2H_2O \rightleftharpoons H_3O^+ + OH^-$$

The only cations present are hydrogen ions, and they are present in considerable quantity in dilute sulphuric acid. The hydrogen ions will migrate to the cathode, and the production of hydrogen gas at this electrode, is accounted for, very simply, by their discharge:

CATHODE REACTION:

$$2H^+ + 2e^- \rightarrow H_2$$

Sulphate ions, together with the few hydroxide ions,

will migrate to the anode. The requirements of the anode reaction are that it produces oxygen and electrons. Furthermore, it should be remembered that sulphuric acid is not lost during the electrolysis, and yet the discharge of hydrogen ions at the cathode amounts to a consuming of acid. Hence a further requirement of the anode reaction is that it produces hydrogen ions in sufficient number to make good this consumption at the cathode. Thus the requirements of the anode reaction are the same as those for the anode reaction in electrolysis of sodium sulphate solution, and the reaction can be represented:

ANODE REACTION:
$$2H_2O \rightarrow O_2 + 4H^+ + 4e^-$$
The sulphate ions which arrive at the anode by migration, will partner the hydrogen ions which are produced at the anode.

In order to show the cathode reaction consuming the same number of electrons as are produced at the anode, the cathode reaction is multiplied by two:

CATHODE REACTION:
$$4H^+ + 4e^- \rightarrow 2H_2$$
It can now be seen that the theory gives an explanation of two further observational facts, viz., that the volume of hydrogen is twice the volume of oxygen (if measured at the same temperature and pressure), and that there is no net loss of hydrogen ions.

The ions conducting the current through the solution are mainly hydrogen ions and sulphate ions.

6. Caustic Soda Solution with Unreactive Electrodes

OBSERVATIONS:
Two volumes of hydrogen are evolved at the cathode for each one volume of oxygen produced at the anode (at the same temperature and pressure). The caustic soda is not consumed during the process.

EXPLANATION:
The entities present in the solution are indicated by the equations:
$$NaOH \rightarrow Na^+ + OH^-$$
$$2H_2O \rightleftharpoons H_3O^+ + OH^-$$
The explanation of the cathode reaction is the same as for electrolysis of sodium chloride solution or sodium sulphate solution:

CATHODE REACTION:
$$2H_2O + 2e^- \rightarrow H_2 + 2OH^-$$
The sodium ions which migrate to the cathode, partner the hydroxide ions produced in the cathode reaction.

Hydroxide ions are the only anions present, and they are present in large numbers. The hydroxide ions migrate to the anode, and the most direct explanation of the anode reaction is that the hydroxide ions discharge:

ANODE REACTION:
$$4OH^- \rightarrow O_2 + 2H_2O + 4e^-$$
If the cathode reaction is multiplied by two:
$$4H_2O + 4e^- \rightarrow 2H_2 + 4OH^-$$
some further points of explanation become clear. There is no net consumption of hydroxide ions (i.e. the alkali is not used up), and there are two volumes of hydrogen for each one volume of oxygen.

7. Copper(II) Sulphate Solution with Copper Electrodes

OBSERVATIONS:
During this electrolysis, no gas is evolved. Copper deposits on the cathode while the copper anode slowly becomes smaller (see fig. 13.4).

EXPLANATION:
The particles or entities in solution are the ions from copper(II) sulphate, together with water molecules

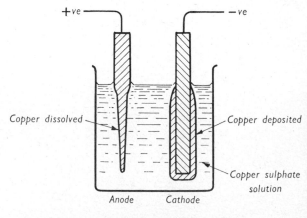

Fig. 13.4. Electrolysis of copper(II) sulphate solution using copper electrodes.

and a relatively few hydrogen and hydroxide ions:

$$CuSO_4 \rightarrow Cu^{2+} + SO_4^{2-}$$
$$2H_2O \rightleftharpoons H_3O^+ + OH^-$$

The deposition of copper on the cathode can be explained by the migration of the copper(II) ions to the cathode, followed by their discharge:

CATHODE REACTION:

$$Cu^{2+} + 2e^- \rightarrow Cu$$

Sulphate ions, together with the few hydroxide ions, will migrate to the anode. However, the discharge of sulphate ions would appear unlikely, just as it is unlikely in the electrolysis of dilute sulphuric acid. This is supported by the fact that it is possible to show by analysis that there is no loss of sulphate ions during the electrolysis. Anode reactions involving hydroxide ions of water seem unlikely, because as shown earlier, they would lead to evolution of a gas.

The most likely interpretation of the observations at the anode, is that copper atoms which comprise the metal of the anode yield up two electrons each, and go into solution as copper(II) ions:

ANODE REACTION:

$$Cu \rightarrow Cu^{2+} + 2e^-$$

The copper(II) ions formed at the anode are partnered by sulphate ions arriving by migration.

The ions conducting the current in solution are mainly copper(II) ions and sulphate ions.

The net result of this electrolysis is the transference of copper metal from the anode to the cathode. This electrolysis has industrial applications in copper plating and in purification of crude copper.

INDUSTRIAL APPLICATIONS OF ELECTROLYSIS

Electrolysis is widely employed industrially. The applications include:

1. Production of Sodium

Sodium metal is produced by electrolysis of the molten chloride. Production of magnesium and calcium metals are similar. Chlorine is a valuable by-product in each process.

2. Electrolysis of Brine

Caustic soda, together with hydrogen and chlorine, is produced by the electrolysis of a concentrated solution of sodium chloride.

3. Production of Hydrogen and Oxygen

Hydrogen and oxygen gases are produced by the electrolysis of dilute sulphuric acid, or preferably by the electrolysis of a solution of an alkali. This latter method gives purer samples of gas.

4. Electroplating Metals with Copper

This is widely employed to form a coating of copper over a base metal, prior to plating with other metals such as nickel or chromium.

The article to be copper plated is first cleaned of rust and grease and is then made the cathode in an electrolytic cell. The electrolyte is copper(II) sulphate solution containing a little sulphuric acid. The sulphuric acid improves the electrolytic conductance and prevents rough coatings. If firm deposits are to be formed, careful attention must be given to the concentration of the electrolyte, the current used, the size of the cathode and the temperature of the bath (see fig. 13.5). Electroplating with other metals is similar in principle.

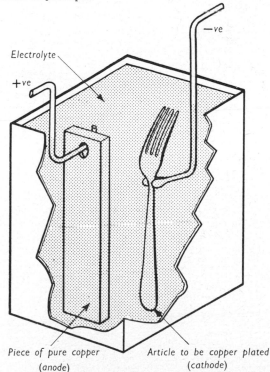

Fig. 13.5. Electroplating with copper.

5. Purification of Copper

In the industrial extraction of copper from its ores, the final stage is the refining of impure copper metal by electrolysis. The impure (crude) copper is made the anode and the electrolyte is copper(II) sulphate solution acidified with dilute sulphuric acid. The cathode is a thin sheet of pure copper.

During the electrolysis, copper dissolves at the anode and deposits at the cathode. If a suitable voltage is maintained pure copper is deposited. The impurities either drop off the anode as 'anode mud' or go into solution at the anode, but do not deposit at the cathode. The anode mud is usually a valuable source of silver.

6. Anodizing

Aluminium metal owes its resistance to corrosion, to a thin adherent film of oxide. This film may be thickened electrolytically, thereby increasing the resistance of the metal to corrosion. The aluminium is made the anode in an electrolytic cell. Dilute sulphuric acid is one of several suitable electrolytes. The oxygen produced on the anode thickens the existing film of oxide.

The freshly formed oxide film is able to absorb dyestuffs very strongly producing an attractive finish.

7. Other Uses

Other important industrial uses of electrolysis include the production of aluminium metal and the purification of zinc metal.

SUMMARY

The common feature of **cathode reactions** is that **electrons are consumed.** The various cathode reactions discussed are shown in the table, fig. 13.6.

The common feature of **anode reactions** is that they **yield electrons.** The various anode reactions discussed are shown in the table, fig. 13.7.

CATHODE REACTIONS	COMMENTS
$Na^+ + e^- \rightarrow Na$	*Occurs with fused salt, but not in aqueous solutions.*
$2H_2O + 2e^- \rightarrow H_2 + 2OH^-$	*Occurs in aqueous solutions of Na^+ salts.*
$2H^+ + 2e^- \rightarrow H_2$	*Occurs in aqueous solutions of acids.*
$Cu^{2+} + 2e^- \rightarrow Cu$	*Occurs in aqueous solutions of Cu^{2+} salts.*

Fig. 13.6

ANODE REACTIONS	COMMENTS
SO_4^{2-}	*Sulphate ions do not discharge in aqueous solutions.*
$2Cl^- \rightarrow Cl_2 + 2e^-$	*Occurs in concentrated aqueous solutions or with fused chlorides.*
$2H_2O \rightarrow O_2 + 4H^+ + 4e^-$	*Occurs in dilute aqueous solutions of Cl^-, SO_4^{2-}.*
$4OH^- \rightarrow 2H_2O + O_2 + 4e^-$	*Occurs in aqueous solutions of alkalis.*
$Cu \rightarrow Cu^{2+} + 2e^-$	*Occurs if the anode is copper.*

Fig. 13.7

STUDY QUESTIONS

1. Electrolysis of molten magnesium chloride produces magnesium at the cathode and chlorine at the anode.

Write equations to illustrate the probable electrode reactions.

2. Electrolysis of an aqueous solution of sodium nitrate with platinum electrodes, yields two volumes of hydrogen at the cathode for each one volume of oxygen at the anode. In addition, the liquid around the cathode becomes alkaline and the liquid around the anode becomes acidic.

Give a full explanation of these observations.

3. Electrolysis of aqueous copper(II) sulphate using platinum electrodes, causes deposition of copper on the cathode and evolution of oxygen at the anode. The solution around the anode becomes acidic.

Write equations to illustrate the probable electrode reactions which occur.

4. Electrolysis of dilute sulphuric acid with copper electrodes produces hydrogen at the cathode. The copper anode slowly dissolves and a blue solution forms.

Write equations to illustrate the probable electrode reactions.

5. Electrolysis of zinc sulphate solution containing dilute sulphuric acid, will produce zinc metal at a zinc cathode and oxygen gas at a lead anode.

Write equations to illustrate the probable electrode reactions which occur.

6. It is possible to make certain generalizations concerning electrode reactions most likely to occur in any given electrolysis. Test your appreciation of the generalizations by attempting to predict the cathode and anode reactions which occur during the electrolysis of each of the following electrolytes, using unreactive electrodes:

(a) concentrated hydrochloric acid;
(b) dilute hydrochloric acid;
(c) fused potassium chloride;
(d) fused sodium hydroxide;
(e) aqueous potassium sulphate;
(f) aqueous silver nitrate.

7. Write an equation for the overall reaction which occurs during the electrolysis of:

(a) dilute sulphuric acid;
(b) caustic soda solution;
(c) copper(II) sulphate solution.

8. Iodide ions are more readily discharged than chloride ions, at a platinum anode.

Write equations for the electrode reactions which occur during electrolysis of an aqueous solution of potassium iodide using platinum electrodes.

9. Under what circumstances could it be expected that sulphate ions would be discharged in an electrolytic cell?

14 : Ionic Equations

Reactions in which gases are involved require that the equations show the molecular formulae of the gases. Thus combustion of carbon monoxide in air or oxygen is represented:

$$2CO + O_2 \rightarrow 2CO_2$$

However, it has been seen already that many aqueous solutions are electrolytes and contain ions. In fact, in dilute aqueous solutions, all salts, alkalis and strong acids exist entirely as ions. Evidently the reactions of these substances involve reactions of ions and are best represented by ionic equations.

REACTIONS INVOLVING ASSOCIATION OF IONS

1. Formation of a Precipitate

This is the formation of an insoluble product, and it may occur on mixing two solutions.

EXAMPLES

● (*a*) *Reaction between silver nitrate solution and sodium chloride solution.*

This reaction was discussed in Chapter 10. It was seen that a white curdy precipitate forms and that this precipitate can be identified as silver chloride. The reaction was represented by the equation:

$$AgNO_3 + NaCl \rightarrow AgCl(s) + NaNO_3 \qquad \text{(i)}$$

However, all the substances except silver chloride, are present as ions in solution. Thus the equation may be re-written:

$$Ag^+ + NO_3^- + Na^+ + Cl^-$$
$$\rightarrow AgCl(s) + Na^+ + NO_3^- \qquad \text{(ii)}$$

This equation emphasizes that the NO_3^- and NA^+ ions do not react. They can be described as '*spectator ions*', and they can be omitted from the equation. If this is done, the equation becomes:

$$Ag^+ + Cl^- \rightarrow AgCl(s) \qquad \text{(iii)}$$

This equation is called the ionic equation for the reaction.

Equations (i) and (iii) both describe the same reaction. Each equation has its value. Thus equation (i) would be preferred in (say) calculating the mass of sodium chloride required for complete reaction with a given mass of silver nitrate. In other cases, equation (iii) may be preferred, because it is a better representation of the reaction and is easier to write. It is also more general, in that it represents the reaction which occurs when a solution of a silver salt is mixed with a solution of any ionic chloride.

● (*b*) *Reaction between barium chloride solution and sodium sulphate solution.*

This reaction has also been discussed earlier. A white precipitate forms and the precipitate can be identified as barium sulphate. The reaction was represented by the equation:

$$BaCl_2 + Na_2SO_4 \rightarrow BaSO_4(s) + 2NaCl$$

It will be seen that this equation may be written as either:

$$Ba^{2+} + 2Cl^- + 2Na^+ + SO_4^{2-}$$
$$\rightarrow BaSO_4(s) + 2Na^+ + 2Cl^-$$
$$\text{or:} \quad Ba^{2+} + SO_4^{2-} \rightarrow BaSO_4(s)$$

● (*c*) *Reaction between copper(II) sulphate solution and sodium hydroxide solution.*

This reaction results in the formation of a blue gelatinous precipitate of copper(II) hydroxide and the reaction can be represented:

$$CuSO_4 + 2NaOH \rightarrow Cu(OH)_2(s) + Na_2SO_4$$

Alternatively it may be represented as:

$$Cu^{2+} + SO_4^{2-} + 2Na^+ + 2OH^-$$
$$\rightarrow Cu(OH)_2(s) + 2Na^+ + SO_4^{2-}$$
$$\text{or:} \; Cu^{2+} + 2OH^- \rightarrow Cu(OH)_2(s)$$

NOTE: Each of the three reactions discussed above occurs to a very considerable extent. However, none of them is absolutely complete, and in some circumstances it may be desirable to emphasize this, by showing that each reaction is reversible to some extent, e.g.:

$$Ag^+ + Cl^- \rightleftharpoons AgCl(s)$$

2. Acid-Alkali Reactions: Neutralization

Neutralization occurs when a solution of an acid is mixed with a solution of an alkali.

$$\text{acid} + \text{alkali} \rightarrow \text{salt} + \text{water}.$$

A similar reaction occurs if an acid is mixed with an insoluble hydroxide such as copper(II) hydroxide, but the discussion will be restricted, for the present, to soluble hydroxides like sodium hydroxide.

● *Neutralization of dilute hydrochloric acid with caustic soda solution.*

This reaction can be represented by the equation:

$$HCl + NaOH \rightarrow NaCl + H_2O$$

It can be shown by the addition of an indicator that a reaction occurs when the two solutions are mixed. However, all the substances present except water, exist as ions in solution. Hence the reaction must be a reaction producing water:

$$H^+ + Cl^- + Na^+ + OH^-$$
$$\rightarrow Na^+ + Cl^- + H_2O$$
or: $\quad H^+ + OH^- \rightarrow H_2O$

This is consistent with the evidence from conductivity experiments on water, which suggest that water is largely a molecular compound, i.e., it is ionized to only a very small extent.

In a similar way, *the neutralization of any strong acid with an alkali*, can be shown to be simply *the formation of water.*

This conclusion is supported by evidence relating to heats of neutralization. It is found that if an acid is neutralized with an alkali, heat is evolved. The quantity of heat can be measured, and the results of experiments are shown in fig. 14.1.

ACID		ALKALI		HEAT EVOLVED
(Amounts quoted in mole)				*(kJ)*
1	HCl	1	NaOH	57.35
1	HCl	1	KOH	57.35
1	HNO₃	1	NaOH	57.35
2	HCl	1	Ca(OH)₂	114.8
2	HCl	1	Ba(OH)₂	114.7

Fig. 14.1. Heats of neutralization of strong acids and alkalis.

The table shows that the heat evolved, when 1 mole of the hydroxide ion is neutralized, is about 57.3 kJ in each case. This supports the conclusion that the reaction is the same in all cases, and is simply:

$$H^+ + OH^- \rightarrow H_2O$$

3. Formation of Other Weak Electrolytes

It can be seen that the reaction occurring during neutralization, can be accounted for because water is a very weak electrolyte. In the chapter on conductivity of electrolytes, it was seen that there are other weak electrolytes, e.g., H_2CO_3, H_2SO_3, and ammonia solution. Hence reactions can probably occur to produce these weak electrolytes.

EXAMPLES

(a) *If dilute hydrochloric acid is added to a solution of sodium carbonate, or solid sodium carbonate,* effervescence occurs, and the gas evolved can be identified as carbon dioxide. The reaction may be represented:

$$Na_2CO_3 + 2HCl \rightarrow 2NaCl + H_2CO_3$$
$$H_2CO_3 \rightarrow H_2O + CO_2$$

The first of these two equations may be written as:

$$2Na^+ + CO_3^{2-} + 2H^+ + 2Cl^- \rightarrow$$
$$2Na^+ + 2Cl^- + H_2CO_3$$
or: $CO_3^{2-} + 2H^+ \rightarrow H_2CO_3$

Evidently the association of carbonate ions and hydrogen ions takes place because the carbonate ion is a strong base, i.e. a good proton acceptor.
The carbonic acid then largely decomposes:

$$H_2CO_3 \rightarrow H_2O + CO_2 \text{ (g)}$$

The overall reaction may be represented:

$$2H^+ + CO_3^{2-} \rightarrow H_2CO_3$$
$$H_2CO_3 \rightarrow H_2O + CO_2 \text{ (g)}$$

(b) *Similarly, if either dilute hydrochloric or sulphuric acid is added to sodium sulphite,* an effervescence occurs, especially on warming. The gas evolved can be identified as sulphur dioxide. The observations can be explained by suggesting that hydrogen ions (from the acid) and sulphite ions (from the salt) associate because the sulphite ion is a strong base.

$$2H^+ + SO_3^{2-} \rightarrow H_2SO_3$$

The sulphurous acid then decomposes, particularly if warmed:

$$H_2SO_3 \rightarrow H_2O + SO_2 \text{ (g)}$$

(c) *The addition of an acid to sodium sulphide* produces a gas which can be identified, partly by odour, as hydrogen sulphide. The reaction is similar to those above, except that the acid itself is volatile:

$$2H^+ + S^{2-} \rightarrow H_2S \text{ (g)}$$

REACTIONS INVOLVING OXIDATION AND REDUCTION — REDOX REACTIONS

Before discussing ionic equations for some redox reactions, it is desirable to make a general review of earlier concepts of oxidation and reduction. The simple concepts of oxidation and reduction are not to be regarded as being replaced by the concept of ionic reactions. Thus for gas reactions such as the combustion of carbon monoxide, ions are not involved. It is simply that the concept of redox reactions is extended to include some other reactions. The following is a brief review of the elementary concepts.

Elementary Concepts of Oxidation

(i) *Oxidation is the addition of oxygen.*

$$2CO + O_2 \rightarrow 2CO_2$$
$$2Mg + O_2 \rightarrow 2MgO$$

Carbon monoxide and magnesium are each oxidized by the addition of oxygen.

(ii) *Oxidation is the removal of hydrogen.*

If hydrogen sulphide gas is burnt in a gas jar a yellow deposit, identifiable as sulphur, forms on the walls of the gas jar. The reaction can be represented:

$$2H_2S + O_2 \rightarrow 2H_2O + 2S$$

A similar reaction occurs if gas jars of hydrogen sulphide gas and chlorine gas are mixed. A yellow deposit, identifiable as sulphur, again forms. Moreover, a choking gas is produced which forms white clouds with ammonia gas. The choking gas is evidently hydrogen chloride and the reaction can be represented:

$$H_2S + Cl_2 \rightarrow 2HCl + S$$

The first reaction is clearly oxidation of hydrogen sulphide because oxygen has been added. Another way of viewing this, is to emphasize the removal of

hydrogen from the H_2S. The second reaction of H_2S is very similar, and although no oxygen is involved in the reaction, it is convenient to regard the reaction as a redox reaction.

(iii) *Oxidation is an increase in valency of the cation.*

Iron(II) oxide burns in air to form iron(III) oxide.

$$4FeO + O_2 \rightarrow 2Fe_2O_3$$

This reaction is oxidation of the iron(II) oxide (addition of oxygen).

If iron(II) chloride solution is treated with a saturated solution of chlorine, the iron(II) chloride solution changes colour from pale green to yellow. If caustic soda solution is added to the product, a red-brown precipitate forms (evidently iron(III) hydroxide), whereas the original iron(II) salt forms a green precipitate with sodium hydroxide and potash solution (iron(II) hydroxide).

Hence chlorine water changes the iron(II) salt to an iron(III) salt and the reaction may be represented:

$$2FeCl_2 + Cl_2 \rightarrow 2FeCl_3$$

This latter reaction is so similar to the oxidation of iron(II) oxide, that it would be desirable to regard it as oxidation of iron(II) chloride.

Similarly for the reaction:

$$2FeS + S \rightarrow Fe_2S_3$$

The three reactions of oxidation of iron(II) compounds are similar in that they show the addition of oxygen or some other non-metal. The reactions also suggest that in general, it is convenient to regard an increase of valency of the cation as oxidation.

Elementary Concepts of Reduction

(i) *Reduction is the removal of oxygen.*

$$CuO + H_2 \rightarrow Cu + H_2O$$
$$H_2O + Mg \rightarrow MgO + H_2$$

In these reactions the copper(II) oxide and the steam are reduced by removal of oxygen.

(ii) *Reduction is the addition of hydrogen.*

$$H_2S + Cl_2 \rightarrow 2HCl + S$$

This reaction is oxidation of H_2S by Cl_2. The Cl_2 is therefore reduced in forming HCl. Hence the reaction

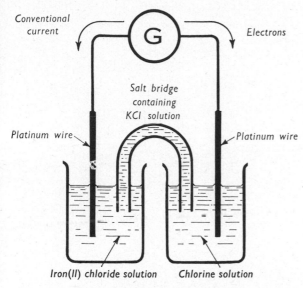

Fig. 14.2. Electron transfer during redox reactions in solution.

illustrates that the addition of hydrogen can be regarded as reduction.

(iii) *Reduction is a decrease of the valency of the cation.*

$$CuO + H_2 \rightarrow Cu + H_2O$$

The copper oxide has been reduced, and at the same time the valency of the cation, copper(II), has decreased from 2 to zero (uncombined copper).

ELECTRON TRANSFER DURING REDOX REACTIONS IN SOLUTION

1. Reaction of iron(II) chloride solution and chlorine water.

$$2FeCl_2 + Cl_2 \rightarrow 2FeCl_3$$

If the salts are written in ionic form, the equation becomes

$$2Fe^{2+} + 4Cl^- + Cl_2 \rightarrow 2Fe^{3+} + 6Cl^-$$
$$\text{or:} \quad 2Fe^{2+} + Cl_2 \rightarrow 2Fe^{3+} + 2Cl^-$$

From earlier chapters it can be seen that the change of *an iron(II) ion to an iron(III) ion involves the removal of one electron.* Likewise, the change of *a chlorine atom to a chloride ion involves the addition of one electron.* It therefore appears that *electrons are transferred* from the iron(II) ions to the chlorine molecules. This theory suggests that an experiment

be devised to test the theory that electrons are transferred.

If a transfer of electrons is to be detected, it is necessary to separate the solution of the oxidizer from the solution of the reducer, and provide an external circuit through metallic wires. The external circuit must contain a current-detecting device such as a galvanometer. The arrangement is shown in the diagram fig. 14.2. The circuit is completed between the two beakers by using a salt bridge containing an electrolytic conductor such as potassium chloride. The platinum wire shown in fig. 14.2 can be replaced by the more elaborate electrode shown in fig. 14.3, or carbon electrodes may be used.

When the circuit is completed, it is observed that the galvanometer needle is deflected. The direction of deflection indicates that conventional current flows from the chlorine water to the iron(II) salt solution along the wires. It can therefore be inferred that *electrons flow along the wires from the iron(II) salt solution to the chlorine water.* This supports the conclusion that iron(II) ions are donating electrons and the chlorine molecules are accepting electrons:

$$Fe^{2+} \rightarrow Fe^{3+} + e^- \tag{i}$$
$$\text{and } Cl_2 + 2e^- \rightarrow 2Cl^- \tag{ii}$$

In order to balance the number of electrons in the half reactions (i) and (ii) above, the first half reaction must be multiplied by 2.

$$[Fe^{2+} \rightarrow Fe^{3+} + e^-] \times 2$$
$$Cl_2 + 2e^- \rightarrow 2Cl^-$$

Adding these two, gives the equation:

$$2Fe^{2+} + Cl_2 \rightarrow 2Fe^{3+} + 2Cl^-$$

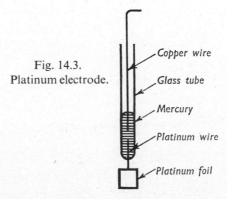

Fig. 14.3.
Platinum electrode.

Note on the Salt Bridge

It will be seen that the loss of electrons from the solution containing the iron(II) salt, will result in a net overall positive charge developing in the solution. Similarly, the gain of electrons by the chlorine water will result in a net negative charge developing. These charges will stop the reactions almost immediately, unless they are neutralized.

The electrolytic conductor in the salt bridge contains cations, K^+, and anions, Cl^-. These ions can migrate in opposite directions so that the potassium ions can partner the chloride ions produced in the chlorine water, and the chloride ions of the salt bridge can help to partner the iron(III) ions being produced in the solution of the iron(II) salt.

Summing Up

In the half reactions shown above it can be seen that:
the oxidizer (Cl_2) is an electron acceptor:

$$Cl_2 + 2e^- \rightarrow 2Cl^-$$

the reducer (Fe^{2+}) is an electron donor:

$$Fe^{2+} \rightarrow Fe^{3+} + e^-$$

This suggests that other oxidizers and reducers may react in this way, and this can be shown to be true experimentally. Furthermore, certain reactions, which are not regarded as redox reactions in more elementary chemistry, can also be shown to involve electron transfer.

2. Reaction of zinc metal and copper(II) sulphate solution.

If zinc metal is placed in copper(II) sulphate solution a dark brown deposit slowly forms on the zinc and the blue solution slowly becomes colourless. The brown deposit can be identified as copper and the solution can be shown to contain zinc sulphate. The reaction may be represented:

$$Zn + CuSO_4 \rightarrow ZnSO_4 + Cu$$

Writing the salts in ionic form, the equation becomes:

$$Zn + Cu^{2+} + SO_4^{2-} \rightarrow Zn^{2+} + SO_4^{2-} + Cu$$
$$\text{or} \qquad Zn + Cu^{2+} \rightarrow Zn^{2+} + Cu$$

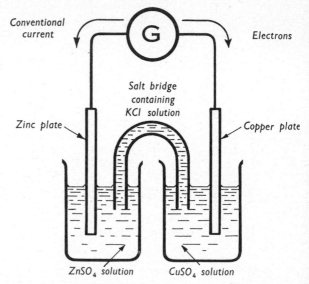

Fig. 14.4. Electron transfer during redox reactions in solution.

Hence it appears that the reaction consists of transfer of electrons from the zinc metal to the copper(II) ions. The flow of electrons can be detected in a manner similar to that described above for the iron(II) salt and chlorine. The apparatus is shown in fig. 14.4.

The direction of deflection of the galvanometer needle, shows that conventional current flows along the wires from the copper electrode to the zinc electrode. Hence electrons flow from the zinc to the copper. This supports the conclusion that the zinc metal is an electron donor and the copper(II) ions are electron acceptors.

$$Zn \rightarrow Zn^{2+} + 2e^-$$
$$Cu^{2+} + 2e^- \rightarrow Cu$$

Addition of these half reactions gives the equation:

$$Zn + Cu^{2+} \rightarrow Zn^{2+} + Cu$$

In this reaction *the zinc metal can be regarded as a reducer (the electron donor) and the copper(II) ions can be regarded as the oxidizer (the electron acceptor).* This is consistent with the view that the copper(II) cation undergoes reduction by decrease of valency from $+2$ to zero while the zinc metal undergoes oxidation by increase of valency from zero to $+2$.

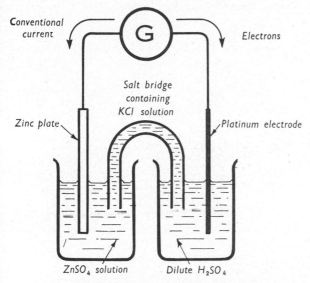

Fig. 14.5. Electron transfer during redox reactions in solution.

3. Reaction of zinc metal with dilute sulphuric acid

This reaction has been described earlier, and it can be represented:

$$Zn + H_2SO_4 \rightarrow ZnSO_4 + H_2$$

or $Zn + 2H^+ + SO_4^{2-} \rightarrow Zn^{2+} + SO_4^{2-} + H_2$

or $\qquad Zn + 2H^+ \rightarrow Zn^{2+} + H_2$

The reaction appears to involve electron transfer and this conclusion can be supported by evidence from experiments similar to those already described. The apparatus is shown in fig. 14.5.

The direction of deflection of the galvanometer needle shows that the conventional current flows in the wires from the platinum electrode to the zinc electrode, and hence electrons flow in the wires from the zinc electrode to the platinum electrode.

In addition, a gas which can be identified as hydrogen is evolved at the platinum electrode and the zinc metal slowly dissolves.

The observations are consistent with the representation:

$$Zn \rightarrow Zn^{2+} + 2e^-$$
$$2H^+ + 2e^- \rightarrow H_2$$

Adding these equations for the half reactions, gives:

$$Zn + 2H^+ \rightarrow Zn^{2+} + H_2$$

In this reaction *the zinc metal can be regarded as a reducer (electron donor) and the hydrogen ions can be regarded as an oxidizer (electron acceptor).* This is consistent with saying that the zinc is oxidized by having its valency increased from 0 to +2.

Summing Up

Generalizing from all the electron-transfer reactions discussed, it is reasonable to extend the notions of oxidizers and reducers to the following:

A substance which accepts electrons is an oxidizer.
A substance which donates electrons is a reducer.

It should be noted that an oxidizer can function as such only if a reducer is present. The relationship between an oxidizer and a reducer can be seen in fig. 14.6.

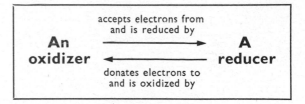

Fig. 14.6.

The table also emphasizes the following:

The addition of electrons to a substance is reduction of the substance.
The removal of electrons from a substance is oxidation of the substance.

Further Redox Reactions

Many other redox reactions will be encountered later, but for the present only one further redox reaction will be considered. This example will be used to illustrate the method of building up an overall ionic equation, by consideration of partial ionic equations.

Reaction of copper and concentrated nitric acid.

If concentrated nitric acid is poured on copper a brisk effervescence occurs and a brown gas is evolved. This brown gas can be identified as nitrogen dioxide (NO_2). At the same time the liquid turns green and later turns blue (particularly if some water is added at the end of the reaction). The blue colour suggests the presence of hydrated copper(II) ions, and appropriate tests confirm the presence of these ions in the solution.

The presence of copper(II) ions suggests that one half reaction occurring is:

$$Cu \rightarrow Cu^{2+} + 2e^-$$

The production of nitrogen dioxide is evidently due to the reduction of nitrate ions. Electrons are also evidently consumed:

$$NO_3^- + e^- \rightarrow NO_2$$

It now appears that some positive ions are also consumed, otherwise there is a lack of charge-balance in the 'equation'. The positive ions are presumably the hydrogen ions of the acid and it is these which finish up with the oxygen atom so far not accounted for.

Hence the half reaction can be represented:

$$NO_3^- + 2H^+ + e^- \rightarrow NO_2 + H_2O$$

If the second half reaction is multiplied by two, the number of electrons in each half reaction is the same. Thus:

$$[NO_3^- + 2H^+ + e^- \rightarrow NO_2 + H_2O] \times 2$$
$$Cu \rightarrow Cu^{2+} + 2e^-$$

adding ───────────────────

$$2NO_3^- + 4H^+ + Cu \rightarrow 2NO_2 + 2H_2O + Cu^{2+}$$

This last equation is the required overall ionic equation for the reaction.

NOTE:

As for the equations written in chapter 10, the following points apply:

(i) the reactants and products must be known before the equation can be written;

(ii) the number of atoms or ions on each side of the equation must balance.

There is, however, the additional factor of charge to be accounted for. Each equation must show a balance of charge, i.e., conservation of charge as well as conservation of atoms must apply.

Finally it should be noted that *ionic equations do not necessarily represent the mechanism of a reaction.* This was illustrated for the equations written in chapter 10, and it applies with equal force to ionic equations.

Redox Reactions in Electrolysis

The electrode reactions occurring during electrolysis are reactions which consume or produce electrons and are therefore redox reactions.

Thus, in the electrolysis of dilute sulphuric acid using unreactive electrodes, the electrode reactions can be represented:

CATHODE REACTION:

$$2H^+ + 2e^- \rightarrow H_2$$

ANODE REACTION:

$$2H_2O \rightarrow O_2 + 4H^+ + 4e^-$$

The cathode reaction is one in which the hydrogen ions accept electrons and the reaction is therefore reduction of the H^+.

● *Cathode reactions always consume electrons and are therefore reduction reactions.*

Conversely, the anode reaction shown above, is one in which water donates electrons and the reaction is therefore oxidation of the water.

● *Anode reactions always produce electrons and are therefore oxidation reactions.*

SUMMARY

When solutions are mixed, reactions may take place between the ions and/or molecules present in the mixture. Reactions may occur as follows:

(i) the ions of an insoluble substance combine and form a precipitate of the substance;

(ii) the ions of a weak electrolyte combine and form molecules of the weak electrolyte;

(iii) atoms, ions or molecules capable of donating and accepting electrons may react by transferring electrons, and a redox reaction takes place.

STUDY QUESTIONS

1. Write ionic equations for the reactions which occur when the following pairs of substances are mixed:
   (a) silver nitrate solution and dilute hydrochloric acid;
   (b) silver sulphate solution and potassium chloride solution;
   (c) barium chloride solution and dilute sulphuric acid;
   (d) caustic soda solution and iron(III) chloride solution;
   (e) caustic potash solution and copper(II) sulphate solution;
   (f) dilute sulphuric acid and sodium hydroxide and potash solution;
   (g) dilute nitric acid and calcium hydroxide solution;
   (h) dilute nitric acid and sodium carbonate;
   (i) dilute hydrochloric acid and iron(II) sulphide.

2. For each of the following reactions, name the oxidizer and give a reason for your answer:
   (a) $NO_2 + SO_2 \rightarrow SO_3 + NO$.
   (b) $H_2 + Cl_2 \rightarrow 2HCl$.
   (c) $CuO + H_2 \rightarrow Cu + H_2O$.
   (d) $2Na + Cl_2 \rightarrow 2Na^+ + 2Cl^-$.
   (e) $2Fe^{3+} + Sn^{2+} \rightarrow 2Fe^{2+} + Sn^{4+}$.
   (f) $2I^- + Br_2 \rightarrow 2Br^- + I_2$.
   (g) $Cl_2 + CH_4 \rightarrow CH_3Cl + HCl$.
   (h) $2Fe(OH)_2 + H_2O_2 \rightarrow 2Fe(OH)_3$.
   (i) $CO_2 + C \rightarrow 2CO$.

3. If clean iron nails are placed in copper(II) sulphate solution, copper is deposited on the iron and a solution of iron(II) sulphate forms.
   Write the ionic equation for the reaction.

4. Write ionic equations for the reactions which occur when:
   (a) magnesium dissolves in dilute sulphuric acid;
   (b) iron dissolves in dilute sulphuric acid;
   (c) zinc dissolves in dilute hydrochloric acid.

5. When dilute nitric acid is poured on to copper, a colourless gas (nitrogen monoxide, NO) is evolved and a blue solution (copper(II) ions) forms.
   Write partial ionic equations for the reactions which occur and hence derive the overall ionic equation.

6. Neutralization of a variety of strong acids with sodium hydroxide solution always results in the evolution of very nearly the same amount of heat per mole of water formed. What conclusions can be drawn from this?

7. Draw and label a diagram of the apparatus which could be used to demonstrate that the reaction between iron(II) sulphate solution and chlorine solution involves electron transfer.
   Indicate the expected direction of electron flow and write the equations for the reactions which occur.

8. Explain why electrode reactions occurring during electrolysis are regarded as redox reactions.

15 : Displacement Reactions of Metals

Although the term 'metal' is widely used to classify certain elements it is not possible to give an exact definition of the term. In the early chapters in this book, the elements which can easily act as electron donors, i.e. elements which are reducers, were called metals. If an atom of an element is capable of giving up some of its electrons easily, it must have only a few electrons in its outer shell or have many electron shells so that the outer shell electrons are not strongly attracted by the positive charge of the nucleus. Atoms which have both of these features will give up their electrons most easily and these elements will be the most 'metallic' of the elements. Thus the metals are found towards the left hand side (few outer shell electrons) and towards the bottom (many electron shells) of the periodic classification. This is shown in the diagram which also includes the symbols of the twelve metals which will be described in this chapter (fig. 15.1).

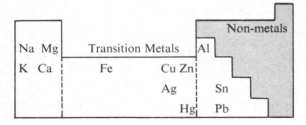

Fig. 15.1.

The twelve metals chosen for study are common, have many important applications and cover a range of metals of the main groups and of the transition metals.

If the relationship between electronic structure and metallic properties were a simple one, the order of reactivity of the metals, listing the most reactive first, would be K, Na, Ca, Mg, Fe, Ag, Cu, Hg, Zn, Al, Pb and Sn. However, as will be shown, this is not the correct order although parts of it are correct. The periodic table cannot be used to predict chemical reactivity. Instead, experimental investigations of the metals must be made first. Then, when the order of reactivity is known, it may be possible to derive some useful correlation with atomic structure.

The periodic table can be used as a guide to finding general trends in properties. Thus, elements with high atomic numbers have relatively heavy atoms and so their densities should be higher. Approximate values for melting points, boiling points and densities for the elements listed in order of increasing atomic number are given in the table (fig. 15.2).

METAL	ATOMIC NUMBER	Approximate Values for:		
		M.P. (°C)	B.P. (°C)	DENSITY (g cm⁻³)
sodium	11	97	880	0.97
magnesium	12	650	1100	1.7
aluminium	13	660	2060	2.7
potassium	19	62	760	0.87
calcium	20	845	1240	1.5
iron	26	1540	3000	7.9
copper	29	1080	2340	8.9
zinc	30	420	905	7.1
silver	47	960	1950	10.5
tin	50	230	2270	7.3
mercury	80	−39	360	13.6
lead	82	330	1620	11.0

Fig. 15.2.

These metals have lustrous surfaces when freshly cut or scraped and they are silvery coloured except for copper which is pink. They all tarnish in air except tin and mercury. Aluminium does not appear

Fig. 15.3. The surface of a piece of aluminium sheet which has been rubbed with mercury.
Fibres of aluminium oxide form, and in less than an hour a thick deposit forms.

to tarnish but this is because it is covered with a very thin transparent film of aluminium oxide. This film protects the metal from further corrosion unless the oxide film is removed. This can be demonstrated by rubbing some mercury on the surface of a piece of aluminium. White powdery aluminium oxide forms (see fig. 15.3).

Potassium and sodium tarnish very rapidly in air and, for this reason, they are usually stored under a layer of oil.

REDUCTION OF WATER BY METALS
Sodium and Cold Water

When a small piece of sodium is cautiously placed on cold water it reacts vigorously. It is advisable to carry out this test in a shallow, stout walled vessel such as a pneumatic trough, and to cover the vessel with a sheet of glass.

OBSERVATIONS:

The sodium melts, and the molten bead of metal moves rapidly around on the surface of the water and gradually becomes smaller. A fizzing sound can be be heard while the sodium is reacting (see fig. 15.4).

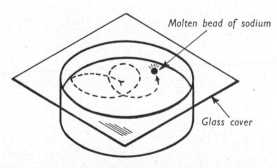

Fig. 15.4. Sodium on water.

If an indicator is added to the water before the reaction, the colour change indicates that the solution is becoming alkaline during the reaction.

The fizzing sound and the rapid movement of the sodium suggest that a gas is being evolved underneath the bead where it is in contact with the water. This can be verified by dropping a cone of steel wire mesh over the bead. This forces it below the water and holds it steady. A stream of bubbles of a colourless gas,

rises from the sodium and can be collected in an inverted test tube full of water. The gas burns when ignited and is apparently hydrogen (see fig. 15.5).

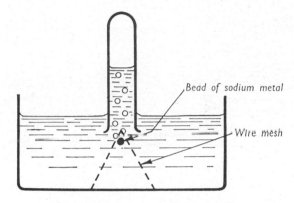

Fig. 15.5. Collection of hydrogen.

EXPLANATION:

The observations suggest that the reaction produces hydrogen and a solution of sodium hydroxide.

$$2Na + 2H_2O \rightarrow 2NaOH + H_2 \qquad (i)$$

Because NaOH is completely ionized, a better representation is:

$$2Na + 2H_2O \rightarrow 2Na^+ + 2OH^- + H_2 \qquad (ii)$$

But the formation of a sodium ion from a sodium atom has already been shown to involve the removal of the one valency electron of the sodium atom:

$$Na \rightarrow Na^+ + e^- \qquad (iii)$$

In the discussion of electrolysis it was shown that water can accept electrons when it takes part in the cathode reaction:

$$2H_2O + 2e^- \rightarrow 2OH^- + H_2 \qquad (iv)$$

Equation (iii) shows the oxidation of sodium. Equation (iv) shows the reduction of water. If equation (iii) is doubled and added to equation (iv), equation (ii) is obtained. Thus it is reasonable to interpret the reaction as one in which *sodium acts as an electron donor (a reducer) and water acts as an electron acceptor (an oxidizer)*.

The equation (ii) summarizes the reaction. Some of the other observations may also be explained. The sodium melts because heat is evolved. The sodium gradually becomes smaller because it is consumed in the reaction.

Potassium and Cold Water

This is very similar to the reaction of sodium. However, the reaction is more violent and a lilac coloured flame is seen.

● *Caution: Potassium metal sometimes explodes after being placed on water, and the same protective device should be used as with sodium.*

The hydrogen being evolved is ignited by the hot potassium. This suggests that the reaction between the potassium and water produces a higher temperature than the reaction of sodium with cold water. The colour of the flame is due to burning potassium (see chapter 3).

$$(K \rightarrow K^+ + e^-) \times 2 \text{ (oxidation)}$$
$$2H_2O + 2e^- \rightarrow 2OH^- + H_2 \text{ (reduction)}$$
$$\overline{2K + 2H_2O \rightarrow 2K^+ + 2OH^- + H_2}$$

reducer   oxidizer

Calcium and Cold Water

The calcium sinks to the bottom and dissolves fairly rapidly. A gas is evolved which can be shown to be hydrogen. The solution becomes alkaline and then, after a time, a white precipitate begins to form.

The reaction is similar in principle to the above examples.

$$Ca \rightarrow Ca^{2+} + 2e^- \quad \text{(oxidation)}$$
$$2H_2O + 2e^- \rightarrow 2OH^- + H_2 \quad \text{(reduction)}$$
$$\overline{Ca + 2H_2O \rightarrow Ca^{2+} + 2OH^- + H_2}$$

reducer   oxidizer

The solution becomes alkaline because calcium hydroxide solution is formed. Calcium hydroxide is not very soluble in water, and the white precipitate is apparently solid calcium hydroxide which forms when the solution becomes saturated. This precipitation can be represented by the equation:

$$Ca^{2+} + 2OH^- \rightleftharpoons Ca(OH)_2 \text{ (s)}$$

Magnesium and Water

If a short strip of clean magnesium is placed in cold water, little apparent reaction takes place. However, if phenolphthalein is added to the water, a red colour, which indicates an alkaline solution, will slowly appear around the magnesium.

Evidently the reaction of water with magnesium is similar to the reaction with calcium, but is very much slower and proceeds only to a limited extent. With hot water, the reaction is somewhat faster but is still quite slow.

Aluminium and Water

Aluminium does not react with water. However, if mercury is rubbed over the surface of the aluminium, the metal slowly displaces hydrogen even from cold water. The lack of reactivity of aluminium can therefore be attributed to the thin layer of oxide (see earlier).

REDUCTION OF STEAM BY METALS

The following metals do not react readily with water and their reactions with steam will be studied.

Magnesium and Steam

A hole is blown in the end of a test tube and the test tube is set up as shown in fig. 15.6. The water is boiled so that the steam will expel all the air from the test tube. The magnesium is heated strongly until it

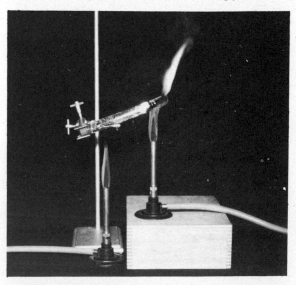

Fig. 15.6. Combustion of magnesium in steam. The burning magnesium glows brightly and the hydrogen issuing from the hole in the end of the tube burns in the air.

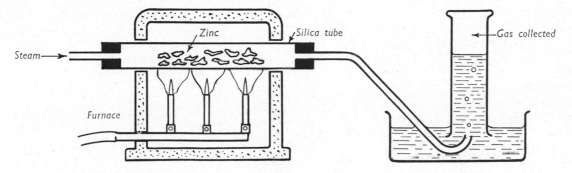

Fig. 15.7. Zinc and superheated steam.

begins to glow and then heating of the magnesium is discontinued. The magnesium continues to burn in the steam with a brilliant white light.

● *Caution: It is dangerous to look straight at the burning magnesium as the light will damage the eyes.*

A white powdery residue is left and a flame is observed if the gas leaving the tube is ignited. The gas is apparently hydrogen.

If the powdery residue is removed from the tube and placed in water, it forms a slightly alkaline solution although very little of it dissolves. The white residue after combustion is apparently magnesium oxide (and not magnesium hydroxide, because the hydroxide is easily decomposed to the oxide on heating).

$$Mg + H_2O \rightarrow MgO + H_2$$
reducer oxidizer

Zinc and Superheated Steam

Zinc does not burn in steam, but if steam is passed over zinc heated in a furnace to 400 °C, reaction takes place slowly. A gas is evolved which can be shown to be hydrogen (see fig. 15.7). The zinc becomes coated with a powdery residue which is yellow when hot and white when cold. This colour change is a property of zinc oxide, and apparently this is the residue.

$$Zn + H_2O \rightarrow ZnO + H_2$$
reducer oxidizer

Iron and Superheated Steam

This is similar to zinc but requires a temperature of about 700 °C. The iron becomes coated with a hard black substance which can be shown to be Fe_3O_4. This is called magnetic iron oxide or iron(II, III) oxide, because it contains iron with both valencies: $Fe_3O_4 = (Fe^{2+})(Fe^{3+})_2(O^{2-})_4$.

$$3Fe + 4H_2O \rightleftharpoons Fe_3O_4 + 4H_2$$
reducer oxidizer

A reversible sign is used here because the reverse reaction can also occur. Thus if hydrogen is passed over hot Fe_3O_4, the oxide is reduced to iron and the hydrogen is oxidized to steam.

Tin, Lead, Copper, Mercury and Silver

These metals do not react with water or with steam even at fairly high temperatures.

Summing Up

The order in which the metals are tested is such that the first metals are the more powerful reducers. Potassium, sodium and calcium reduce cold water. Magnesium, zinc and iron reduce steam. Aluminium appears to be out of order, but its apparent lack of reactivity can be explained in terms of the coating of oxide on its surface. This protects it from water and steam.

REDUCTION OF ACIDS BY METALS

For the present only reactions with hydrochloric acid and dilute sulphuric acid will be considered. The reactions of nitric acid and concentrated sulphuric acid will be considered in the next chapter.

The reactions of potassium, sodium and calcium will not be considered, because they react with water; if hydrogen were evolved no useful conclusion could be drawn as it could not be clearly said whether the metal had reacted with the water or with the acid. Furthermore, it would be extremely dangerous to place any of these metals in an acid solution as the reaction could be extremely violent.

Magnesium

If magnesium is placed in dilute hydrochloric acid or in dilute sulphuric acid, it dissolves rapidly with vigorous evolution of a colourless gas which can be shown to be hydrogen.

Reactions of this type were discussed in the previous chapter and it was shown that the reaction can be interpreted as a transfer of electrons from the metal (the reducer) to the hydrogen ions of the acid (the oxidizer).

$$\frac{\begin{array}{ll} Mg \rightarrow Mg^{2+} + 2e^- & \text{(oxidation)} \\ 2H^+ + 2e^- \rightarrow H_2 & \text{(reduction)} \end{array}}{\underset{\text{reducer oxidizer}}{Mg + 2H^+ \rightarrow Mg^{2+} + H_2}}$$

If the acid used is hydrochloric, two chloride ions remain unused for each magnesium ion formed. These are spectator ions and partner the magnesium ion in solution. Thus the solution contains magnesium chloride. If sulphuric acid is used, the sulphate ions are the spectator ions and the solution contains magnesium sulphate.

$$Mg + 2H^+ + 2Cl^- \rightarrow Mg^{2+} + 2Cl^- + H_2$$
$$Mg + 2H^+ + SO_4^{2-} \rightarrow Mg^{2+} + SO_4^{2-} + H_2$$

Aluminium

There appears to be no reaction at first but after a time, particularly if the mixture is warmed, the aluminium begins to react. The reaction with dilute hydrochloric acid becomes quite vigorous after a time, with rapid evolution of a colourless gas which can be shown to be hydrogen. The reaction with dilute sulphuric acid is much slower.

The delay of the start of the reaction can be attributed to the protective layer of oxide on the surface of the aluminium. This oxide apparently dissolves fairly slowly in the acid. Once the aluminium is exposed it dissolves and reduces the acid to hydrogen.

$$\frac{\begin{array}{ll} (Al \rightarrow Al^{3+} + 3e^-) \times 2 & \text{(oxidation)} \\ (2H^+ + 2e^- \rightarrow H_2) \times 3 & \text{(reduction)} \end{array}}{\underset{\text{reducer oxidizer}}{2Al + 6H^+ \rightarrow 2Al^{3+} + 3H_2}}$$

Zinc

The reaction is similar to the reaction of magnesium but is slower: the hydrogen is produced at a steady rate, without frothing. Because of this, the reaction is often used as a method for preparing hydrogen (see chapter 22).

$$\frac{\begin{array}{ll} Zn \rightarrow Zn^{2+} + 2e^- & \text{(oxidation)} \\ 2H^+ + 2e^- \rightarrow H_2 & \text{(reduction)} \end{array}}{\underset{\text{reducer oxidizer}}{Zn + 2H^+ \rightarrow Zn^{2+} + H_2}}$$

Iron

This is similar to zinc but the reaction is slower. Often the acid must be heated to obtain a reasonably rapid evolution of gas. The solution formed as the iron dissolves is pale green, which suggests that the iron is oxidized to the iron(II) state and not the iron(III) state (iron(II) ions are green, iron(III) ions are yellow-brown). The addition of excess caustic soda solution to the green solution, causes formation of a green gelatinous precipitate. It was seen in the previous chapter that iron(II) hydroxide is green and hence the precipitate is evidently iron(II) hydroxide. This confirms that the reaction of iron with the acids, produces an iron(II) salt:

$$\frac{\begin{array}{ll} Fe \rightarrow Fe^{2+} + 2e^- & \text{(oxidation)} \\ 2H^+ + 2e^- \rightarrow H_2 & \text{(reduction)} \end{array}}{\underset{\text{reducer oxidizer}}{Fe + 2H^+ \rightarrow Fe^{2+} + H_2}}$$

Tin

Granulated tin dissolves very slowly in cold dilute hydrochloric acid. If the mixture is heated, or if hot concentrated hydrochloric acid is used, the tin dissolves more rapidly forming a colourless solution and with evolution of a colourless gas.

The colourless solution can be shown to contain tin(II) ions and the gas can be shown to be hydrogen.

$$Sn \rightarrow Sn^{2+} + 2e^- \quad \text{(oxidation)}$$
$$2H^+ + 2e^- \rightarrow H_2 \quad \text{(reduction)}$$
$$\overline{Sn + 2H^+ \rightarrow Sn^{2+} + H_2}$$
reducer oxidizer

Lead

Lead slowly becomes coated with a white layer of an insoluble substance when placed in cold hydrochloric or sulphuric acid. Since lead(II) chloride and lead(II) sulphate are both insoluble white substances, these are apparently formed on the surface. These insoluble layers slow down the reactions. However, lead dissolves fairly readily in hot concentrated hydrochloric acid.

Copper, Mercury and Silver

There is no apparent reaction if these metals are placed in hydrochloric or dilute sulphuric acid. Thus they do not reduce the hydrogen ions in aqueous solutions of these acids.

Summing Up

The order in which the metals are tested is such that the first metals are the more powerful reducers as shown by the readiness with which they reduce acids.

REDUCTION OF SALT SOLUTIONS BY METALS

Potassium, sodium and calcium will not be discussed because they react with water. Aluminium will be omitted because its oxide film protects it from reaction with the solution. The discussion will start with zinc.

Reduction of Salt Solutions by Zinc

Copper(II) sulphate solution is often prepared by dissolving copper(II) sulphate in water and then adding dilute sulphuric acid to produce a clear solution. If zinc metal is added to such a solution,

hydrogen gas is evolved, and therefore *aqueous* copper(II) sulphate solution should be used in the following experiment.

The displacement of copper from aqueous copper(II) sulphate solution by zinc was discussed in chapter 14.

$$Zn \rightarrow Zn^{2+} + 2e^- \quad \text{(oxidation)}$$
$$Cu^{2+} + 2e^- \rightarrow Cu \quad \text{(reduction)}$$
$$\overline{Zn + Cu^{2+} \rightarrow Zn^{2+} + Cu}$$
reducer oxidizer

This reaction can be interpreted as being due to the readiness with which zinc forms positive ions as compared to copper forming copper(II) ions. It seems likely then, that zinc would displace all metals which are poorer reducers, from solutions of their salts.

● *In the following tests it is assumed that the salt solution is prepared by dissolving the salt in water without the addition of any acid.*

Zinc in iron(II) sulphate solution slowly becomes coated with a black deposit which is apparently metallic iron.

$$Zn \rightarrow Zn^{2+} + 2e^- \quad \text{(oxidation)}$$
$$Fe^{2+} + 2e^- \rightarrow Fe \quad \text{(reduction)}$$
$$\overline{Zn + Fe^{2+} \rightarrow Zn^{2+} + Fe}$$
reducer oxidizer

Zinc in tin(II) chloride solution becomes coated with a spongy grey deposit of tin.

$$Zn \rightarrow Zn^{2+} + 2e^- \quad \text{(oxidation)}$$
$$Sn^{2+} + 2e^- \rightarrow Sn \quad \text{(reduction)}$$
$$\overline{Zn + Sn^{2+} \rightarrow Zn^{2+} + Sn}$$
reducer oxidizer

Zinc in lead(II) nitrate solution becomes coated with a dark grey crystalline deposit of metallic lead.

$$Zn \rightarrow Zn^{2+} + 2e^- \quad \text{(oxidation)}$$
$$Pb^{2+} + 2e^- \rightarrow Pb \quad \text{(reduction)}$$
$$\overline{Zn + Pb^{2+} \rightarrow Zn^{2+} + Pb}$$
reducer oxidizer

Zinc in mercury(II) chloride solution slowly becomes coated with a grey deposit of metallic mercury.

$$Zn \rightarrow Zn^{2+} + 2e^- \quad \text{(oxidation)}$$
$$Hg^{2+} + 2e^- \rightarrow Hg \quad \text{(reduction)}$$
$$\overline{Zn + Hg^{2+} \rightarrow Zn^{2+} + Hg}$$
reducer oxidizer

Zinc in silver nitrate solution slowly becomes coated with a black deposit of finely divided metallic silver.

$$Zn \rightarrow Zn^{2+} + 2e^- \quad \text{(oxidation)}$$
$$\underline{(Ag^+ + e^- \rightarrow Ag) \times 2 \quad \text{(reduction)}}$$
$$\underset{\text{reducer} \quad \text{oxidizer}}{Zn + 2Ag^+ \rightarrow Zn^{2+} + 2Ag}$$

Thus it can be seen that zinc displaces all of the metals which are poorer reducers, from solutions of their salts.

Reduction of Salt Solutions by Other Metals

It would be expected that any metal should reduce the ions of all of the metals which are poorer reducers. The observations which are obtained from such tests are summarized in the table (fig. 15.8). Detailed explanations will not be given as the discussion of the reactions with zinc should be sufficient guide.

Summing Up

The reduction of salt solutions by metals further confirms the order of reactivity with water and acids.

The most useful reactions are those which allow the least reactive metals to be arranged in order. Previously it had only been possible to say that none of them react with water or acids.

THE ACTIVITY SERIES

The sequence of metals arranged in order of the readiness with which they react with water or steam is the same as their sequence of reactivity with dilute hydrochloric or sulphuric acid and salt solutions. Not all of the metals were tested with each of the reagents listed, but in all cases, those tested fell into practically the same sequence. Occasionally other factors obscure the order, such as the lack of reactivity of aluminium with water because of its protective oxide film, but the generalization is still worthwhile.

The similarity in sequence suggests a similarity in the reactions of the metals in each case. This has been illustrated by showing that each reaction can be interpreted as donation of electrons by atoms of the metals during formation of positive ions.

METAL	Observations obtained with a solution of:				
	$SnCl_2$	$Pb(NO_3)_2$	$CuSO_4$	$HgCl_2$	$AgNO_3$
Fe	black deposit	grey deposit (also a white substance)	red-brown deposit (1)	dark grey coating	black deposit
Sn		little apparent change	red-brown deposit (1)	white coating	black deposit
Pb			little apparent change	white coating	grey deposit
Cu		*NO APPARENT CHANGE*		dark grey coating (2)	black deposit (2)
Hg					grey crystalline deposit

Some reactions are very slow and require several days.

Note (1) Blue colour of solution fades. (2) Solution slowly becomes blue.

Fig. 15.8. Reduction of salt solutions by metals.

● *The sequence is called the activity series or the displacement series of the metals and is shown in fig. 15.9.*

It has been suggested that the activity series is the order of ease of formation of positive ions in solution. This seems to be justified by the reasons given but further experimental work is desirable to test the validity of the statement. An experiment can be devised to do this and it will now be described.

EXPERIMENT: *To test whether metals high on the activity series can donate electrons more easily than metals low on the activity series.*

To observe that electrons can be donated by a metal, it is suitable to use cells such as those described in chapter 14, in the discussion of redox reactions in terms of electron transfer. So that comparisons can be made for a number of metals it is desirable to keep all factors constant, except the nature of the metal used as the electrode.

A cell which can be used is represented in the diagram (fig. 15.10). The voltmeter measures the

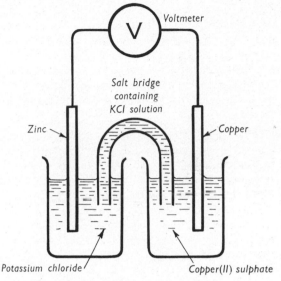

Fig. 15.10. Voltage between zinc and copper electrodes.

METAL	ACTIVITY WITH COLD WATER	ACTIVITY WITH STEAM	ACTIVITY WITH ACIDS	ACTIVITY WITH SOLUTIONS OF METALLIC SALTS
K				
Na	displace hydrogen from cold water			
Ca				
Mg				
Al	do not displace hydrogen from cold water	displace hydrogen from steam	displace hydrogen from hydrochloric and dilute sulphuric acids	displace lower metals from solutions of their salts
Zn				
Fe				
Sn	do not displace hydrogen from cold water or steam			
Pb				
Cu	do not displace hydrogen from cold water, steam or acids			
Hg				
Ag				

Fig. 15.9. The Activity Series.

difference in electrical potential between the two electrodes. It indicates that for the particular conditions (size of electrode, concentration of solution, etc.) the cell develops a potential of 1.0 volt and the zinc electrode is negative relative to the copper electrode. Thus electrons are flowing from the zinc electrode to the copper electrode, suggesting that the reaction of the zinc is:

$$Zn \rightarrow Zn^{2+} + 2e^-$$

The voltage gives a measure of the ease with which electrons can be removed from the zinc compared with copper. The potassium chloride solution is used so that the electrical circuit can be completed.

A number of other metals should now be used instead of zinc. A suitable set is: magnesium, aluminium, iron and lead. Each electrode should be of the same size and should be freshly cleaned by rubbing with emery paper.

A typical set of results is shown in the table (fig. 15.11). It will be seen that the voltage generated is greatest for magnesium and least for lead. The voltage obtained using aluminium is lower than might be expected and this is probably due to the surface film of oxide. It is easy to show that the voltage is much less for a piece of aluminium which has not been rubbed with emery paper and this lends support to the suggestion made concerning the oxide film.

ELECTRODE	VOLTMETER READING
magnesium	1.5
aluminium	0.7
zinc	1.0
iron	0.3
lead	0.2

Fig. 15.11.

The experiment shows that the reactions:

$$Mg \rightarrow Mg^{2+} + 2e^-$$
$$Zn \rightarrow Zn^{2+} + 2e^-$$
$$Fe \rightarrow Fe^{2+} + 2e^-$$
$$Pb \rightarrow Pb^{2+} + 2e^-$$

are in the order of the best electron donor first, and the poorest last. This provides further support for the

postulate *that the activity series lists the metals in order of the ease of formation of positive ions in solution.*

If the copper electrode is examined at the end of the experiment it will be seen that a fresh deposit of copper has formed on it. The electrons produced by the more active metals are apparently consumed at the copper electrode by the reaction:

$$Cu^{2+} + 2e^- \rightarrow Cu$$

Commercial Applications

Portable electrically operated appliances use a variety of power sources commonly called dry cells, batteries or powerpacks. Several different chemical systems are at present in use but batteries using electrodes of *carbon* and *zinc* are the most widely used, the cheapest, the most reliable, and the most readily available.

Carbon-Zinc Batteries

The most usual shape is cylindrical with a carbon rod down the centre. This is surrounded by a solid mixture of MnO_2 (s), NH_4Cl (s) and powdered C (s). This mixture is wet with an electrolyte consisting of a solution containing $ZnCl_2$ and NH_4Cl. Because NH_4^+ is an acid, the electrolyte solution is acidic. The outer cylindrical container is made of zinc. See fig. 15.12.

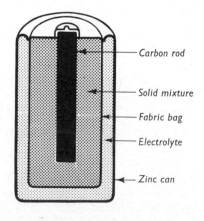

Fig. 15.12. A carbon-zinc dry cell.

The electricity produced by the cell is generated by chemical reactions of the Zn and the MnO_2. The Zn dissolves and releases electrons and hence generates a negative charge on the zinc container:

$$Zn \rightarrow Zn^{2+} + 2e^-$$

The MnO_2 consumes H^+ ions from the electrolyte and electrons from the carbon rod. Hence a positive charge is generated on the carbon rod:

$$MnO_2 + 4H^+ + 2e^- \rightarrow Mn^{2+} + 2H_2O$$

One disadvantage of this arrangement is that the zinc container develops holes as the zinc dissolves and this allows the electrolyte to leak out. An alternative design uses a pressed carbon outer container in a steel can. Several strips of zinc are placed near the centre of the battery. This insures much more efficient use of the zinc and also makes the battery leak-proof.

Alkaline Batteries

These are essentially the same as carbon-zinc batteries but use an alkaline solution of KOH as electrolyte. Because the electrolyte is not acidic, steel can be used as the positive terminal instead of carbon and the whole steel container can be sealed. These batteries are leak-proof. The reactions which generate the electricity in an alkaline battery are:

$$Zn + 4OH^- \rightarrow Zn(OH)_4^{2-} + 2e^-$$
$$MnO_2 + 2H_2O + e^- \rightarrow Mn(OH)_3 + OH^-$$

Rechargeable Batteries

Many electronic devices are powered by rechargeable batteries based on compounds of nickel and cadmium. The electrodes in these batteries are made by heating a layer of powdered nickel with nickel gauze to produce a porous, flexible layer of the metal. The battery is then assembled from three layers:

(i) a porous nickel sheet soaked in a solution of a nickel salt;

(ii) a sheet of absorbent paper soaked in potassium hydroxide solution;

(iii) a porous nickel sheet soaked in a solution of a cadmium salt.

The sheets are rolled into a tight cylinder and sealed in a steel container. This is fully leak-proof.

The KOH reacts with the salts and precipitates insoluble hydroxides in the pores of each metal plate: thus one plate is filled with $Ni(OH)_2(s)$ and the other with $Cd(OH)_2(s)$. The cell is then charged by passing an electric current through it. The $Ni(OH)_2(s)$ plate is made the anode and oxidation occurs at this electrode:

$$Ni(OH)_2 (s) + 2OH^- \rightarrow NiO_2 (s) + 2H_2O + 2e^-$$

The $Cd(OH)_2$ (s) plate is the cathode and reduction occurs at this electrode:

$$Cd(OH)_2 (s) + 2e^- \rightarrow Cd (s) + 2OH^-$$

The electrolyte is not consumed, no gas is evolved and, because all of the nickel and cadmium compounds are solids, they remain trapped in the pores of the electrodes.

When the cell is used to generate electricity, the reverse reactions take place. The $NiO_2(s)$ consumes electrons and so generates a positive charge. The Cd (s) releases electrons and so generates a negative charge. The insoluble hydroxides are reformed in the pores of the electrodes.

The operation of the cell can be represented by the equation:

$$NiO_2 (s) + Cd (s) + 2H_2O$$

charging ⇧   discharging ⇩

$$Ni(OH)_2 (s) + Cd(OH)_2 (s)$$

Several hundreds of cycles of charging and discharging can be achieved giving a long service life to the battery.

THE ELECTROCHEMICAL SERIES

The potential differences between the potentials of various electrodes and the potential of the Cu^{2+}/Cu electrode were measured (fig. 15.10). By convention, electrode potentials are listed relative to the standard hydrogen electrode (see fig. 15.13). E.M.F.'s measured in such cells are treated mathematically to predict what the value would be in an imaginary ideal one molar solution. These imaginary solutions are idealized in the sense that the mathematical procedure removes the experimentally observed effect which the ions in a solution have on each other. The sign given to the computed standard electrode potential is the sign of the charge developed by the electrode relative to the standard hydrogen electrode. Thus, the value of -0.76 V for the Zn^{2+}/Zn electrode means that

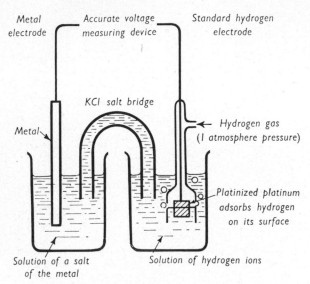

Fig. 15.13. Electrode potential relative to the standard hydrogen electrode.

HALF-REACTION	STANDARD ELECTRODE POTENTIAL
oxidizer + electrons $\rightleftharpoons$ reducer	(Volt)
$F_2 + 2e^- \rightleftharpoons 2F^-$	+ 2.87
$H_2O_2 + 2H^+ + 2e^- \rightleftharpoons 2H_2O$	+ 1.77
$MnO_4^- + 8H^+ + 5e^- \rightleftharpoons Mn^{2+} + 4H_2O$	+ 1.52
$Cr_2O_7^{2-} + 14H^+ + 6e^- \rightleftharpoons 2Cr^{3+} + 7H_2O$	+ 1.36
$Cl_2 + 2e^- \rightleftharpoons 2Cl^-$	+ 1.36
$MnO_2 + 4H^+ + 2e^- \rightleftharpoons Mn^{2+} + 2H_2O$	+ 1.28
$Br_2 + 2e^- \rightleftharpoons 2Br^-$	+ 1.07
$HNO_2 + H^+ + e^- \rightleftharpoons NO + H_2O$	+ 0.99
$NO_3^- + 3H^+ + 2e^- \rightleftharpoons HNO_2 + H_2O$	+ 0.94
$Hg^{2+} + 2e^- \rightleftharpoons Hg$	+ 0.85
$Ag^+ + e^- \rightleftharpoons Ag$	+ 0.80
$Fe^{3+} + e^- \rightleftharpoons Fe^{2+}$	+ 0.77
$O_2 + 2H^+ + 2e^- \rightleftharpoons H_2O_2$	+ 0.68
$I_2 + 2e^- \rightleftharpoons 2I^-$	+ 0.54
$Cu^{2+} + 2e^- \rightleftharpoons Cu$	+ 0.35
$SO_4^{2-} + 4H^+ + 2e^- \rightleftharpoons H_2SO_3 + H_2O$	+ 0.20
$S + 2H^+ + 2e^- \rightleftharpoons H_2S$	+ 0.14
$2H^+ + 2e^- \rightleftharpoons H_2$ (assigned)	**0.00**
$Pb^{2+} + 2e^- \rightleftharpoons Pb$	− 0.12
$Sn^{2+} + 2e^- \rightleftharpoons Sn$	− 0.14
$Ni^{2+} + 2e^- \rightleftharpoons Ni$	− 0.25
$Fe^{2+} + 2e^- \rightleftharpoons Fe$	− 0.44
$2CO_2 + 2H^+ + 2e^- \rightleftharpoons (COOH)_2$	− 0.49
$Cr^{3+} + 3e^- \rightleftharpoons Cr$	− 0.71
$Zn^{2+} + 2e^- \rightleftharpoons Zn$	− 0.76
$Mn^{2+} + 2e^- \rightleftharpoons Mn$	− 1.05
$Al^{3+} + 3e^- \rightleftharpoons Al$	− 1.67
$Mg^{2+} + 2e^- \rightleftharpoons Mg$	− 2.34
$Na^+ + e^- \rightleftharpoons Na$	− 2.71
$Ca^{2+} + 2e^- \rightleftharpoons Ca$	− 2.87
$Ba^{2+} + 2e^- \rightleftharpoons Ba$	− 2.90
$K^+ + e^- \rightleftharpoons K$	− 2.92

the electrode is negative with respect to the standard hydrogen electrode. Similarly, the Cu^{2+}/Cu electrode is positive (value $+0.35$ V) with respect to the standard hydrogen electrode. A tabulation of standard electrode potentials is given in fig. 15.14.

In the table, fig. 15.8, it is recorded that Pb undergoes little apparent change with $CuSO_4$ solution. From fig. 15.14 it would appear that the potential difference between a Pb^{2+}/Pb electrode and a Cu^{2+}/Cu electrode should be 0.47 v. The actual difference which was measured (see fig. 15.11) was only 0.2 V. These contradictions are pointed out to emphasize the danger of attempting to predict chemical behaviour from a table of standard electrode potentials. The activity series developed in this chapter was based on experimental observations. This approach cannot lead to incorrect predictions. The electrochemical series is given only because it is a widely used device in more advanced chemistry but extreme care must be exercised in using it to make predictions.

Fig. 15.14. Table of Standard Electrode Potentials.

STUDY QUESTIONS

1. Write ionic equations to represent the reaction between iron and:
(a) tin(II) chloride solution;
(b) copper(II) sulphate solution;
(c) mercury(II) chloride solution;
(d) silver nitrate solution.

2. (i) Using the principles outlined in this chapter, write down the expected observations, if freshly scraped magnesium is placed in:
(a) copper(II) sulphate solution;
(b) silver nitrate solution;
(c) sodium chloride solution.
   (ii) Carry out these tests and modify your ideas about these reactions if the actual observations are not as you predicted.

3. If granulated zinc is placed in an acidified iron(III) chloride solution, a colourless odourless gas is evolved and the solution changes colour from yellow to pale green.
Discuss the inferences which can be drawn from these observations and write equations for each half reaction which is apparently taking place.

4. Write ionic equations to represent the reaction of cold water with:
(a) sodium;
(b) calcium;
(c) magnesium.
Indicate the oxidizer and reducer in each reaction.

5. Write partial ionic equations and hence develop ionic equations for the overall reaction, between dilute hydrochloric acid and:
(a) magnesium;
(b) zinc;
(c) iron.
For each overall equation, identify the oxidizer and the reducer.

6. Write partial ionic equations to represent the reaction between zinc and:
(a) iron(II) sulphate solution;
(b) copper(II) sulphate solution;
(c) silver nitrate solution.
For each partial equation, state whether the reaction is oxidation or reduction.

7. Rewrite the equations in questions 4, 5 and 6 using empirical formulae for all reactants and products.

8. A simple battery (known as a Daniell Cell) can be made by using a zinc electrode in dilute sulphuric acid and a copper electrode in copper(II) sulphate solution. Write equations for the reactions which occur in this cell. Which electrode develops a positive charge?

9. In aqueous solutions, MnO_4^- ions are purple, Mn^{2+} ions are colourless, Fe^{2+} ions are pale green and Fe^{3+} ions are very pale yellow. If a green solution of iron(II) sulphate is added to a purple solution of potassium permanganate acidified with dilute sulphuric acid, a very pale yellow solution forms. Develop partial ionic equations and hence write the ionic equation for the overall reaction. Rewrite the equation for the overall reaction using empirical formulae.

10. Use the information given in the table, fig. 15.14, to write the equation for the reaction between an acidified solution of potassium dichromate, $K_2Cr_2O_7$, and a solution of sulphurous acid, H_2SO_3.

11. The most common type of battery used in automobiles is made with:
(a) plates packed with PbO_2 (s);
(b) H_2SO_4 solution as the electrolyte;
(c) plates made of Pb (s).
PbO_2 (s) and Pb (s) are both insoluble in sulphuric acid. As the battery is discharged, both plates become coated with $PbSO_4$(s) which is also insoluble in sulphuric acid.
Write equations to show how this chemical system generates electricity. Which plate develops a positive charge? Is the electrolyte consumed during discharge of the battery?
Which plate should be made the anode during charging of the battery? Why is it important that Pb, PbO_2 and $PbSO_4$ are all insoluble in H_2SO_4 solution?

16 : Further Reactions of Metals

OXIDATION OF METALS BY OXYGEN

In most cases, when a metal is heated in air or oxygen, it burns.

POTASSIUM AND SODIUM

If these metals are heated in air they burn forming a solid product. Potassium burns with a lilac flame and sodium with a yellow flame. Analysis of the product shows that a mixture of oxides is formed.

Sodium forms a mixture of sodium oxide and sodium peroxide, depending on the air supply. In excess air the product is mainly the peroxide:

$$4Na + O_2 \rightarrow 2Na_2O \qquad \text{(sodium monoxide)}$$
$$2Na + O_2 \rightarrow Na_2O_2 \qquad \text{(sodium peroxide)}$$

CALCIUM

If clean pieces of calcium are dropped on to an iron tray which is being heated strongly, the calcium ignites and burns very rapidly with a red coloured flame. The residue is a white solid which turns moist litmus blue and is apparently calcium oxide:

$$2Ca + O_2 \rightarrow 2CaO$$

MAGNESIUM

If a piece of clean magnesium is heated in air, it ignites and burns with an intense white light.

● *Caution: It is dangerous to look at the burning magnesium because the light will damage the eyes.*

The residue is a white solid which is mostly the oxide (some magnesium nitride is also formed).

$$2Mg + O_2 \rightarrow 2MgO$$

ALUMINIUM

Aluminium does not burn in air but can be burnt in oxygen. Photographic flash bulbs use aluminium foil in oxygen gas. When the aluminium is heated by an electric current, it ignites and burns with an intense white light forming aluminium oxide:

$$4Al + 3O_2 \rightarrow 2Al_2O_3$$

ZINC

Zinc is very difficult to burn in air although the vapour of zinc is burnt industrially. Thin zinc foil will burn in air if heated fairly strongly. The product is a powder which is yellow when hot and white when cold.

$$2Zn + O_2 \rightarrow 2ZnO \qquad \text{(zinc oxide)}$$

IRON

The reaction can be most easily observed by plunging a bundle of red hot pieces of fine iron wire into a gas jar of oxygen. The iron ignites and burns with a shower of sparks. A black residue is formed which is evidently Fe_3O_4. Fine steel wool (steel is an alloy containing mainly iron) burns readily if heated in air, forming a black residue:

$$3Fe + 2O_2 \rightarrow Fe_3O_4 \qquad \text{(iron(II, III) oxide)}$$

TIN

Tin is very difficult to burn in air although it is burnt industrially at temperatures above 1500 °C:

$$Sn + O_2 \rightarrow SnO_2 \qquad \text{(tin(IV) oxide)}$$

LEAD

Molten lead oxidizes fairly rapidly on the surface forming a layer of yellow or orange oxide called litharge (PbO):

$$2Pb + O_2 \rightarrow 2PbO \qquad \text{(lead(II) oxide)}$$

COPPER

Copper oxidizes slowly on the surface when heated. At fairly high temperatures, black copper(II) oxide (CuO) forms, but at very high temperatures red copper(I) oxide (Cu_2O) is the product.

$$2Cu + O_2 \rightarrow 2CuO$$
$$4Cu + O_2 \rightarrow 2Cu_2O$$

MERCURY

Mercury oxidizes slowly when heated just below its boiling point, forming a layer of red oxide called mercury(II) oxide:

$$2Hg + O_2 \rightleftharpoons 2HgO$$

This is a reversible reaction because at higher temperatures the oxide decomposes.

SILVER

Silver does not burn or oxidize on its surface when heated in air.

Summing Up

The metals high on the activity series are oxidized by oxygen more readily than metals low on the series.

The most reactive metals, potassium and sodium, can also form peroxides. These peroxides contain O_2^{2-} ions whereas the ordinary oxides of these metals contain O^{2-} ions. The different properties of these compounds are described in chapter 17.

OXIDATION OF METALS BY OTHER NON-METALS

Only a few typical examples will be discussed.

OXIDATION OF SODIUM BY CHLORINE

Sodium burns in chlorine gas with a yellow flame forming a white solid. The reaction has been discussed in detail in earlier chapters:

$$2Na + Cl_2 \rightarrow 2NaCl$$
$$\text{reducer} \quad \text{oxidizer}$$

Sodium (the reducer) donates electrons to chlorine (the oxidizer).

OXIDATION OF MAGNESIUM BY NITROGEN

When magnesium burns in air, a white solid residue is formed which is mostly the oxide. However, if a little water is dropped on to the residue while it is still hot, the characteristic odour of ammonia can be detected. The ammonia can also be detected by its effect on litmus and its reaction with hydrogen chloride.

Since ammonia is evolved, the product of combustion must have contained a compound of nitrogen. Thus some reaction occurred with nitrogen from the air while the metal was burning:

$$3Mg + N_2 \rightarrow Mg_3N_2 \qquad \text{(magnesium nitride)}$$
$$\text{reducer} \quad \text{oxidizer}$$

Figs. 16.1 and 16.2.
(left) Concentrated nitric acid is added to copper and after a time the jar is inverted in water (right). The water forms a blue solution and rises almost to the top of the jar.

OXIDATION OF IRON BY SULPHUR

If clean iron filings and powdered sulphur are mixed together and heated in a hard glass test tube, the mixture begins to glow and continues to glow when the burner is removed. If the test tube is broken, a stick of a black solid is obtained. This substance can be shown to be iron(II) sulphide, e.g., by its reaction with acids.

$$Fe + S \rightarrow FeS$$
$$\text{reducer} \quad \text{oxidizer}$$

REDUCTION OF NITRIC ACID BY METALS

Nitric acid is commonly used at three different concentrations in chemical reactions:
(i) commercial concentrated acid, about 70% HNO_3;
(ii) 35% nitric acid, obtained by mixing equal volumes of concentrated acid and water;
(iii) dilute nitric acid, obtained by mixing one volume of concentrated acid with seven volumes of water, about 9%.

No matter which concentration is used, the HNO_3 is dissolved in a considerable amount of water, and since nitric acid is a strong electrolyte, it can be regarded as being ionized:

$$HNO_3 + H_2O \rightarrow H_3O^+ + NO_3^-$$

The reduction of nitric acid by metals will be discussed first for copper and silver, because it has been established that these metals do not reduce hydrogen ions.

COPPER AND CONCENTRATED NITRIC ACID

This can be conveniently demonstrated in the following way. Place some copper pellets in the bottom of a glass jar fitted with a cork stopper and an outlet tube leading down the sink. Pour in some concentrated nitric acid and replace the stopper immediately (see fig. 16.1).

After a short time there is a very brisk effervescence of a brown gas, and the copper dissolves forming a green solution.

● *Caution: The brown gas formed is very poisonous and has a corrosive action on the tissues of the throat and lungs.*

When the jar is full of the brown gas, invert it in a trough of water and remove the stopper. It will be observed that the water rises rapidly almost to the top of the jar, and the solution changes colour from green to blue (see fig. 16.2).

The brown gas which is very soluble in water is nitrogen dioxide (NO_2). The blue colour is due to the hydrated copper(II) ions, $Cu(H_2O)_4^{2+}$.

The reaction can be represented:

$$Cu \rightarrow Cu^{2+} + 2e^-$$
$$\underline{(NO_3^- + 2H^+ + e^- \rightarrow NO_2 + H_2O) \times 2}$$
$$Cu + 2NO_3^- + 4H^+ \rightarrow Cu^{2+} + 2NO_2 + 2H_2O$$

NOTE. In nitric acid solution the number of hydrogen ions is equal to the number of nitrate ions. However, in this reaction, *only two nitrate ions are used for each four hydrogen ions used.* The other two nitrate ions are *spectator* ions. They partner the copper(II) ion formed and the solution therefore contains two nitrate ions for each copper(II) ion present. Although the copper(II) ions and nitrate ions do not associate, the solution is called copper(II) nitrate solution.

COPPER AND 35% NITRIC ACID
Use the same apparatus as before (fig. 16.1). After a short time there is a brisk effervescence of a gas. However, the colour of the gas evolved is not nearly as intense as when concentrated nitric acid is used. The copper dissolves forming a blue solution.

If the jar is inverted in water, the water rises only a small way up the jar and all of the brown gas dissolves (see fig. 16.3).

Fig. 16.3.
Dilute nitric acid is added to copper as in fig. 16.1) and the jar is inverted in water. A blue solution forms but the water rises only a small way up the jar.

If the gas jar is lifted from the water momentarily, so that some air enters, the colourless gas in the jar turns brown. This reaction identifies the gas as nitrogen monoxide (NO), because when nitrogen monoxide is mixed with air, it is oxidized to brown nitrogen dioxide by the oxygen of the air:

$$\underset{\text{colourless}}{2NO} + O_2 \rightarrow \underset{\text{brown}}{2NO_2}$$

Thus the reaction differs from the reaction with concentrated nitric acid in that NO is the main product instead of NO_2:

$$(Cu \rightarrow Cu^{2+} + 2e^-) \times 3$$
$$\underline{(NO_3^- + 4H^+ + 3e^- \rightarrow NO + 2H_2O) \times 2}$$
$$3Cu + 2NO_3^- + 8H^+ \rightarrow \underset{\text{blue}}{3Cu^{2+}} + \underset{\text{colourless}}{2NO\ (g)} + 4H_2O$$

NOTE. For every 8 H^+ ions used, only 2 NO_3^- ions are used. The other 6 NO_3^- ions are spectator ions and partner the 3 Cu^{2+} ions in the solution. The solution therefore contains copper(II) nitrate.

The reduction of nitric acid by metals usually yields a mixture of reduction products (e.g. NO_2 and NO) as described above. However, only the main products will be described below.

SILVER AND CONCENTRATED NITRIC ACID
If concentrated nitric acid is added to a small piece of silver in a test tube, there is a slow effervescence which becomes more rapid on heating. A poisonous brown gas with a pungent irritating odour is evolved and the silver dissolves forming a colourless solution (see fig. 16.4).

The brown gas is nitrogen dioxide. If dilute hydrochloric acid is added to the colourless solution, a white curdy precipitate forms which turns purple in sunlight. The precipitate is apparently silver chloride and thus the colourless solution must have contained silver ions.

$$Ag \rightarrow Ag^+ + e^-$$
$$\underline{NO_3^- + 2H^+ + e^- \rightarrow NO_2 + H_2O}$$
$$Ag + NO_3^- + 2H^+ \rightarrow Ag^+ + NO_2\ (g) + H_2O$$

NOTE. One NO_3^- ion is a spectator ion. The solution contains silver nitrate.

SILVER AND 35% NITRIC ACID
If 35% nitric acid is added to a small piece of silver in a test tube, there is a slow effervescence which becomes rapid on heating. A colourless gas is evolved which forms a poisonous brown gas with a pungent

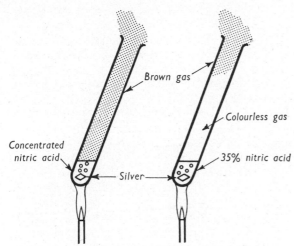

Fig. 16.4. Reduction of nitric acid by silver.

irritating odour as it mixes with air. The silver dissolves forming a colourless solution (see fig. 16.4).

The colourless gas is nitrogen monoxide. The colourless solution can be shown to contain silver ions as described above:

$$(Ag \rightarrow Ag^+ + e^-) \times 3$$
$$NO_3^- + 4H^+ + 3e^- \rightarrow NO + 2H_2O$$
$$\overline{3Ag + NO_3^- + 4H^+ \rightarrow 3Ag^+ + NO \text{ (g)} + 2H_2O}$$

The nitrogen monoxide is oxidized by the oxygen of the air:

$$2NO + O_2 \rightarrow 2NO_2$$

NOTE. Three NO_3^- ions are spectator ions. The solution contains silver nitrate.

Reduction of Nitric Acid by Other Metals

Generally only concentrated nitric acid will be discussed here because the reactions with dilute nitric acid are often very complex.

IRON AND CONCENTRATED NITRIC ACID
There is a slight reaction at first and then the reaction ceases.

It is found that after such treatment with concentrated nitric acid, the iron becomes unusually unreactive. Thus it no longer displaces copper from copper(II) sulphate solution nor does it react with dilute hydrochloric acid. The iron is said to have been *rendered passive*.

Careful experimental investigation of passive iron has shown that it is coated with an oxide film and it is even possible to remove the film of oxide for examination. The composition of such oxide films varies with the thickness of the film.

ZINC AND CONCENTRATED NITRIC ACID
There is a violent reaction producing a brown gas (NO_2) and a colourless solution (which can be identified as zinc nitrate).

$$Zn \rightarrow Zn^{2+} + 2e^-$$
$$\underline{(NO_3^- + 2H^+ + e^- \rightarrow NO_2 + H_2O) \times 2}$$
$$Zn + 2NO_3^- + 4H^+ \rightarrow Zn^{2+} + 2NO_2 + 2H_2O$$
$$\qquad\qquad\qquad\quad \text{colourless} \quad \text{brown}$$

Two NO_3^- ions are spectator ions and partner the Zn^{2+} ion.

ALUMINIUM AND CONCENTRATED NITRIC ACID
There is no reaction. Apparently the surface film of oxide is insoluble in concentrated nitric acid.

MAGNESIUM AND CONCENTRATED NITRIC ACID
The reaction is very similar to the reaction with zinc.

$$Mg + 2NO_3^- + 4H^+ \rightarrow Mg^{2+} + 2NO_2 + 2H_2O$$
$$\qquad\qquad\qquad \text{colourless} \quad \text{brown}$$

MAGNESIUM AND VERY DILUTE NITRIC ACID
When dilute nitric acid is mixed with an equal volume of water and poured on to magnesium in a test tube, there is an effervescence of a colourless, odourless gas which burns when brought near a flame. As the reaction proceeds, some brown gas may be evolved but this is only formed in small amounts and is not the main product of the reaction.

The main gaseous product is hydrogen.

$$Mg \rightarrow Mg^{2+} + 2e^- \quad \text{(oxidation)}$$
$$\underline{2H^+ + 2e^- \rightarrow H_2 \qquad\quad \text{(reduction)}}$$
$$Mg + 2H^+ \rightarrow Mg^{2+} + H_2$$
$$\text{reducer} \;\; \text{oxidizer}$$

Magnesium is the only metal which will displace hydrogen from nitric acid, and then only if the acid is very dilute.

The small amount of brown nitrogen dioxide gas which is produced is due to reduction of some nitrate ions.

REDUCTION OF CONCENTRATED SULPHURIC ACID BY METALS

Commercial sulphuric acid contains about 98% H_2SO_4 and only about 2% H_2O. Because there is so little water present the acid is only slightly ionized. It will be regarded as virtually un-ionized in the following reactions although some H^+ and HSO_4^- ions are present. The reactions will be limited to the oxidation of copper and silver.

COPPER AND CONCENTRATED SULPHURIC ACID

When the mixture is heated, there is an effervescence. The gas evolved fumes in air and has an irritating choking odour and turns potassium dichromate from orange to green. These are the properties of sulphur dioxide, which is apparently the gas evolved:

$$Cu \rightarrow Cu^{2+} + 2e^-$$
$$\underline{H_2SO_4 + 2H^+ + 2e^- \rightarrow SO_2 + 2H_2O}$$
$$Cu + H_2SO_4 + 2H^+ \rightarrow Cu^{2+} + SO_2 + 2H_2O$$

The mixture often becomes cloudy for a short time and then turns black. These changes are due to other reactions which will not be discussed, beyond the fact that the black residue consists of a mixture of copper(I) and copper(II) sulphides.

Sulphur dioxide is the only gaseous product and this reaction can be used as a convenient method for preparing dry sulphur dioxide.

SILVER AND CONCENTRATED SULPHURIC ACID

The reaction is similar to the reaction with copper. Sulphur dioxide is the gaseous product.

$$(Ag \rightarrow Ag^+ + e^-) \times 2$$
$$\underline{H_2SO_4 + 2H^+ + 2e^- \rightarrow SO_2 + 2H_2O}$$
$$2Ag + H_2SO_4 + 2H^+ \rightarrow 2Ag^+ + SO_2 + 2H_2O$$

REDUCTION OF CAUSTIC ALKALI SOLUTIONS BY METALS

The common caustic alkalis are sodium hydroxide (caustic soda) and potassium hydroxide (caustic potash). Most metals do not react with alkalis but aluminium, zinc and tin do react. These reactions will now be discussed.

ALUMINIUM AND CAUSTIC SODA SOLUTION.

This reaction can be safely and conveniently demonstrated in the following way. Add caustic soda solution to aluminium powder and place the test tube in a rack on a tray. A gas is evolved, very slowly at first but the effervescence soon becomes very vigorous and the mixture may froth out of the tube. The gas is evolved so rapidly that when it is lit it continues to burn at the top of the tube. The gas evolved is apparently hydrogen. The mixture becomes very hot, and eventually a colourless solution remains.

If this solution is evaporated to dryness, a white residue is obtained. Analysis has shown that the empirical formula of this solid is $NaAlO_2$. This is called sodium aluminate.

Experimental work has shown that the *solution* contains Na^+ ions, but not AlO_2^- ions. The aluminate ion is hydrated in solution and can be represented as $AlO_2.(H_2O)_2^-$ or $Al(OH)_4^-$. This latter formula will be regarded as the best representation of the aluminate ion in solution.

The equation for the formation of $Al(OH)_4^-$ from Al can now be developed:

$$Al \qquad \rightarrow Al(OH)_4^-$$

Four OH^- ions are needed on the left hand side to achieve mass balance:

$$Al + 4OH^- \rightarrow Al(OH)_4^-$$

Electrons must now be included to achieve charge balance:

$$\underset{\text{reducer}}{Al} + 4OH^- \rightarrow Al(OH)_4^- + 3e^- \text{ (oxidation)}$$

This partial equation shows aluminium acting as a reducer (an electron donor) and is thus consistent with all of the reactions of metals so far discussed. Evidently there must be an oxidizer (an electron acceptor). Furthermore, the formation of hydrogen is still to be explained. Both of these requirements are met by regarding the water as the electron acceptor.

$$\underset{\text{oxidizer}}{2H_2O} + 2e^- \rightarrow 2OH^- + H_2 \text{ (reduction)}$$

Thus the reaction can be represented as follows:

$$(Al + 4OH^- \rightarrow Al(OH)_4^- + 3e^-) \times 2$$
$$\underline{(2H_2O + 2e^- \rightarrow 2OH^- + H_2) \times 3}$$
$$\underset{\text{reducer}}{2Al} + 2OH^- + \underset{\text{oxidizer}}{6H_2O} \rightarrow 2Al(OH)_4^- + 3H_2 \text{ (g)}$$

NOTE. The Na^+ ions are spectator ions and partner the $Al(OH)_4^-$ ions. The solution contains sodium aluminate. On evaporation to dryness, water is lost:

$$Na^+ + Al(OH)_4^- \rightarrow NaAlO_2 + 2H_2O \text{ (g)}$$

ZINC AND CAUSTIC SODA SOLUTION

If a concentrated solution of caustic soda is poured

on to granulated zinc, a very slow effervescence is observed. The reaction becomes faster if the mixture is heated but is still very slow.

● *Caution: Hot concentrated alkali solutions are very dangerous.*

The gas evolved can be identified as hydrogen if the mixture is allowed to react for long enough to allow a reasonable volume of gas to be collected. Analysis has shown that the residue on evaporation is Na_2ZnO_2 (sodium zincate). In solution, the zincate ion is hydrated, similar to the aluminate ion. The formula of the zincate ion in solution will be taken to be $Zn(OH)_4^{2-}$.

$$Zn + 4OH^- \rightarrow Zn(OH)_4^{2-} + 2e^-$$
$$2H_2O + 2e^- \rightarrow 2OH^- + H_2$$
$$Zn + 2OH^- + 2H_2O \rightarrow Zn(OH)_4^{2-} + H_2$$

NOTE. Two Na^+ ions (spectator ions) partner each $Zn(OH)_4^{2-}$ ion. The solution contains sodium zincate. On evaporation to dryness, water is lost:

$$2Na^+ + Zn(OH)_4^{2-} \rightarrow Na_2ZnO_2 + 2H_2O \text{ (g)}$$

TIN AND CAUSTIC SODA SOLUTION

This is similar to the reaction of caustic soda solution with zinc, and forms the stannate(II) anion, $Sn(OH)_3^-$.

$$Sn + OH^- + 2H_2O \rightarrow Sn(OH)_3^- + H_2$$

Summing Up

The reactions studied in this chapter can be briefly summarized although the reactions with nitric acid, concentrated sulphuric acid and caustic alkalis were limited to selected examples only and no generalizations should be made concerning their reactions with other metals (see fig. 16.5).

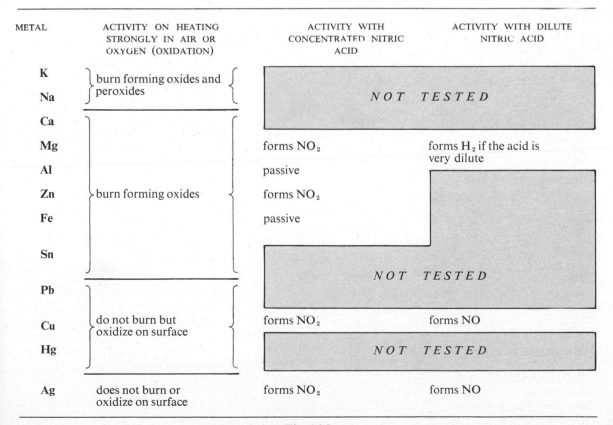

METAL	ACTIVITY ON HEATING STRONGLY IN AIR OR OXYGEN (OXIDATION)	ACTIVITY WITH CONCENTRATED NITRIC ACID	ACTIVITY WITH DILUTE NITRIC ACID
K	burn forming oxides and peroxides	NOT TESTED	
Na			
Ca	burn forming oxides		
Mg		forms NO_2	forms H_2 if the acid is very dilute
Al		passive	
Zn		forms NO_2	
Fe		passive	
Sn		NOT TESTED	
Pb	do not burn but oxidize on surface		
Cu		forms NO_2	forms NO
Hg		NOT TESTED	
Ag	does not burn or oxidize on surface	forms NO_2	forms NO

Fig. 16.5

Note for More Advanced Students

The reaction of concentrated sulphuric acid as an oxidizer with the metals copper and silver, has been represented by the partial equation:

$$H_2SO_4 + 2H^+ + 2e^- \rightarrow SO_2 + 2H_2O$$

The hydrogen ion is better represented by the symbol H_3O^+ rather than H^+. The equation can be rewritten:

$$H_2SO_4 + 2H_3O^+ + 2e^- \rightarrow SO_2 + 4H_2O$$

Furthermore, it is unlikely that the water shown in this reaction could remain as a final product. It would almost certainly react with the H_2SO_4 which is present:

$$H_2SO_4 + H_2O \rightarrow H_3O^+ + HSO_4^-$$

The reaction is better represented by the sum of these last two reactions:

$$H_2SO_4 + 2H_3O^+ + 2e^- \rightarrow SO_2 + 4H_2O$$
$$\underline{(H_2SO_4 + H_2O \rightarrow H_3O^+ + HSO_4^-) \times 4}$$
$$5H_2SO_4 + 2e^- \rightarrow SO_2 + 2H_3O^+ + 4HSO_4^-$$

Hence, oxidation of copper by concentrated sulphuric acid can be represented:

$$Cu \rightarrow Cu^{2+} + 2e^-$$
$$\underline{5H_2SO_4 + 2e^- \rightarrow SO_2 + 2H_3O^+ + 4HSO_4^-}$$
$$Cu + 5H_2SO_4 \rightarrow SO_2 + Cu^{2+} + 2H_3O^+ + 4HSO_4^-$$

Similarly, the reaction with silver is:

$$2Ag + 5H_2SO_4 \rightarrow SO_2 + 2Ag^+ + 2H_3O^+ + 4HSO_4^-$$

These equations are better representations of the reactions than those used earlier in this chapter, because they show reaction of the metals with molecules of H_2SO_4 and, as conductivity measurements show, these are the main species present in concentrated sulphuric acid. On the other hand, the equations used earlier are simpler to construct. The choice between these different forms of the equation depends on what the equations are required to represent.

STUDY QUESTIONS

1. Explain why the reaction between sodium metal and chlorine gas is regarded as a redox reaction.

2. Write equations for the reaction of magnesium with:
  (a) chlorine;
  (b) nitrogen;
  (c) sulphur.
Indicate the oxidizer and reducer in each equation.

3. Write partial ionic equations for the reaction of concentrated nitric acid with magnesium. Hence develop the overall ionic equation.

4. Silver nitrate is prepared commercially by dissolving silver in 35% nitric acid, rather than in concentrated nitric acid. Explain why the more dilute acid is used.

5. (a) Write partial ionic equations and hence develop the overall ionic equation for the reaction between copper and 35% nitric acid. Show that the spectator ions are in sufficient number to 'partner' the ions formed in the reaction.
  (b) Rewrite the equation for the overall reaction using empirical formulae.

6. What is the usual gaseous product of oxidation of metals with:
  (a) concentrated nitric acid;
  (b) dilute nitric acid;
  (c) concentrated sulphuric acid.

7. How can the gases, nitrogen monoxide and nitrogen dioxide be identified and distinguished from each other?

8. Describe any changes which can be observed and write the equation for the reaction between silver and hot concentrated sulphuric acid.

9. When concentrated sulphuric acid reacts with zinc, hydrogen sulphide gas (H_2S) is evolved. Build up a partial ionic equation for the reduction of H_2SO_4 to H_2S, and hence derive the overall ionic equation for the reaction.

10. Develop the partial ionic equations and hence write the overall ionic equation for the reaction between aluminium and caustic soda solution.

11. Reaction between iron and oxygen produces a black product, said to be Fe_3O_4.
  How could it be established that the formula of this black substance is Fe_3O_4?

12. Describe how the truth of the following statement could be tested: iron and sulphur react to form iron(II) sulphide.

13. You are provided with a sample of calcium metal. Devise an experiment to show that combustion of calcium in air produces CaO and not CaO_2.

14. How could you show that magnesium combines with nitrogen but that calcium does not combine with nitrogen?

17 :   Oxides and Hydroxides of Metals

The discussion of oxides and hydroxides is restricted to the metals studied in previous chapters.

OXIDES

A summary of the preparation and specific properties of individual oxides is given in the table fig. 17.1.

Preparation of Oxides

Oxides of sodium may be prepared by burning the metal in air, e.g.:

$$2Na + O_2 \rightarrow Na_2O_2$$

Oxides of less reactive metals can frequently be prepared by burning the metal in air, but it is usually more convenient and more economical to decompose an oxy-compound, e.g.:

$$CaCO_3 \rightarrow CaO + CO_2$$

A convenient method of preparing the oxides of mercury(II) and silver is to add caustic soda solution to solutions of the salts. In these reactions, the oxide is formed because the hydroxides do not exist under ordinary conditions:

$$Hg^{2+} + 2OH^- \rightarrow HgO(s) + H_2O$$
$$2Ag^+ + 2OH^- \rightarrow Ag_2O(s) + H_2O$$

Stability of Oxides

Most oxides are stable to heat. However, when a metal has several oxides, one of the oxides may be more stable than the others.

Mercury(II) oxide and silver oxide decompose at moderate temperatures, e.g.:

$$2HgO \rightarrow 2Hg + O_2$$

Reactions with Water

Oxides of the more reactive metals react with water to form alkaline solutions. Oxides of sodium, potassium and calcium react vigorously with water and magnesium oxide reacts slightly with water. In these reactions, the oxide ion, O^{2-}, functions as a strong base, i.e. it is a very good proton acceptor.

$$O^{2-} + HOH \rightarrow OH^- + OH^-$$

The oxides of other metals, e.g. CuO, are generally insoluble in water and therefore do not react with water.

Peroxides, such as sodium peroxide, Na_2O_2, contain the peroxide ion, O_2^{2-}, which also functions as a base.

$$O_2^{2-} + 2HOH \rightarrow H_2O_2 + 2OH^-$$

The hydrogen peroxide formed will decompose if the mixture is warm.

$$2H_2O_2 \rightarrow 2H_2O + O_2$$

Thus, the overall reaction of sodium peroxide with warm water is:

$$2Na_2O_2 + 2H_2O \rightarrow 4Na^+ + 4OH^- + O_2$$

Reactions with Acids

Most oxides of metals dissolve in acids to form a salt. Thus most oxides of metals have a basic character, e.g.:

$$CuO + 2H^+ \rightarrow Cu^{2+} + H_2O$$

A few oxides of metals dissolve in acids with difficulty, especially if the oxides have been prepared at high temperatures, e.g. Fe_2O_3.

Iron(II, III) oxide is unusual in that it forms a mixture of salts when dissolved in acids.

$$Fe_3O_4 + 8H^+ \rightarrow Fe^{2+} + 2Fe^{3+} + 4H_2O$$

Peroxides react with acids to form hydrogen peroxide.

$$O_2^{2-} + 2H^+ \rightarrow H_2O_2$$

The hydrogen peroxide formed will decompose if the mixture is warm and the overall reaction of peroxides with acids is:

$$2O_2^{2-} + 4H^+ \rightarrow 2H_2O + O_2$$

Reactions with Alkalis

The oxides of some metals, in addition to dissolving in acids, will also dissolve in alkalis.

$$Al_2O_3 + 3H_2O + 2OH^- \rightarrow 2\,Al(OH)_4^-$$
$$ZnO + H_2O + 2OH^- \rightarrow Zn(OH)_4^{2-}$$
$$SnO_2 + 2H_2O + 2OH^- \rightarrow Sn(OH)_6^{2-}$$
$$PbO + H_2O + OH^- \rightarrow Pb(OH)_3^-$$

Such oxides therefore display both acidic and basic characters and are said to be amphoteric.

If aluminium oxide is heated to very high temperatures, a hard dense form of the oxide is obtained which is practically insoluble in acids and alkalis.

Reactions with Ammonia Solutions

Silver oxide dissolves readily in ammonia solution forming diamminesilver(I) ions:

$$Ag_2O + 4NH_3 + H_2O \rightarrow 2\,Ag(NH_3)_2^+ + 2\,OH^-$$

Redox Reactions

Oxides of reactive metals e.g. MgO, are difficult to reduce, whereas oxides of less reactive metals are easy to reduce, e.g.:

$$CuO + H_2 \rightarrow Cu + H_2O$$

METALLIC OXIDES	PREPARATION	COLOUR	EFFECT OF HEAT	REACTION WITH WATER
Na_2O_2	Burn Na in excess O_2	Pale yellow	↑	Forms alkali and H_2O_2 (cold) or O_2 (warm)
Na_2O	Burn Na in limited O_2	White		Forms alkali solution, Na^+OH^-
CaO (quicklime)	Heat $CaCO_3$	White		Forms slightly soluble $Ca(OH)_2$
MgO	Heat the hydroxide.	White		Very slight reaction, weakly alkaline solution
Al_2O_3 (alumina)	These oxides are also formed by heating certain salts	White		
ZnO		Yellow when hot, white when cold	Stable	
FeO		Black		
Fe_2O_3	Heat $Fe(OH)_3$	Red-brown		
Fe_3O_4	Heat Fe in air or steam	Black		
SnO		Black		
SnO_2	Occurs naturally as cassiterite	White		Practically no reaction
PbO (litharge)	Heat Pb in air	Yellow or orange	↓	
PbO_2		Dark brown	Decompose to litharge at high temperatures	
Pb_3O_4 (red lead)		Red		
CuO	Heat $Cu(OH)_2$	Black	Stable (up to 1000 °C)	
Cu_2O		Red	Stable	
HgO	Add NaOH solution to solutions of Hg^{2+} and Ag^+ salts	Red, sometimes yellow Black when hot	Decompose to the metal plus oxygen	
Ag_2O		Brown		↓

Fig. 17.1. Comparison

REACTION WITH ACIDS	REACTION WITH CAUSTIC ALKALIS	CLASSIFICATION	REDOX REACTIONS
Forms salt and H_2O_2 (cold) or O_2 (warm)	Do not react with alkalis but will react with the water in which the alkali is dissolved	*Peroxide*	Cannot be reduced by carbon, carbon monoxide or hydrogen
↑ Dissolves forming corresponding salt ↓		*Basic*	
		Basic	
	Insoluble	*Basic*	
	Dissolves forming aluminate $Al(OH)_4^-$	*Amphoteric*	
	Dissolves forming zincate $Zn(OH)_4^{2-}$	*Amphoteric*	Can be reduced by carbon and carbon monoxide
Dissolves with difficulty forming Fe^{3+} salt	Insoluble	*Basic*	↑
Dissolves forming Fe^{2+} and Fe^{3+} salts	Insoluble	*Compound* $(Fe_3O_4 \equiv FeO.Fe_2O_3)$	
Dissolves with difficulty	Dissolves forming stannate(IV) $Sn(OH)_6^{2-}$	*Amphoteric*	
Dissolves forming Pb^{2+} salt	Dissolves forming plumbate(II) $Pb(OH)_3^-$	*Amphoteric*	Can be reduced by carbon, carbon monoxide and hydrogen
Dissolves forming corresponding salt	Insoluble	*Basic*	
Dissolves forming Cu^{2+} salt and Cu	Insoluble	*Basic*	
Dissolves forming corresponding salt	Insoluble	*Basic*	
Dissolves forming corresponding salt	(Soluble in NH_3 forming diamminesilver(I) $Ag(NH_3)_2^+$)	*Basic*	↓

of the metallic oxides

METALLIC HYDROXIDES	PREPARATION	PHYSICAL PROPERTIES	SOLUBILITY IN WATER	EFFECT OF HEAT	REACTION WITH ACIDS
NaOH (caustic soda)	Electrolysis of conc. NaCl solution or Na and water	White crystalline	Soluble	Stable to heat	
Ca(OH)$_2$ (slaked lime)	CaO (quicklime) and water	White powder	Slightly soluble		
Mg(OH)$_2$	Excess NaOH to solution of an Mg^{2+} salt	White gelatinous	Very slightly soluble		
Al(OH)$_3$	Excess NH$_3$ solution to a solution of an Al^{3+} salt	White gelatinous			
Zn(OH)$_2$	Limited addition of NaOH solution to a Zn^{2+} salt solution	White gelatinous			All hydroxides react with acids to form the corresponding salt and water
Fe(OH)$_2$	Excess NaOH to a Fe^{2+} salt solution	Green gelatinous		Decompose to the oxide and water	
Fe(OH)$_3$	Excess NH$_3$ or NaOH to an Fe^{3+} salt solution	Red-brown gelatinous			
Sn(OH)$_2$	Excess NH$_3$ to a Sn^{2+} salt solution	White gelatinous	Insoluble		
Sn(OH)$_4$	Excess NH$_3$ to a Sn^{4+} salt solution	White gelatinous			
Pb(OH)$_2$	Excess NH$_3$ to a Pb^{2+} salt solution	White gelatinous			
Cu(OH)$_2$	Excess NaOH to a Cu^{2+} salt solution	Blue gelatinous		Cu(OH)$_2$ is the least stable of these hydroxides	

Fig. 17.2. Comparison of the metallic hydroxides

REACTION WITH CAUSTIC ALKALIS	REACTION WITH AMMONIA SOLUTION
Do not react with alkalis but will dissolve in the water of the alkali solution. See under solubility in water	
Insoluble	Insoluble
Dissolves forming aluminate $Al(OH)_4^-$ *AMPHOTERIC*	Insoluble
Dissolves forming zincate $Zn(OH)_4^{2-}$ *AMPHOTERIC*	Dissolves forming tetramminezinc(II) $Zn(NH_3)_4^{2+}$
Insoluble (Oxidized by air to form brown $Fe(OH)_3$)	Insoluble
Insoluble	Insoluble
Dissolves forming stannate(II) $Sn(OH)_3^-$ *AMPHOTERIC*	Insoluble
Dissolves forming stannate(IV) $Sn(OH)_6^{2-}$ *AMPHOTERIC*	Insoluble
Dissolves forming plumbate(II) $Pb(OH)_3^-$ *AMPHOTERIC*	Insoluble
Dissolves in concentrated solutions forming cuprate(II) $Cu(OH)_4^{2-}$ *AMPHOTERIC*	Dissolves forming tetramminecopper(II) $Cu(NH_3)_4^{2+}$

HYDROXIDES

A summary of the preparation and properties of individual hydroxides is given in the table fig. 17.2.

Preparation of Hydroxides

Most hydroxides have slight to very slight solubility in water. Such hydroxides can be prepared by treating a solution of the salt with a suitable base, e.g.

$$Mg^{2+} + 2OH^- \rightarrow Mg(OH)_2 (s)$$

Calcium hydroxide is usually prepared by adding water to calcium oxide.

Sodium hydroxide is readily soluble in water and is prepared industrially by electrolysis of concentrated solutions of sodium chloride.

Stability of Hydroxides

Apart from sodium and potassium hydroxides, the hydroxides of metals usually decompose on heating to form the corresponding oxide, e.g.

$$Cu(OH)_2 \rightarrow CuO + H_2O$$

Reactions with Acids

All hydroxides of metals show a basic character by dissolving in acids to form a salt and water, e.g.

$$Fe(OH)_3 + 3H^+ \rightarrow Fe^{3+} + H_2O$$

Reactions with Alkalis

A few hydroxides, in addition to showing a basic character, can also show an acidic character by reacting with alkalis, e.g.

$$Al(OH)_3 + OH^- \rightarrow Al(OH)_4^- \text{ aluminate anion}$$
$$Zn(OH)_2 + 2OH^- \rightarrow Zn(OH)_4^{2-} \text{ zincate anion}$$

Hydroxides which show this dual character are said to be amphoteric (or amphiprotic).

In preparing these amphoteric hydroxides it is clearly necessary to avoid adding excess sodium hydroxide solution, because the precipitate of the insoluble hydroxide would dissolve in excess as shown above.

Reactions with Ammonia Solution

Zinc hydroxide and copper(II) hydroxide dissolve readily in ammonia solution forming complex ions.

$$Zn(OH)_2 + 4NH_3 \rightarrow Zn(NH_3)_4^{2+} + 2OH^-$$
$$\text{tetramminezinc cation}$$
$$Cu(OH)_2 + 4NH_3 \rightarrow Cu(NH_3)_4^{2+} + 2OH^-$$
$$\text{tetramminecopper(II) cation}$$

In preparing these hydroxides it is necessary to avoid adding excess ammonia solution.

STUDY QUESTIONS

1. Describe the tests which would be used to decide whether an oxide or hydroxide is basic or amphoteric.

2. Write equations for the reactions of caustic soda solution with:
(a) iron(III) chloride solution;
(b) aluminium sulphate solution;
(c) zinc nitrate solution.

3. Write equations for the reactions of ammonia solution with each of the substances named in question two.

4. Each of the following hydroxides is to be prepared by precipitation from a solution of the appropriate salt:
(a) $Al(OH)_3$;
(b) $Cu(OH)_2$;
(c) $Fe(OH)_3$.
Explain, with reasons, whether caustic soda solution or ammonia solution should be used.

5. Write equations for the dissolution of each of the following oxides in dilute nitric acid:
(a) MgO;
(b) ZnO;
(c) Ag_2O.

6. Write equations for the dissolution of each of the following oxides in caustic soda solution:
(a) ZnO;
(b) Al_2O_3.

7. Describe and explain the effect of heat on:
(a) $Cu(OH)_2$;
(b) $Fe(OH)_3$;
(c) HgO;
(d) Ag_2O.

8. Describe and explain the changes which occur when iron(II) hydroxide is exposed to air.

9. Write the equations for the reactions which occur if water is added to:
(a) Na_2O_2;
(b) CaO.

10. Write the equations for the reactions which occur when caustic soda solution is added to:
(a) tin(II) chloride solution;
(b) tin(IV) chloride solution.

11. Write the equations for the reactions of lead(II) hydroxide with dilute nitric acid and with caustic soda solution.

12. A mixture of aluminium metal and iron(III) oxide reacts vigorously after the reaction has been started. What reaction occurs? How can this reaction be correlated with the positions of these metals in the activity series?

13. What are the chemical names of Fe_3O_4 and Pb_3O_4?

14. Barium is a metal which forms an oxide, BaO, as well as a peroxide, BaO_2. What chemical tests could be used to distinguish between these two compounds?

15. 10.00 g of hydrated copper(II) sulphate is dissolved in water. When an excess of dilute caustic soda solution is added, a blue gelatinous precipitate is formed. When the mixture is boiled, the precipitate changes to a black residue. Write equations for the reactions and calculate the mass of the black residue.
($Cu = 63.5$, $H = 1.0$, $Na = 23.0$, $O = 16.0$, $S = 32.1$)

16. Calculate the volume of oxygen, expressed at s.t.p., which is evolved when 1.00 g of sodium peroxide is dissolved in warm sulphuric acid.
($Na = 23.0$, $O = 16.0$)

17. Calculate the mass of sodium hydroxide which is just needed to dissolve 1 000 kg of aluminium oxide to form a solution of sodium aluminate. If the solution of sodium aluminate is diluted it decomposes to aluminium hydroxide and sodium hydroxide. How much sodium hydroxide is recovered?
($Al = 27.0$, $H = 1.0$, $Na = 23.0$, $O = 16.0$)

Answers to Numerical Problems

15. 3.18 g.
16. 144 cm^3.
17. 784 kg, 784 kg.

18 : Metals and Their Uses

The Occurrence of Metals

Very few metals are found 'free' or uncombined amongst the minerals which occur in the Earth's crust. Gold occurs free as very small particles in quartz veins or in river gravels. Metallic silver, mercury and copper are also sometimes found free, but these metals occur mainly as sulphides. It is only the least reactive metals which are found free in nature. Most metals occur as compounds with non-metals (see fig. 18.1).

Chlorides	Carbonates	Oxides	Sulphides
KCl	$MgCO_3$	Al_2O_3	ZnS
$NaCl$	$CaCO_3$	Fe_2O_3	PbS
$MgCl_2$		SnO_2	Cu_2S
			HgS
			Ag_2S

Fig. 18.1. Most Commonly Occurring Minerals of Metals.

The naturally occuring minerals and ores must be *reduced* in order to extract the metals. For example, the conversion of Fe_2O_3 to Fe is a reduction process.

A study of the methods of extracting metals from their ores, together with the processing of these metals, is called *metallurgy*, and this is an important branch of science, because of the practical and industrial uses of metals. Some of the properties of metals which make them particularly useful will be briefly reviewed.

● *Malleability.*
A malleable substance is capable of being beaten or rolled into thin sheets.

● *Ductility.*
A ductile substance is capable of being drawn out into a rod or wire.

● *Hardness.*
A hard substance is firm and unyielding. It is difficult to scratch its surface.

● *Toughness.*
A tough substance is difficult to break because it is flexible and not brittle.

● *Electrical conductivity.*
Metals are good conductors of electricity and the metals near the bottom of the activity series are mostly very good conductors. Copper is used in electrical circuits because it is the cheapest of these metals. Silver is sometimes used for special purposes such as contact points.

● *Reducing properties.*
Metals near the top of the activity series are powerful reducers and some of their uses are dependent on this fact, e.g., magnesium is used for removing traces of gases from radio valves and sodium is used to purify molten metals

The Extraction of Metals

The reducing properties of metals are due to the fact that they form positive ions by loss of electrons. The reverse process of reducing the positive ions leads to the formation of the metal. Compounds of metals high on the activity series are difficult to reduce to the metals, whereas the compounds of metals low on the activity series are relatively easy to reduce.

The production of a few metals will be described in detail, in order to illustrate the general principles of extraction of metals.

PRODUCTION OF SODIUM

The production of sodium is described as an example of the production of a metal high on the activity series.

The commercial process starts with sodium chloride. Large quantities of sodium chloride are obtained by evaporation of sea water, or by dissolving underground rock salt deposits in water and extracting the salt by evaporation. Some purification of the residue is necessary to remove other salts.

To obtain metallic sodium from sodium chloride it is necessary to reduce the sodium ions:

$$Na^+ + e^- \rightarrow Na$$

Because sodium is high on the activity series this reduction is fairly difficult to bring about, and is usually carried out by electrolysis. The electrolyte used is fused sodium chloride. It is unsuitable to use an aqueous solution, because if water is present, hydrogen gas is produced at the cathode instead of sodium.

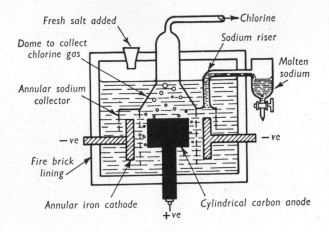

Fig. 18.2. The Downs Cell.

The Downs Cell

The electrolysis of fused sodium chloride is carried out in the *Downs cell* (fig. 18.2). Sodium chloride fuses at 801 °C but the electrolyte used in the cell is molten at about 600 °C. The lower operating temperature is obtained by using a mixture of about 40% NaCl and 60% $CaCl_2$ as the electrolyte.

It is important to keep the temperature as low as possible because of the fairly low boiling point of sodium. If the temperature rises much above 600 °C, the sodium disperses through the electrolyte as a metallic fog instead of forming as molten droplets. The heat necessary to maintain the temperature of the cell is produced by the electric current, but the Downs cell is essentially an *electrolytic cell* and not an electric furnace.

Chlorine gas is liberated at the carbon anode and is collected under the dome. It escapes under its own pressure, is led off and stored.

Sodium metal is liberated at the annular cathode and rises upwards, because it is less dense than the molten salt. The two metal screens help to prevent the sodium from mixing with the chlorine. The sodium passes up the riser and is collected in the closed container at the side of the cell. Some calcium is also formed and this is filtered from the molten sodium.

Fresh salt is added at intervals so that the cell operates continuously. A typical cell operates at about 8 volt and draws a current of up to 20 000 ampere.

The electrode reactions occurring in the cell are:

CATHODE REACTION:
$$Na^+ + e^- \rightarrow Na$$

ANODE REACTION:
$$2Cl^- \rightarrow Cl_2 + 2e^-$$

Uses of Sodium

(i) Manufacture of sodium peroxide (Na_2O_2).

(ii) Manufacture of sodamide ($NaNH_2$) and sodium cyanide (NaCN), which are used in the production of dyes.

(iii) Manufacture of an alloy with lead which corresponds to the proportions NaPb. This alloy is used for the production of tetra-ethyl lead, an anti-knock additive for petrol.

(iv) The reduction of compounds of metals to the pure metal. This is sometimes used to purify a molten metal by removing traces of compounds of the metal.

(v) Sodium vapour lamps.

(vi) Heat transfer. Molten sodium has been used for transferring the heat of nuclear reactors to steam boilers. It has also been used inside the stems of valves in petrol engines, because of its high thermal conductivity.

(vii) Sodium is used in small proportions to improve the properties of some alloys.

(viii) Pure caustic soda (NaOH) can be prepared by dissolving sodium cautiously in pure water.

PRODUCTION OF IRON

The production of iron is described because it is a very important industrial process. It is also an example of the production of a metal fairly low on the activity series.

Occurrence of Iron

Iron ores occur in most parts of the world and the main ores are:

(i) haematite: Fe_2O_3;

(ii) limonite: Fe_2O_3, xH_2O;

(iii) magnetite: Fe_3O_4.

Iron also occurs as the carbonate and the sulphide.

Iron Monarch, South Australia. More than 100 000 million kg of iron ore have been quarried in the Iron Knob area and most of this has come from Iron Monarch.

All the ore deposits contain a fair proportion of impurities, e.g., sand and clay. In metallurgy such materials are called gangue (pronounced 'gang'). Australia is fortunate in possessing large deposits of high grade haematite.

● *The production of iron is essentially the chemical reduction of iron(III) oxide to the metal*

The ore is usually given a preliminary roasting to dry it and convert any other iron compounds to the oxide. Because iron is fairly low on the activity series, the reduction of its oxide is fairly easily accomplished and can be brought about by heating it with carbon monoxide.

● *At the same time the opportunity is taken to remove some of the impurities*

Limestone is used to do this because it is basic in character and the impurities are acidic, e.g., silica. The reactions are carried out in the blast furnace.

The Blast Furnace

A blast furnace is a tapered steel tower lined with fire brick (fig. 18.3). It is not unusual for the furnace to be about 30 metres high and *consume* about 1 700 000 kg of iron ore, 850 000 kg of coke, 350 000 kg of limestone and 4000 tons of hot air every 24 hours, *producing* about 1 million kg of pig iron, 400 000 kg of slag and 5 500 000 kg of furnace gas.

The ore, coke and limestone are carried in separate loads to the top of the furnace by a hoist and tipped

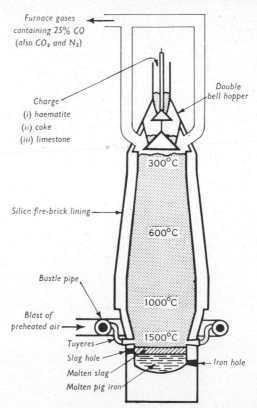

Fig. 18.3. The blast furnace.

Top: No. 4 Blast Furnace, Port Kembla, New South Wales. This furnace can produce 1 700 000 kg of pig iron each day. Bottom: the raw materials for the production of pig iron. Left to right: limestone: haematite; coke.

into the bell hoppers. The bell hoppers at the top of the furnace prevent the escape of gas during addition of the charge. The hoppers are operated separately and are rotated as they are lowered so that the charge is spread evenly.

The air used in the blast furnace is preheated to 550-950 °C and is blown into the bottom of the furnace from small pipes called tuyeres (pronounced 'tweers') which lead off from a large circular pipe called the bustle pipe. It is from the air blast that the furnace takes its name.

The blast furnace is designed for continuous operation because this is generally cheaper for large scale industrial processes. The raw materials are added and the products are removed at suitable intervals. The furnace is stopped only when the lining is burnt out. To maintain the blast furnace in operation, a great deal of ancillary equipment is necessary. Ore, coal and limestone must be mined and transported to the works. The coal is processed to make coke, plus a number of valuable by-products, including coke-oven gas. The charge has to be crushed, graded, weighed, and hoisted to the top of the furnace. Blowers are needed to compress the air and blow it through heating stoves before it is blasted into the furnace. The molten iron and slag from the furnace are collected in enormous ladles and moved about by heavy trucks or cranes. It is not intended to attempt to describe the complex of processes involved in the production of iron. The discussion will be limited to the chemical changes in the blast furnace.

The Reduction of Iron(III) Oxide in the Blast Furnace

At the bottom of the furnace the coke, which is essentially carbon, burns in the blast of pre-heated air:

$$C + O_2 \rightarrow CO_2$$

This reaction also produces heat and the temperature in this part of the furnace is as high as 1500 °C. At this high temperature the carbon dioxide is reduced by the hot carbon:

$$CO_2 + C \rightarrow 2CO$$

The carbon monoxide so formed reduces the iron (III) oxide to iron in the middle and upper parts of the furnace. The reaction probably involves a number of steps, but the overall reduction, which is a reversible reaction, can be regarded as being due to the carbon monoxide:

$$Fe_2O_3 + 3CO \rightleftharpoons 2Fe + 3CO_2$$

The reaction is driven to completion by using a large excess of carbon monoxide in the blast. This also sweeps away the carbon dioxide and minimizes the reverse reaction.

Although the reaction has been represented as being due to the carbon monoxide, it is also possible that part of the reduction is due to the carbon.

The iron forms as a spongy mass at first but later melts as it moves down the furnace. As the molten iron runs down the furnace it dissolves a fair amount of carbon and other substances from the coke. The impure molten iron collects as a layer at the bottom of the furnace.

Removal of Impurities

In the central part of the furnace the limestone reacts with any silica, SiO_2, and other acidic oxides, present in the ore, forming calcium silicate or *slag*:

$$CaCO_3 + SiO_2 \rightarrow CaSiO_3 + CO_2$$

This is molten at the temperature of the furnace and runs down to form a layer on top of the molten iron. The molten slag serves the useful purpose of protecting the layer of molten iron from oxidation by the air blast.

Products of the Blast Furnace

The hot gases leaving the top of the furnace contain about 25% carbon monoxide. The furnace gases are freed from dust, mixed with air and burnt to provide the heat for preheating the air blast and for generating steam to operate some of the ancillary equipment.

The slag is run off about every two hours. It solidifies as a black glassy solid which can be used as a filling material in construction works or crushed for use in road making or as an aggregate in ready-mixed concrete.

The iron is tapped about every five hours and the molten iron is run into huge ladles which may hold up to 100 000 kg of metal. The molten iron is either cast into large slabs called 'pigs' or used immediately for the manufacture of steel.

Properties of Cast Iron or Pig Iron

Cast iron has a granular crystalline structure. This can be observed at a surface exposed by breaking a piece of cast iron. Close investigation has shown that as the metal solidifies, much of the dissolved carbon forms at the boundaries of the growing crystals of metal. It solidifies with the remaining iron as a brittle glassy carbide of iron which corresponds to the formula Fe_3C, and is called cementite.

Briefly, the properties of cast iron are:

(i) It contains 4 to 5% carbon together with smaller amounts of silicon, phosphorus and sulphur and still smaller amounts of other elements.

(ii) The melting point is about 1200 °C which is considerably below the melting point of pure iron (1530 °C).

Slag.

Top: Molten slag from blast furnaces and open hearth furnaces being poured on a slag heap to solidify. Bottom: Pieces of slag. After molten slag has solidified and been crushed, black glassy pieces such as these are obtained.

A section of an iron pig.

The pig has been broken at the left-hand end, showing the brittleness of pig iron.

(iii) Compared with pure iron, cast iron is very hard and brittle.

(iv) As molten cast iron solidifies, it expands slightly. This makes it particularly useful for casting the metal into various shapes.

Uses of Cast Iron

Most of the iron produced by the blast furnace is used for the manufacture of steel, and only a small proportion is used in foundry work for the production of cast iron objects.

Cast iron is used mainly as a structural material where brittleness is unimportant. It is used for bed-plates for machinery, fire-grates and other objects not subjected to great strain. A mould is made by stamping a special mixture of clay and sand firmly around a wooden model. The model is removed and cast iron is poured into the mould. As the iron solidifies, it forms a very good copy of the original shape. It is generally cheaper to cast such pieces, rather than to start with pieces of metal and machine them down to the required shape.

PRODUCTION OF STEEL
The Open Hearth Process

The production of steel from cast iron is essentially the oxidation of impurities to bring about their removal. The charge for the furnace is mainly molten iron with a little haematite to help oxidize the impurities.

In addition, it is also possible to use up the ends of steel cut off from billets, bars and sheets, etc., produced in the steel rolling mills. Lumps of limestone ($CaCO_3$) and dolomite ($MgCO_3,CaCO_3$) are included with the charge.

A typical furnace produces about 250 000 kg of steel per 'heat'. Such a furnace would be about 29 metres long, 9 metres wide and about 5 metres high. The hearth of such a furnace would be about 1 metre deep with a hearth area of about 9 square metres. Large furnaces can make between 350 000 and 500 000 kg of steel per heat. The charge is added to the furnace through doors in the working face (see fig. 18.4).

The charge is heated by the combustion of coke-oven gas, fuel oil or tar, in preheated air. Then a stream of oxygen gas is played on to the surface of the molten charge through a water-cooled pipe called an 'oxygen lance'. Large furnace use more than one oxygen lance. A typical furnace uses about 200 000 kg of oxygen per day.

The carbon, sulphur, phosphorus and silicon are oxidized by the oxygen and iron(III) oxide. The carbon is mainly oxidized to carbon dioxide and the sulphur to sulphur dioxide. These gases escape from the molten mixture together with carbon dioxide from the decomposing limestone and dolomite. The escaping gases cause the mixture to bubble fairly rapidly. The oxides of phosphorus and silicon combine with the limestone and dolomite forming a slag of calcium

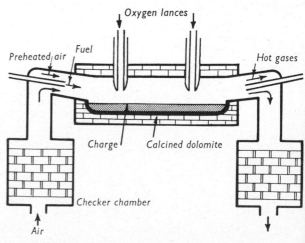

Fig. 18.4. The open hearth furnace.
Left: charging a furnace with molten metal. Right: diagram showing section through the length of the furnace.

and magnesium phosphates and silicates.

The furnace is lined with a layer of calcined dolomite, i.e., dolomite which has been heated very strongly to decompose it to CaO.MgO. This is to ensure that the impurities are slagged off.

The hot gases leaving the furnace pass over fire bricks set in an open checkered pattern. The gases pass through the labyrinth of spaces and heat the bricks. After a time the direction in which the gases pass through the furnace is reversed so that the incoming air is preheated by passing through the set of hot bricks. The other set of bricks is reheated by the exit gases. The direction of gas flow is reversed every 15 to 20 minutes.

When the impurities have been removed, the furnace is tapped and the molten steel is run out into a large ladle. The ladle has a side spout to allow the slag to overflow. Each 'heat' takes 5 to 10 hours.

The molten steel is run into moulds from a hole in the bottom of the ladle to avoid mixing it with the slag. The moulds are removed, leaving large blocks of hot steel called ingots. The ingots are held in a furnace to attain an even temperature throughout the ingot and then rolled between very heavy rollers. These knead the hot metal, strengthening it, shaping it and producing suitably sized pieces of metal, which are then further processed into the many different shapes in which steel is used.

Basic Oxygen Steelmaking

A basic oxygen steel furnace is a large cylindrical vessel lined with refractory bricks (see fig. 18.5).

The charge of burnt lime (CaO), steel scrap and molten iron (from the blast furnace) is delivered into the open top of the furnace while the furnace is tilted. The furnace is then turned upright. A water-cooled oxygen lance is lowered until it is just above the surface of the charge and a stream of oxygen gas is played on the charge. The oxygen oxidizes the impurities which are converted to a slag by the calcium oxide.

The oxygen lance is then withdrawn and the furnace is tilted to pour the steel from a taphole below the mouth of the vessel. Alloying materials are added to the steel in the ladle. Finally the slag is poured from the furnace.

The basic oxygen steel furnace produces a batch of steel in about one hour.

Steel ingots.
Molten steel is allowed to solidify in molds which are then removed. The photo shows a white hot ingot being picked up by a crane fitted with crab-like fingers.

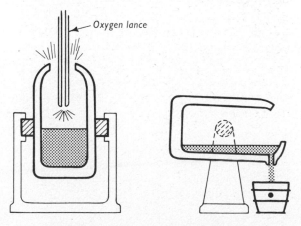

Fig. 18.5. A basic oxygen steel furnace.

Injecting oxygen into an electric furnace, to improve the quality and speed up the production of steel.

Other methods for the production of steel.
Special high quality steels are manufactured in electric furnaces and in induction furnaces. These processes begin with mild steel and produce small batches of steel which have very carefully controlled composition.

Properties of Mild Steel
Steel has a very fine grained structure because very little carbon remains in it. It is usually called mild steel and its properties are as follows:

(i) It contains about 0.15% carbon.

(ii) It is much softer than cast iron but is tough and ductile.

Uses of Mild Steel
Mild steel is used for structural girders, angles, channels, beams, plates, etc. It is used for the manufacture of railway lines, parts of trains, cars, refrigerators, washing machines, tools, tractors and thousands of other articles.

General note on the properties of steel
The properties of a particular piece of steel depend largely on three factors.

● *The carbon content*
Steel containing about 0.15% carbon is mild steel and is relatively soft and tough. A steel containing up to 1.5% carbon is called cast steel and is much harder than mild steel but comparatively brittle.

● *Heat treatment*
If a piece of steel is heated to red heat (about 750 °C),

it can be allowed to cool slowly (called annealing) or plunged into cold water (called quenching).

Annealed steel is as tough as possible, depending on its carbon content, but is relatively soft.

Quenched steel is extremely hard, depending on its carbon content, but it is relatively brittle, the hardness being obtained at the expense of ductility and elasticity.

If a steel in between these two extremes is required, the quenched steel can be *tempered*. This is done by heating it carefully to a certain temperature (between 200 and 300 °C) and then allowing it to cool slowly. By careful selection of the temperature used for tempering, the brittleness can be largely removed while still retaining the hardness.

● *Alloying*
The addition of varying amounts of other elements to steel gives it special properties (see fig. 18.6).

ELEMENTS ADDED	SPECIAL PROPERTIES OF ALLOY
Carbon	High strength. Used for machine parts, axles, bolts, etc.
Silicon	Good electrical properties and resistance to corrosion. Used in electrical equipment and for making chemical reaction vessels.
Manganese	High strength. Used for machine parts.
Manganese and molybdenum	High impact strength. Used for crushing machinery and for tram and rail crossovers and points.
Tungsten, vanadium and chromium	High strength even when hot. Used for high speed cutting tools.
Chromium	Resists corrosion. Used in stainless steel for equipment associated with food preparation and for surgical instruments.
Chromium and nickel	Resists corrosion. Used in stainless steel.
Aluminium, cobalt and nickel	Good magnetic properties. Called Alnico and used in magnets.

Fig. 18.6.

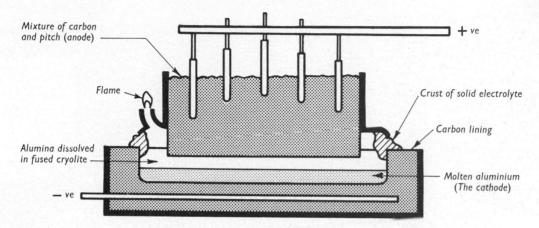

Fig. 18.7. Cell with continuous self-baked anode. As the anode burns away, more of the mixture of carbon and pitch is added. This mixture is baked hard by the heat from the furnace.

PRODUCTION OF OTHER METALS
Potassium, Calcium and Magnesium

Potassium occurs mainly as the chloride and is produced by a method similar to the extraction of sodium. Calcium and magnesium occur as the carbonates. These are converted to the chlorides, which are then electrolyzed.

Aluminium

The main source of aluminium is bauxite which is a mineral consisting mainly of Al_2O_3,xH_2O with some Fe_2O_3 and SiO_2. Bauxite is essentially a clay which is very low in silica content.

The first stage of the process is the purification of the aluminium oxide. To do this, the bauxite is dissolved in hot concentrated caustic soda solution under pressure.

$$Al_2O_3 + 3H_2O + 2OH^- \rightarrow 2Al(OH)_4^-$$

The sodium ions partner the aluminate ions formed and the solution obtained is therefore a solution of sodium aluminate.

The iron(III) oxide does not dissolve and only a very small amount of silica dissolves. When the solution is cooled, diluted, and seeded with freshly precipitated aluminium hydroxide, most of the aluminate changes to aluminium hydroxide, which precipitates in a coarsely crystalline form and is easily filtered.

$$Al(OH)_4^- \rightleftharpoons Al(OH)_3 \text{ (s)} + OH^-$$

The precipitate is filtered off. The filtrate is concentrated by evaporation and used to dissolve the next batch of bauxite.

Some of the aluminium hydroxide is kept to seed the next batch of solution and the rest is heated to decompose it to aluminium oxide. The residue obtained is a white powder which is very pure aluminium oxide and is called *alumina*:

$$2Al(OH)_3 \rightarrow Al_2O_3 + 3H_2O$$

The second stage of the process consists of the electrolysis of the alumina. The alumina is dissolved in a molten mineral called cryolite (Na_3AlF_6), producing an electrolytic conductor. The electrolytic reduction is carried out in a cell represented in the diagram (fig. 18.7). Aluminium is formed at the cathode and collects at the bottom of the cell. Oxygen produced at the anode attacks the carbon forming mainly carbon dioxide. Some carbon monoxide and volatile products from the pitch are also formed, and these are burnt as they escape from the cell. From time to time, liquid metal is tapped from the cell by suction.

Zinc

The main source of zinc is zinc blende, which is found mixed with large amounts of other minerals such as lead sulphide (PbS), pyrites (FeS_2) and silica, as well as a great deal of gangue. The ore is finely ground and the useful components are separated out by selective flotation (see chapter 7). The zinc

The main Electrolytic Zinc cell room of the works at Risdon, Tasmania, showing cathodes lifted from a cell prior to stripping of the zinc.

sulphide is then burnt in air to form zinc oxide and sulphur dioxide.

$$2ZnS + 3O_2 \rightarrow 2ZnO + 2SO_2$$

The zinc oxide can be reduced to zinc in one of two ways.

(i) *Reduction by Carbon*

A mixture of zinc oxide and carbon (e.g. coke) is heated strongly in a retort:

$$ZnO + C \rightleftharpoons Zn + CO$$

The zinc distils over and is condensed.

(ii) *Reduction by Electrolysis*

The zinc oxide is dissolved in dilute sulphuric acid:

$$ZnO + 2H^+ \rightarrow Zn^{2+} + H_2O$$

Zinc dust is then added to the solution to displace all metals below zinc on the activity series. These may be present from impurities in the ore, e.g.:

$$Zn + Cu^{2+} \rightarrow Zn^{2+} + Cu(s)$$

The solution is electrolyzed using a lead anode (not attacked by oxygen or acids) and an aluminium cathode. The electrode reactions are:

CATHODE REACTION:

$$Zn^{2+} + 2e^- \rightarrow Zn$$

ANODE REACTION:

$$2H_2O \rightarrow 4H^+ + O_2 + 4e^-$$

Thus the zinc ions are replaced by hydrogen ions and the solution contains sulphuric acid. This is used to dissolve the next batch of zinc oxide. The zinc is peeled from the aluminium cathodes when the deposit has reached a suitable thickness.

Tin

Tin occurs mainly as cassiterite which is impure tin (IV) oxide (SnO_2). Metallic tin is produced by reduction of the oxide by carbon.

Lead

Lead is produced from the sulphide ore, which is concentrated by flotation. A number of different processes are used to extract the metal, one of which is to convert the sulphide to the oxide, which is then reduced by carbon.

Copper

The sulphide ore is roasted and then reduced to impure copper in a converter. The impure copper is refined by electrolysis (see chapter 13).

Mercury

The main source is cinnabar which is mainly mercury(II) sulphide (HgS). This is heated with oxygen, which converts it to the oxide. The oxide is unstable to heat, and decomposes giving mercury.

STUDY QUESTIONS

1. Sodium has many uses including the purification of molten metals, its use as a heat transfer medium and for the production of pure caustic soda. What are the properties of sodium which make it suitable for these purposes?

2. Aluminium is used for the manufacture of kitchenware, alloys such as duralium used in aircraft construction and for electrical transmission cables. Explain why the physical properties of aluminium make it suitable for these purposes.

3. Aluminium powder is used in the thermite process for welding. A mixture of iron(III) oxide and aluminium is ignited. It becomes extremely hot and molten iron flows from the mixture and can be used to join two pieces of steel.
Write the equation for the reaction and give an explanation of the reaction with reference to the properties of aluminium.

4. Aluminium is the most abundant metal in the Earth's crust, mostly combined in the form of clay. Why is it that iron is so widely used and aluminium only in much smaller quantities?

5. Strontium is a metal high on the activity series. It has many uses including removal of traces of oxide from copper during casting, and in making special alloys. Suggest a likely method for the extraction of strontium from the mineral strontianite ($SrCO_3$).

6. What are the products of electrolysis of:
(a) dilute sodium chloride solution;
(b) concentrated sodium chloride solution;
(c) molten sodium chloride?

7. The lead(II) sulphide obtained from naturally occurring ores is concentrated by flotation. This produces a very finely ground powder. This powder is burnt in air and the residue is mixed with carbon and heated. Write equations for the chemical changes which occur during this process.

8. What are the principal ores of iron? What are the reactions occurring in the blast furnace and in a steel furnace? What are the differences between cast iron and mild steel?

9. A solution contains a mixture of zinc sulphate and copper(II) sulphate. How can this solution be made to yield separate samples of metallic zinc and copper? What would the solution contain at the end of this treatment?

10. Which metals occur free in nature? What position do these metals occupy:
(a) on the activity series;
(b) in the periodic table?

11. What methods are generally used to extract metals which are:
(a) very high on the activity series;
(b) fairly low on the activity series?

12. Describe how the properties of steel can be modified by controlling the carbon content and by heat treatment of the metal.

19 : Corrosion of Iron and Steel

Iron and steel are used so extensively that their corrosion is of considerable importance. When unprotected steel structures are exposed, they corrode and become coated with a red-brown deposit called rust. In order to devise effective methods to protect steel, it is important to know as much as possible about the process of its corrosion.

The first step towards such an understanding is to analyse rust and determine its composition. Analysis has shown that rust can be represented by the formula $Fe_2O_3.xH_2O$ where x is variable. Thus rust is *hydrated iron(III) oxide*, and if $x = 3$, then rust could be regarded as iron(III) hydroxide:

$$Fe_2O_3,3H_2O \equiv 2Fe(OH)_3$$

However, analysis has shown that x is usually less than three and therefore the formation of rust probably involves formation of iron(III) hydroxide from the iron followed by loss of water to the atmosphere.

Conditions necessary for the Corrosion of Iron

By general observations of conditions favourable for corrosion of iron exposed to the atmosphere, it seems that water and air are responsible. By using carefully controlled experiments it is possible to establish more precisely which substances cause the formation of rust.

If a clean iron nail is placed in aerated water it rusts, but if a nail is placed in freshly boiled air-free water in a sealed flask it does not rust. Furthermore, a nail kept in dry air in a desiccator does not rust. Thus *both air and water together are necessary for the corrosion of iron.*

If a small cloth bag of iron filings is suspended over water (fig. 19.1), the iron rusts and after a few days the water rises until approximately one fifth of the original volume of air is used up. The remaining gas is unreactive and hence *oxygen from the air is consumed during rusting.*

Precise experimental observations have shown that if the humidity of the air is below 50%, no corrosion occurs. If the humidity is above 80% the iron rusts very rapidly. This latter figure corresponds to the presence of a complete film of liquid water on the surface of the iron.

If some lime water is placed in a sealed flask containing moist air, any carbon dioxide present in the air is absorbed and precipitated as calcium carbonate.

Fig. 19.1. Absorption of atmospheric oxygen during corrosion of iron filings contained in the cloth bag in the gas jar.

An iron nail suspended in the air in the flask does not rust. Thus it *appears* that one of the requirements for corrosion is the presence of carbon dioxide. The carbon dioxide forms carbonic acid (a weak electrolyte) in the film of liquid water on the surface of the iron.

Other experiments have shown that rust can form in the absence of carbon dioxide provided that some other electrolyte is present. The effect of the presence of other electrolytes is illustrated by the accelerated corrosion of iron exposed to sea spray, i.e. NaCl solution. Thus it can be concluded that though CO_2 is not essential, *some electrolyte must be present for corrosion to take place.*

Irregularities in the exposed surface of the metal greatly influence the rate of corrosion because, in similar conditions, rusting is very much more rapid for steel than for chemically pure iron. It is also more rapid near imperfections in the surface and near areas of strain, e.g., at the point and head of a nail.

Summary

The conditions necessary for the corrosion of iron to form rust are:

  (i)   *Water.*

  (ii) *Oxygen gas.*

  (iii) *An electrolyte.*

  (iv) *Impurities in the iron.*

Mechanism of Corrosion of Iron

The process of corrosion may vary under different conditions. An explanation will be developed for iron exposed to typical atmospheric conditions.

The water film on the surface of the iron is exposed to the atmosphere and will dissolve any soluble substances which may be present. Thus, carbon dioxide of the atmosphere will dissolve forming carbonic acid, and near the sea other electrolytes such as sodium chloride may also be dissolved in the water film. These dissolved substances will make the solution an electrolytic conductor.

A small section of the surface of a piece of iron is represented in the diagram (fig. 19.2).

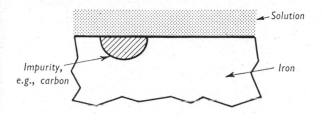

Fig. 19.2. Surface of iron exposed to the atmosphere.

The presence of electrolytes in the solution could allow a current producing cell to be set up between the iron as one electrode and an area of impurity, such as a small crystal of carbon, as the other electrode.

A cell can be set up in the laboratory to examine the behaviour of iron under these conditions. A suitable arrangement of apparatus is shown in the diagram (fig. 19.3). The direction of the electric current detected by the galvanometer suggests that electrons are being produced by the iron and consumed at the carbon electrode.

The production of electrons by iron would involve dissolution of the iron:

$$Fe \rightarrow Fe^{2+} + 2e^- \qquad \text{(oxidation)}$$

The electrons are consumed by a reduction process at the carbon electrode and this could account for the consumption of oxygen gas:

$$2H_2O + O_2 + 4e^- \rightarrow 4OH^- \qquad \text{(reduction)}$$

If iron(II) ions and hydroxide ions are being formed at the electrodes, it should be possible to detect them.

Phenolphthalein is an indicator which is colourless in neutral and acidic solutions but red in alkaline solutions. This indicator can be used to detect the formation of hydroxide ions because they make a solution alkaline and so turn the indicator to its red form.

The compound, $K_3Fe(CN)_6$ is called potassium hexacyanoferrate(III). It reacts with Fe^{2+} ions to form a deep blue coloured precipitate:

$$K^+ + Fe^{2+} + Fe(CN)_6^{3-} + H_2O \rightarrow KFe_2(CN)_6,H_2O(s)$$

This compound can be used to detect the formation of iron(II) ions in solution because these would cause formation of a blue colour.

If a small amount of each of these chemicals is dissolved in distilled water and this is used instead of the tap water in the cell (fig. 19.3) it will be observed that the solution *does* turn blue near the iron and red near the carbon. This verifies that iron(II) ions and hydroxide ions are being formed.

The currents detected in the cells described above are very small because of the small amount of electrolyte in the aqueous solutions used. If the galvanometer is replaced by a milliammeter and the electrolyte is a dilute solution of sodium chloride containing potassium hexacyanoferrate(III) and phenolphthalein, the observations are more striking. The colours develop more rapidly and the current is much greater.

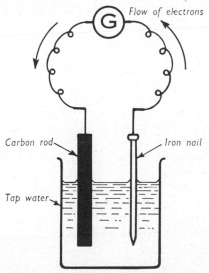

Fig. 19.3. Electrical potential developed between iron and graphite.

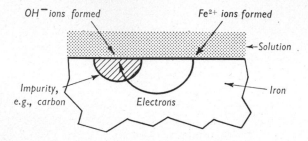

Fig. 19.4. Electron transfer during corrosion of iron.

The observations show that the corrosion of iron is probably due to the transfer of electrons through the metal to areas of impurity (fig. 19.4). The areas of impurity are areas to which the electrons drift, because electrons are not being produced at these positions. Thus the oxidation of the iron to the iron(II) state occurs with simultaneous formation of hydroxide ions.

The result of obtaining a solution containing these ions can be readily seen in the laboratory by adding an alkali to a solution of a iron(II) salt in a test tube. A green gelatinous precipitate of iron(II) hydroxide forms:

$$Fe^{2+} + 2OH^- \rightarrow Fe(OH)_2 \, (s)$$

On standing, the green iron(II) hydroxide is slowly oxidized to brown iron(III) hydroxide where it is in contact with the air:

$$\underset{green}{4Fe(OH)_2} + 2H_2O + O_2 \rightarrow \underset{brown}{4Fe(OH)_3}$$

If the electrode reactions are examined, it will be seen that formation of two iron(II) ions would supply four electrons and that the consumption of these electrons would form four hydroxide ions. Thus the iron(II) ions and hydroxide ions are being formed in the necessary proportions to precipitate iron(II) hydroxide which would then lead to iron(III) hydroxide, or hydrated iron(III) oxide.

Summary

The mechanism outlined above accounts for the composition of rust (hydrated iron(III) oxide) and the conditions necessary for the corrosion of iron (water, oxygen gas, an electrolyte and impurities).

As iron corrodes, it goes into solution as iron(II) ions:

$$Fe \rightarrow Fe^{2+} + 2e^-$$

The electrons are consumed on nearby areas:

$$2H_2O + O_2 + 4e^- \rightarrow 4OH^-$$

Precipitation of iron(II) hydroxide occurs, followed by its oxidation to iron(III) hydroxide:

$$Fe^{2+} + 2OH^- \rightarrow Fe(OH)_2 \, (s)$$

$$4Fe(OH)_2 + 2H_2O + O_2 \rightarrow 4Fe(OH)_3$$

The iron(III) hydroxide loses some of its water to the atmosphere:

$$2Fe(OH)_3 \rightarrow Fe_2O_3, xH_2O + (3 - x)H_2O$$

Note that oxygen is used by the reduction reaction during consumption of electrons and also by the oxidation of the iron(II) hydroxide to iron(III) hydroxide.

PREVENTION OF CORROSION

The corrosion of iron can be greatly diminished or prevented in a number of ways. These include:
(i)   alloying the iron with other elements;
(ii) using a protective coating;
(iii) using electrical protection.

Alloys

If chromium is alloyed with steel the product is called stainless steel because of its resistance to corrosion. The addition of small amounts of molybdenum further improves its resistance. Cast iron alloyed with silicon is very resistant to corrosion but is weak structurally. However, it is useful in building chemical reaction vessels and other articles which require resistance to attack by acids.

Protective Coatings

The most effective method of preventing corrosion of underground steel structures is to completely coat them with an impervious substance. The pipelines laid by the Gas and Fuel Corporation and the Melbourne and Metropolitan Board of Works are protected by a layer of plasticized coal-tar enamel. The coal-tar enamel is melted and the liquid is coated on the pipe. Fibre glass is included to give the enamel greater strength and the layer is covered with asbestos sheet to protect it from mechanical damage. A few inches at each end of each length of pipe are left bare so that the lengths can be welded together, but the exposed parts are then covered by a plastic tape. Thus the entire length of pipe is completely covered by an impervious layer and is protected from corrosion.

Plastic coatings on metal structures are widely used to prevent corrosion. Plastics in use include: polythene, polyvinyl chloride (P.V.C.), epoxy resins, rubber and synthetic rubber. Steel can also be coated with glass. These provide very effective barriers and can prevent corrosion even under very severe conditions, such as in sea water, in underground structures, in marine or industrial atmospheres and where corrosive chemical solutions are being used.

Vitreous enamels and acrylic paints are widely used to protect motor car bodies, refrigerators and washing machines. Special paints are also available which are very useful in protecting steel from atmospheric corrosion. These contain substances which are impervious to air and include: bituminous paints, alkyd resin paints, phenolic paints and metallic pigment paints. A well known example of a metallic pigment paint is 'aluminium paint'. Ordinary oil-based paints are not as effective as these special paints and enamels because the oil film is not impervious to water or to oxygen gas.

For some purposes steel is covered with chromium or aluminium because these metals form protective oxide films on their exposed surfaces. Metallic coatings are applied by electrolytic methods or by using mechanical pressure (called cladding). It is also possible to oxidize the surface of the steel itself by heating it in steam so that a layer of Fe_3O_4 is formed.

Temporary protection of small articles such as tools can be obtained by smearing oil or grease on the surface. Another simple method is to package the tools in air-tight containers and include a drying agent such as silica gel.

Electrical Protection

Steel can be protected by being connected to a metal higher on the activity series. Steel coated with zinc is called *galvanized iron* and this method of protection is often called galvanic protection, or, because the steel is protected by preferential corrosion of the zinc, *sacrificial* protection. Even if the layer of zinc has small imperfections the steel is still protected (see

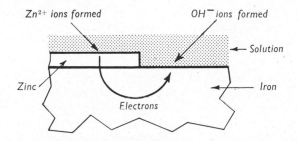

Fig. 19.5. Galvanic protection of iron.

fig. 19.5). The zinc dissolves in preference to the iron because it forms positive ions more readily:

$$Zn \rightarrow Zn^{2+} + 2e^-$$

The electrons are consumed on the iron and so corrosion of the iron is prevented.

Instead of using sacrificial protection, electrons can be supplied by a D.C. generator or a rectifier. This method has proved to be particularly successful in protecting the large steel wharves at Lae, Port Moresby and Darwin. The steel wharf is wired to the negative terminal and the positive terminal is connected to anodes (fig. 19.6). Only about 4 to 5 volt is necessary and a current of a few milliamp per square foot of structure to be protected is maintained. However, for a large wharf this may total up to quite a large current.

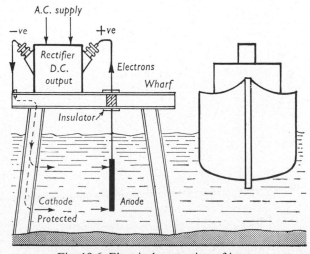

Fig. 19.6. Electrical protection of iron.

The anodes may be lumps of steel, e.g., engine blocks of scrapped earthmoving equipment have been used. In this case the scrap steel is slowly dissolved:

$$Fe \rightarrow Fe^{2+} + 2e^-$$

The corrosion of the anodes poses difficulties because they tend to dissolve away near the point where the electrical connection is made. Silicon-iron anodes are better because they corrode only very slowly but the latest development is to use anodes made of titanium coated with a very thin layer of platinum, in which case there is no corrosion of the anodes.

When these wharves were built the steel was not coated in any way at all. During the first two weeks strips of magnesium were connected to the structure to boost the flow of electrons by galvanic corrosion of the magnesium.

Unreactive Metallic Coatings

Steel is often protected from corrosion by using a layer of tin. This is widely used in the manufacture of food containers although when acidic fruit juices, etc., are to be packaged, it is usually necessary to coat the tin with an impervious layer of lacquer. With tinplate it is essential that the layer of tin has no imperfections,

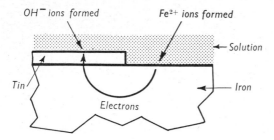

Fig. 19.7. Accelerated corrosion of iron near an imperfection in tin plate.

because tin is below iron on the activity series. If there is a break in the tinplating, however small, corrosion will commence on the exposed iron and will be very rapid (fig. 19.7). The iron dissolves because it forms positive ions more readily than the tin:

$$Fe \rightarrow Fe^{2+} + 2e^-$$

The electrons are consumed on the tin and so the iron corrodes. For special purposes, steel is sometimes coated with gold or platinum and it is essential that the layer has no imperfections because these metals are very low on the activity series.

A similar principle can be used to cause a metal to dissolve more rapidly in an acid. Thus, if pure zinc is placed in dilute sulphuric acid, it dissolves very slowly:

$$Zn + 2H^+ \rightarrow Zn^{2+} + H_2$$

If a few drops of copper(II) sulphate solution are added to the mixture, the zinc dissolves more rapidly. Zinc displaces the copper from solution because copper is below zinc on the activity series:

$$Zn + Cu^{2+} \rightarrow Zn^{2+} + Cu$$

The presence of the patches of copper on the zinc enables current producing cells to be set up. The zinc dissolves:

$$Zn \rightarrow Zn^{2+} + 2e^-$$

The electrons formed are consumed by discharge of hydrogen ions on the surface of the copper:

$$2H^+ + 2e^- \rightarrow H_2$$

The hydrogen gas is evolved on the patches of copper and this leads to more rapid dissolution of the zinc.

Generally the presence of a metal in contact with a more active metal will cause the corrosion of the more active metal to be faster than if it is pure. Similarly an active metal will dissolve more rapidly in acids in the presence of less active metals.

STUDY QUESTIONS

1. Iron fencing wire is a relatively pure form of iron. It does not rust very rapidly. Discuss the effect of the composition of the iron on the rate of rusting.

2. Zinc is used to galvanize iron. What property of zinc makes it suitable for this purpose?

3. It has been found that iron or steel which is deeply immersed in sea water shows little sign of rusting. However, it rusts very rapidly as soon as it is brought to the surface. Explain these observations.

4. The corrosion of water pipes is often most severe where the steel pipe is connected to a copper pipe. What explanation can be offered for this observation?

5. An experimental ship was built with the hull made of monel metal (a copper-nickel alloy). The plates were riveted to a steel frame with iron rivets. The ship was unable to complete even one trip in the sea because of severe corrosion of the rivets which allowed water into the ship. Explain why the rivets corroded so rapidly.

20 : Introduction to the Chemistry of Non-Metals

The term 'non-metal' refers to the elements towards the top right-hand corner of the periodic table (see fig. 20.1).

All elements in the same group, or vertical column, have the same number of electrons in the outer shells of their atoms, except for group O elements.

The *group number* is the number of electrons in the outer shell. All the group O elements have eight electrons in their outer shell, with the exception of the helium atom which has two. The bonding of an atom depends on the number of electrons in the outer shell. Thus there are some similarities in the way in which elements of the same group form bonds in their compounds but there are also many differences. The elements of group VII show group similarities and group trends in their properties and these elements are dealt with as a group. However, there is little similarity between elements such as carbon and silicon (group IV), nitrogen and phosphorus (group V) and oxygen and sulphur (group VI). These elements are dealt with in the order in which they appear in the groups of the periodic table but they are dealt with separately because of their different properties.

Non-metal atoms have vacancies for electrons in their outer shells and, unlike the atoms of metals, they are electron acceptors. They may form covalent bonds by sharing electron pairs with other non-metal atoms. They may also form ionic bonds by accepting electrons from atoms of metals. The non-metal *elements* discussed in this book are listed below. All except the group O elements exist as molecules with two or more atoms of the element per molecule because of the formation of covalent bonds between atoms of the same non-metal:

				He
H_2				
C_n	N_2	O_2	F_2	Ne
Si_n	P_4	S_8	Cl_2	Ar
			Br_2	Kr
			I_2	Xe
				Rn

The values of n for carbon and silicon are very large, because as shown in chapter 4, these elements form giant molecules. The structures of some of the gaseous non-metals are shown in fig. 20.2. These can be compared with the structure of diamond

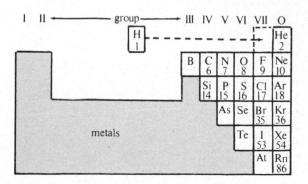

Fig. 20.1. Non-Metals in the Periodic Table.

shown in chapter 4. The structures of the molecules P_4 and S_8 are shown later.

All the above non-metals have structures which are in contrast to the structure of metals discussed in chapter 4.

A major difference between the chemistry of the metals already studied, and the non-metals yet to be studied, is that the *chemistry of the non-metals will include many covalent substances*. This can be illustrated by reference to a few examples:

$NaCl$, MgO and Na_2S are ionic
HCl, H_2S, SO_2, H_2O, PCl_3, CCl_4 are covalent.

The first three of these covalent compounds are gases and the last three are liquids. Because the chemistry of the non-metals will involve the study of gaseous compounds, it is desirable to discuss some of the techniques of handling gases in a laboratory.

Fig. 20.2. Models representing the structure of some molecules. Left to right: H_2; O_2; Cl_2.

Preparation of Gases

Gases are commonly prepared by mixing a solid and a liquid in an apparatus similar to that shown in fig. 20.3.

The liquid can be added, as required, down the funnel. However, with each addition of the liquid, air is pushed into the flask. This could be dangerous when preparing gases which form an explosive mixture with air, particularly if it is intended to burn the gas produced. In such cases, an apparatus similar to that shown in fig. 20.4 can be used. In use, the dropping funnel fitted with the tap is partly filled with liquid, and only then is the tap opened. No air is pushed into the flask by the liquid and the tap is closed as soon as sufficient liquid is added.

A Kipps apparatus is convenient for preparing a gas required for intermittent use (see fig. 20.5). The Kipps apparatus is suitable only if the gas can be prepared by reaction of a liquid with *lumps* of a solid.

It is recommended that a Kipps apparatus *should not be used in preparing hydrogen*. This is because the testing of hydrogen usually involves lighting it. Thus, if air should diffuse into the apparatus while it is not being used, the subsequent testing of the gas could lead to a violent explosion.

Fig. 20.5. A Kipps Apparatus.

Collection of Gases

The method of collecting a gas depends partly on its properties, and partly on whether it is necessary to have a dry sample. Three methods of collection are shown in fig. 20.6, fig. 20.7, and fig. 20.8.

It can be seen that if a gas is to be collected by displacement of air, a knowledge of the density of the gas relative to air is required. Such knowledge will make the method of collection a matter of intelligent planning rather than a matter of trial and error.

Figs. 20.3 (left) and 20.4 (right). Apparatus for preparation of a gas.

Fig. 20.6 (above) 20.7, and 20.8 (right).

CALCULATION OF DENSITY

The density of a gas relative to air can be calculated in the following way (using hydrogen as an example):
The molecular mass of hydrogen = 2.016.

∴ 22.4 dm³ of hydrogen at s.t.p. has a mass of 2.016 g.

The density of air is 1.293 gramme per dm³ at s.t.p.

∴ 22.4 dm³ of air at s.t.p. has a mass of 1.293 × 22.4 g = 29 g (approx.).

Therefore the masses of equal volumes of hydrogen and air are in the ratio of 2:29 (approx). Thus the density of hydrogen is approximately one-fourteenth of the density of air at the same temperature and pressure.

For other gases, the information needed to calculate densities relative to air, are the molecular mass of the gas, and the fact that approximately 29 gramme of air occupies 22.4 dm³ at s.t.p.

Purifying the Gas

It is sometimes necessary to take elaborate measures to purify a gas, although this is not usually necessary. An acid spray is often present in a gas prepared by the reaction of an acid with a solid, as in the laboratory preparation of hydrogen and carbon dioxide. This acid spray can be removed from the gas by bubbling the gas through water, provided that the gas is not very soluble in water. An apparatus similar to that shown in fig. 20.9 is suitable, although the flask would contain water.

Fig. 20.9. Apparatus for drying a gas with concentrated sulphuric acid: (left) a Drechsel bottle and (right) a Woulff's bottle.

Drying a Gas

A moist gas can be dried by passing it through a substance which is a strong absorber of water. The selection of a suitable drying agent requires knowledge of the reactions of the gas and the drying agent.

Some commonly used drying agents are calcium chloride, concentrated sulphuric acid, quicklime and phosphorus pentoxide.

CALCIUM CHLORIDE is used in flakes which have been previously heated (called fused calcium chloride). This drying agent is satisfactory for drying hydrogen, and a suitable apparatus is shown in fig. 20.10.

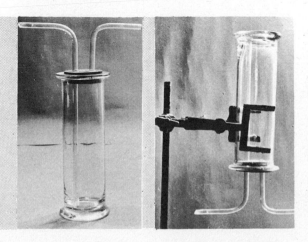

Collection of gases.

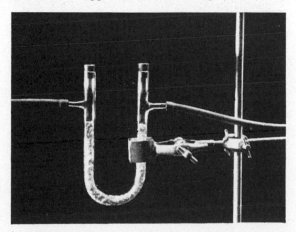

Fig. 20.10 Apparatus for drying hydrogen.

CONCENTRATED SULPHURIC ACID can be used for drying carbon dioxide, and convenient arrangements are shown in fig. 20.9.

QUICKLIME (calcium oxide) is used for drying ammonia gas. Neither concentrated sulphuric acid nor calcium chloride is suitable for this, as the gas is absorbed by each substance. Thus concentrated sulphuric acid reacts with ammonia and forms an ammonium salt, while anhydrous calcium chloride forms $CaCl_2.6NH_3$. Ammonia gas is therefore dried by passing it through a tower packed with lumps of quicklime. This is illustrated in fig. 20.11.

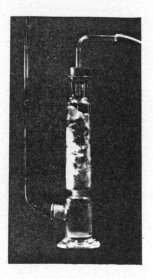

Fig. 20.11. Drying ammonia gas by passing it upwards through a tower packed with lumps of calcium oxide.

PHOSPHORUS PENTOXIDE can be used to dry hydrogen, carbon dioxide and other gases, but not ammonia. The dehydrating property of phosphorus pentoxide is discussed later.

Properties of Gases

A systematic study of the properties of a gas should include:

PHYSICAL PROPERTIES
1. Colour and odour.
2. Density.
3. Solubility in water.
4. Ease of liquefaction.

The molecular mass of a gas is useful information because the molecular masses of gases are in the same ratio as their densities, under similar conditions of temperature and pressure.

CHEMICAL PROPERTIES
1. Stability to heat.
2. Combustibility.
3. Support of combustion.
4. Effect of solution on litmus.
5. Reaction as an oxidizer or reducer.
6. Special reactions.

STUDY QUESTIONS

1. Calculate the relative molecular masses of the following gases and estimate their densities relative to air:

H_2, CO, CO_2, NH_3,
NO_2, SO_2, HCl, Cl_2
($H = 1$, $C = 12$, $N = 14$, $O = 16$, $Cl = 35.5$)

2. What explanation can be offered for the fact that the non-metal elements exist as covalent molecules?

3. Chlorine combines with sodium to form a solid product but chlorine combines with hydrogen to form a gaseous product. Explain why this is so.

4. Describe suitable methods for the collection of:
(a) NO_2 (very soluble in water);
(b) NH_3 (very soluble in water);
(c) CO_2 (not very soluble in water);
(d) H_2 (slightly soluble in water).

21 : Air and the Noble Gases

Air is a mixture. Its composition varies slightly from place to place on the Earth's surface and it varies considerably with altitude.

The various constituents in air can be detected by using a mass spectrometer. The molecules of the different gases have different molecular masses and are detected by their distinctive peaks in the mass spectrum of air. The principal peaks correspond to relative masses of 28 and 32 and are therefore due to N_2 and O_2 molecules. The other constituents of air give smaller but easily detectable peaks. The composition of dry, dust free air near the Earth's surface is shown in the table, fig. 21.1. The composition of air is given in two different ways in the table and these have different numerical values. The values for *composition by mass* depend on the relative molecular masses of the constituents. The *partial pressures* are directly proportional to the number of mole (or molecules) of the constituents.

Pollutants in the air caused by large numbers of automobiles and intense industrial activity are not included in the table (fig. 21.1), because the amounts of these vary from place to place.

Air is a source of oxygen and enormous quantities are separated from the atmosphere for use in chemical processes—particularly in steelmaking. Approximately 20 000 000 000 tons of oxygen are produced each year. To separate the oxygen, the air is first compressed to about 200 atm. This makes the air hot and it is cooled while still under pressure. It is then expanded through a turbine. In this way, some of the energy used to compress the air is recovered. During expansion, the air becomes cold and condenses to a liquid. The liquid is fractionated to separate the oxygen, nitrogen and the noble gases.

THE NOBLE GASES

The noble gases are helium, neon, argon, krypton, xenon and radon. All these gases, except radon, are present in small amounts in the air (fig. 21.2), and can be obtained during processing of liquid air.

GAS	PROPORTION IN AIR BY VOLUME (Parts per million)
helium	5
neon	15
argon	9400
krypton	1
xenon	0.1

Fig. 21.2. Noble gases in the atmosphere.

The separation of the gases depends on their boiling points and these are shown in fig. 21.3.

It can be seen that, at the temperature of liquid air (about $-190\ °C$), helium and neon escape liquefaction. The gases contain a considerable amount of nitrogen which can be removed by chemical means, but the separation of the helium and neon is difficult, and requires the use of liquid hydrogen (b.p. $= -253\ °C$).

The fractionation of the liquid air yields several fractions: first the nitrogen, then the argon mixed with a little oxygen, then the oxygen and finally the krypton and xenon.

Radon is not present in the atmosphere to a detectable extent, but it is produced by the radioactive decay of radium.

Helium is present in the air, but the major commercial source of this gas is the natural gas of some oil wells in the United States.

GAS	PERCENTAGE COMPOSITION BY MASS	PARTIAL PRESSURE (atmospheres)
nitrogen	75.51	0.780 3
oxygen	23.15	0.290 9
noble gases	1.30	0.009 5
carbon dioxide	0.04	0.000 3
	100.00	1.000 0

Fig. 21.1. Composition of air.

GAS	BOILING POINT ($°C$)
helium	−268.9
neon	−245.9
nitrogen	−195.8
argon	−185.7
oxygen	−182.7
krypton	−152.9
xenon	−107.1

Fig. 21.3.

ELEMENT	ATOMIC NUMBER	ATOMIC MASS ($^{12}C = 12$)	ELECTRON CONFIGURATION	MELTING POINT (°C)	BOILING POINT (°C)	DENSITY AT·s.t.p. (g dm^{-3})
helium	2	4.0026	2	−272.2	−268.9	0.177
neon	10	20.179	2, 8	−248.7	−245.9	0.899
argon	18	39.948	2, 8, 8	−189.2	−185.7	1.784
krypton	36	83.80	2, 8, 18, 8	−156.6	−152.9	3.743
xenon	54	131.30	2, 8, 18, 18, 8	−112	−107.1	5.851
radon	86	222	2, 8, 18, 32, 18, 8	−71	−61.8	9.73

Fig. 21.4. The Noble Gases.

Physical Properties of the Noble Gases

It can be seen from the table in fig. 21.4 that the melting points, boiling points and densities of the noble gases show regular increases, related to increases in their atomic masses. Helium has the lowest melting point and boiling point of any known substance.

All the noble gases are colourless and odourless.

Chemical Properties of the Noble Gases

If a mixture of xenon and fluorine is placed in glass vessel and exposed to sunlight, white crystals of xenon difluoride form on the inside of the glass. If xenon is heated with fluorine in a nickel tube, xenon tetrafluoride and xenon hexafluoride can be formed. These are all white, volatile solids composed of molecules (see fig. 21.5).

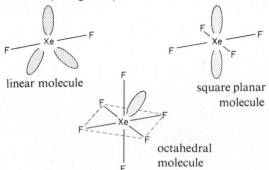

linear molecule

square planar molecule

octahedral molecule

Fig. 21.5. Molecules of xenon fluorides.

SOME REACTIONS OF XENON COMPOUNDS

All of the xenon fluorides are reduced by hydrogen to xenon and hydrogen fluoride, e.g.

$$XeF_2 + H_2 \rightarrow Xe + 2HF$$

XeF_2 is hydrolysed slowly by water, and oxygen is formed.

$$2XeF_2 + 2H_2O \rightarrow 2Xe + 4HF + O_2$$

The hydrolysis of XeF_6 is more complicated. If XeF_6 is mixed with an equimolar amount of water, xenon oxide tetrafluoride is formed:

$$XeF_6 + H_2O \rightarrow XeOF_4 + 2HF$$

However, with excess water, xenon trioxide can be obtained by evaporating off the resulting solution.

$$XeF_6 + 3H_2O \rightarrow XeO_3 + 6HF$$

● *This process is extremely hazardous because XeO_3 is unstable and liable to detonate explosively.*

If, instead of evaporating the solution, an alkali is added, it is possible to obtain white crystalline salts such as:

Ba_3XeO_6 barium xenate
$Na_4XeO_6.8H_2O$ sodium perxenate

Hydrolysis of XeF_4 gives some Xe gas and some of the oxygen compounds. It gives the products obtained from both of the other fluorides. Krypton and radon fluorides have also been prepared.

Uses of the Noble Gases

HELIUM is useful for inflating large balloons because of its low density and because it is not inflammable. It has about 90% of the buoyancy of hydrogen.

Helium is also used, mixed with oxygen, as the gas breathed by divers. This mixture is used in place of air, because helium is much less soluble in the blood than nitrogen. If a diver who is breathing air is suddenly brought to the surface, bubbles of nitrogen gas may form in the blood. This has a crippling effect and is known as 'the bends'.

NEON is used in neon lights.

ARGON is used in incandescent electric light globes, and in discharge tubes. It is also used in radio valves and electrical rectifiers, and in argon arc welding, where the gas provides an inert atmosphere to protect the metal from oxidation during welding.

KRYPTON is used in special large electrical rectifiers.

STUDY QUESTIONS

1. If the oxygen is removed from dry dust-free air, the remaining gas is mainly nitrogen but contains small amounts of the noble gases.

Ramsay absorbed the nitrogen and so obtained a sample of the noble gases, by repeatedly passing the gas over heated magnesium. How does this process remove the nitrogen?

2. A vessel containing a cold liquid is sometimes observed to become frosted on the outside if standing in air. What explanation can be offered for this observation?

3. (a) Why is it unsatisfactory to store solutions of alkalis in open vessels exposed to the air?
   (b) A laboratory bottle contains a substance labelled 'sodium hydroxide'. What other substances are likely to be present in the bottle?

4. Calculate accurately the volume of gas left, if all the oxygen is removed from 1.0000 dm³ of dry dust-free air. (Both volumes expressed at s.t.p.)

5. What property of argon makes it suitable for filling electric light globes?

What other gases are suitable, and why are these gases not used?

6. It has been seen that the percentages of carbon dioxide and oxygen in the atmosphere are fairly constant. However, carbon dioxide is produced and oxygen is consumed during respiration and combustion.

How can these facts be reconciled?

7. Air becomes cold when it is expanded suddenly, i.e., when the molecules move apart. The decrease in temperature of the gas, means that the kinetic energy of the molecules is decreased, and hence the molecules slow down as they move apart.

What forces exist which could explain why the molecules slow down as they move apart?

8. The buoyancy of a balloon can be calculated as follows:

buoyancy = weight of air displaced — weight of gas in the balloon.

Using the densities given below, calculate:
(a) the buoyancy of a balloon filled with hydrogen and having a capacity of 1000 dm³;
(b) the buoyancy of the same balloon filled with helium;
(c) the relative buoyancy of helium as compared to hydrogen.

density of hydrogen = 0.090 g dm⁻³ at s.t.p.
density of helium = 0.177 g dm⁻³ at s.t.p.
density of air = 1.293 g dm⁻³ at s.t.p.

Answers to Numerical Problems

4. 0.7901 dm³.
8. (a) 1203 gf.
   (b) 1116 gf.
   (c) 92.8%.

22 : Hydrogen and Water

HYDROGEN

Hydrogen has the simplest atomic structure of all elements. It is one of the commonest of all elements and ranks in ninth place in order of abundance of the elements in the earth's crust, calculated on the basis of mass. If calculated on the basis of number of atoms, it ranks still higher. Hydrogen appears to be extremely abundant in the stars and is probably the most abundant element in the universe.

Hydrogen does not occur to any significant extent in air, although it occurs widely as compounds, principally water.

Commercial Production of Hydrogen

Hydrogen is produced commercially as a by-product of the electrolysis of a concentrated solution of sodium chloride. Where electrical power is cheap, it can be prepared by the electrolysis of dilute sulphuric acid or sodium hydroxide solution. These methods have been discussed in chapter 13.

Hydrogen is also produced commercially by the reduction of steam with hot iron:

$$3Fe + 4H_2O \rightleftharpoons Fe_3O_4 + 4H_2$$

A pure gas can be obtained.

Reduction of steam with red hot coke is also used to produce hydrogen:

$$C + H_2O \rightarrow CO + H_2$$

It is possible to obtain the hydrogen from the mixture of gases and the procedure used is described later under the 'Haber Process'.

A method of producing hydrogen which is becoming more widely used is to treat hydrocarbons with steam.

Laboratory Preparation and Collection of Hydrogen

Care must be taken with hydrogen, because if the gas is mixed with air and lit, a violent explosion might occur. Hence a small sample of the gas should be collected and burnt. This should be repeated until the gas burns quietly. Only then is it safe to proceed with tests on the gas.

Hydrogen can be prepared in the laboratory by the reduction of dilute hydrochloric acid with zinc.

$$Zn + 2H^+ \rightarrow Zn^{2+} + H_2$$

It may be collected over water as shown in fig. 22.1.

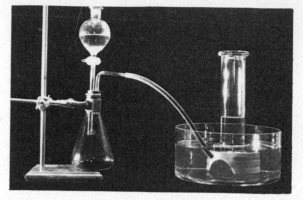

Fig. 22.1. Laboratory preparation and collection of hydrogen: dilute hydrochloric acid is added to zinc.

If the gas is required dry, it should be bubbled through water to remove the acid spray, dried with a suitable drying agent, e.g. calcium chloride, and then collected by downward displacement of air.

If dilute sulphuric acid is used with zinc, the reaction is slower but can be speeded up by adding a few drops of copper(II) sulphate solution (see chapter 19).

The Properties of Hydrogen

It can be observed during the preparation, that hydrogen is a colourless, odourless gas, while its collection over water suggests that it is not very soluble in water.

The molecular mass of hydrogen is 2.016. Thus 22.4 dm³ of hydrogen at s.t.p. weighs 2.016 g, whereas the mass of 22.4 dm³ of air at s.t.p. is about 29 g. Hence the density of hydrogen is about one-fourteenth that of air.

The combustion of hydrogen is a chemical property, i.e., a reaction in which a new substance is formed. If a lighted taper is thrust upwards into an inverted gas jar of hydrogen, the gas burns at the mouth of the gas jar, but the taper goes out when in the gas. Thus the gas does not support combustion, but will burn. A film of droplets of a clear liquid is observable inside the gas jar, suggesting that the product of combustion is water. The fact that water is the sole product of combustion of hydrogen will be discussed later in this chapter.

If hydrogen gas is shaken with water and then tested with litmus, the litmus does not change colour. The remaining chemical properties of hydrogen have been discussed already.

Summary of the Properties of Hydrogen

Physical properties of hydrogen
1. Colourless, odourless gas.
2. The lightest gas known. (The molecular mass of H_2 is about 2.)
3. Very slightly soluble in water. (About 2 cm³ of hydrogen dissolves in 100 cm³ of water at room conditions.)
4. Very difficult to liquefy (b.p. = −252.7 °C).

Chemical properties of hydrogen
1. Stable to heat.
2. Burns quietly with a blue flame if pure, forming water:
$$2H_2 + O_2 \rightarrow 2H_2O$$
The gas explodes if it is mixed with oxygen or air and then lit.
3. If does not support burning.
4. Its solution in water is neutral to litmus.
5. It is a good reducing agent. This has been discussed earlier under the reactions of oxides:
$$CuO + H_2 \rightarrow Cu + H_2O$$
6. It reacts with most non-metals. A mixture of hydrogen and chlorine explodes if exposed to direct sunlight:
$$H_2 + Cl_2 \rightarrow 2HCl$$
Also, a jet of burning hydrogen will continue to burn if introduced into an atmosphere of chlorine.
This reaction of hydrogen with chlorine is a further example of hydrogen acting as a reducing agent.

Test for Hydrogen

Hydrogen is a colourless, odourless gas which burns with a blue flame to form water as the only product. Some other gases are similar but their combustion yields other products in addition to water, e.g., methane, CH_4, produces carbon dioxide and water.

Uses of Hydrogen

1. *In balloons.* Hydrogen has very good buoyancy but there are dangers in its use because of its inflammability.

2. *The conversion of oils to fats.* Oils can be converted to fats by treatment with hydrogen under pressure, in the presence of finely divided nickel as a catalyst.

Thus whale oil can be converted to fat for the manufacture of margarine, and some inedible oils can be converted to fats for use in the production of soap. The process is called *hydrogenation of oils.*

3. *Hydrogenation of coal* can be performed in a manner similar to hydrogenation of oils, and distillation of the product yields a good quality petrol.

4. *Hydrogenation of heavy fuel oils* can be used to produce lighter grade fuel oils.

5. *Production of ammonia.* This is achieved commercially by the combination of nitrogen gas and hydrogen gas—see chapter 28.

6. *Production of methanol.* This is brought about by the combination of carbon monoxide and hydrogen:
$$CO + 2H_2 \rightleftharpoons CH_3OH$$
Methanol is used as a solvent for lacquers, and in the production of other chemicals.

7. *Liquid hydrogen* is currently being used as a fuel for rockets. However, the very low temperature required to store liquid hydrogen, introduces many practical problems.

WATER

Water may be purified by distillation (chapter 7) or by ion-exchange methods (chapter 24). The physical properties of pure water are listed in the table fig. 22.2.

freezing point	0.00 °C at 760 mmHg
boiling point	100.00 °C at 760 mmHg
density of ice at 0 °C	0.916 8 g cm⁻³
density of water at 0 °C	0.999 9 g cm⁻³
at 4 °C	1.000 0 g cm⁻³
at 20 °C	0.998 2 g cm⁻³

Fig. 22.2. Physical properties of water.

The Composition of Water
Composition by volume

The electrolysis of aqueous solutions of electrolytes, suggests that water can be decomposed to give two volumes of hydrogen gas and one volume of oxygen gas, measured at the same temperature and pressure. However, the presence of the electrolyte, e.g. H_2SO_4, may make this conclusion a little doubtful.

The conclusion can be verified by direct combination of hydrogen and oxygen in a eudiometer (see fig. 10.3). About equal volumes of hydrogen and oxygen are mixed in the eudiometer at room temperature and pressure, the volume of each gas being noted. (A mixture of two volumes of hydrogen to one volume of oxygen is dangerously explosive.) The gas mixture is sparked in order to react the hydrogen and oxygen and the mixture is allowed to cool back to room temperature. The water formed condenses to a few droplets and the volume of oxygen left is noted. All volumes are read at the same temperature and pressure. It is found that two volumes of hydrogen combine with one volume of oxygen.

Composition by mass

1. *By Direct Weighing*. In 1895 E. Morley determined the composition by weighing carefully purified samples of hydrogen and oxygen, burning the gases, and then determining the mass of water formed. From his experiments, Morley concluded that 1 part by mass of hydrogen combines with 7.94 parts by mass of oxygen, forming 8.94 parts by mass of water.

2. *By Indirect Weighing*. This method was devised by J. Dumas in 1843, although his results were not published until 1892.

Dumas passed carefully purified and dried hydrogen gas over copper(II) oxide. The oxide was reduced:

$$CuO + H_2 \rightarrow Cu + H_2O$$

The apparatus is illustrated in the diagram fig. 22.3. The right hand U-tube was protected from entry of of moist air, by a further U-tube containing calcium chloride. This further U-tube is not shown in the diagram.

Before the reduction was started two masses were determined:
(i) the mass of the reduction tube and its contents;
(ii) the mass of the right hand U-tube and its contents.

These same two pieces were weighed after the reduction. The decrease in mass of the reduction tube gave the mass of oxygen used. The increase in mass of the right hand U-tube gave the mass of water formed. The mass of hydrogen used was not determined directly, but was calculated by subtracting the mass of oxygen used from the mass of water formed (i.e., by application of the Law of Conservation of Mass).

Dumas repeated the experiment many times in order to obtain reliable results. The following figures illustrate Dumas' results:

loss in mass of copper(II) oxide = 44.22 g
mass of water formed = 49.76 g
∴ mass of hydrogen used = 5.54 g

These results give a ratio of 7.98 gramme of oxygen for every gramme of hydrogen. (More recent figures give the ratio as 7.940 : 1.)

Physical Properties of Water

Water is the most abundant liquid known to man; the oceans cover three-quarters of the earth's surface and it is widely used in industry as a solvent in chemical reactions. The biological fluids in all living things contain large proportions of water.

Although water has been studied for many years, its structure is not yet fully understood. Ice floats on water because solid water is less dense than liquid water at 0 °C—this is most unusual because practically

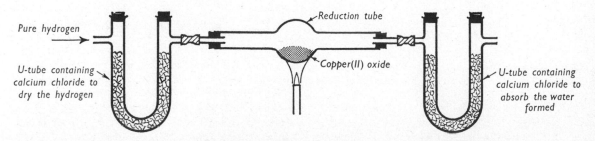

Fig. 22.3. Determination of the composition of water by mass.

all substances are more dense when in their solid state. Ice formed by freezing water under atmospheric pressure is called ice I. This ice can be changed into a denser form, called ice II, by subjecting it to very high pressure.

The crystal structures of both forms of ice are dependent on the electrical charges on water molecules. The hydrogen atoms are slightly positively charged and become attached to the negatively charged electron-pairs on adjacent water molecules (see fig. 22.4). Each oxygen atom is surrounded by four

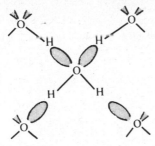

Fig. 22.4. Electrostatic forces of attraction bond a water molecule to other water molecules.

hydrogen atoms which are arranged tetrahedrally and this leads to a hexagonal pattern which is perfectly regular in ice II (see fig. 22.5). The structure of ice I is similar but not as regular; some of the chair shaped

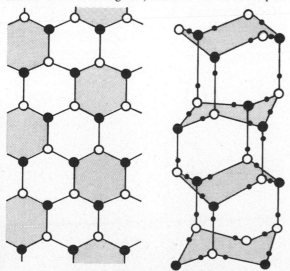

Fig. 22.5. The structure of ice II.

hexagons are stacked with their ends pointing the wrong way and so they are not stacked as efficiently. This is why ice I is less dense than ice II.

The electrostatic forces which are responsible for the structure of ice, are also responsible for the properties of liquid water.

Near 100 °C, water consists of separate molecules in random motion. As the temperature is lowered, the electrostatic forces cause some of the water molecules to cluster together. These clusters form within the bulk of the random molecules. The clusters have the ice I structure and they increase in size and number as the temperature is lowered. Below 60 °C, the clusters begin to group together but, because of imperfect fitting, spaces containing random water molecules form between the clusters. Below 4 °C, very few random water molecules remain and so empty spaces begin to form between the clusters and this is why the liquid *expands* as it is cooled below 4 °C. This is unusual—liquids usually contract when they are cooled.

At 0 °C, the groups of clusters change to a rigid lattice and to do this they must join together. Because the orientation of the columns of hexagonal rings is different in the various groups, they cannot fit perfectly. Many more empty spaces are formed and this is why ice I is less dense than water at 0 °C.

Recently there have been reports of preparations of a heavy syrupy liquid which was claimed to be an unusual form of water. The liquid is made by distilling water under partial vacuum in a closed vessel in which glass capillary tubes are placed. When the ordinary water is allowed to evaporate out of the capillaries, a very small amount of liquid remains, about 10^{-3} cm^3 being typical. The liquid was reported to be 40% denser and 15 times more viscous than ordinary water and to boil at 500°C. This heavy syrupy liquid was thought to be a highly ordered form of water with clusters having the ice II structure and was named 'super water', 'anomalous water' or 'poly water' by various workers.

Indentification of a minute sample is difficult. The results now available indicate that there is very little water in the syrupy liquid; it consists mainly of salts, present as impurities during the preparation. They probably enter the capillaries because of the surface tension of water and are left behind when the water is evaporated. The existence of super water is extremely doubtful.

Chemical Properties of Water

The chemical properties of water have been discussed already in various sections of this book. These properties include:

(a) the acid-base properties of water;

(b) water as a solvent;

(c) the formation of hydrates;

(d) the reduction of water with metals;

(e) the reaction of water with oxides.

STUDY QUESTIONS

1. In most of its compounds hydrogen exhibits a covalency of 1.
Illustrate this statement by reference to the structure of the following molecules:
H_2, HCl, H_2O, NH_3, CH_4.

2. Hydrogen gas can be made to react with sodium metal, forming an ionic compound sodium hydride, Na^+H^-.
What is the electronic arrangement in the hydride ion?

3. Consider each of the following reactions of hydrogen, and classify the hydrogen as an oxidizer or a reducer:

$H_2 + Cl_2 \rightarrow 2HCl$;
$CuO + H_2 \rightarrow Cu + H_2O$;
$2Na + H_2 \rightarrow 2Na^+H^-$.

4. What volume of oxygen can be obtained at s.t.p. by complete electrolysis of one mole of water?

5. Suppose that in a determination of the composition of water by Dumas' method, copper(I) oxide was used instead of copper(II) oxide. Describe and explain what effect this would have on the result.

6. Sketch a molecule of water showing the distribution of electron pairs and indicate the shape of the molecule.

7. Sketch a molecule of water showing the distribution of electrical charge and explain how molecules link together in the formation of ice.

8. At 0 °C and 1 atm pressure, ice is less dense than liquid water. What explanation can be offered for this unusual property?

9. Water shows unusual changes in density as it is cooled. Describe and explain these changes.

10. Give suitable examples to illustrate the reactions of water with:
(a) acids and bases;
(b) metals;
(c) oxides and peroxides.

11. Calculate the volume of hydrogen gas needed to react exactly with:
(a) 50 cm³ of oxygen gas;
(b) 50 cm³ of chlorine gas.
(All volumes measured at the same temperature and pressure.)

12. Calculate the mass of copper formed by complete reduction of 8.0 g of copper(II) oxide.
(Cu = 64, O = 16)

Answers to Numerical Problems

4. 11.2 dm³.
11. (a) 100 cm³. (b) 50 cm³.
12. 6.4 g.

23 :   Carbon, Carbon Dioxide and Carbonates

Carbon occurs mainly in the form of carbonates such as calcium carbonate, although it also occurs as carbon dioxide in the atmosphere, and to a limited extent as the pure element in graphite and diamond. It is not a particularly abundant element and it has been estimated that it comprises about 0.1% of the earth's crust by mass.

Valency of Carbon

Carbon is in group IV of the periodic table. There are four electrons in the outer electron shell of a carbon atom and therefore carbon has a covalency of four, e.g:

$$\underset{\text{methane}}{H-\overset{\displaystyle H}{\underset{\displaystyle H}{C}}-H} \qquad \underset{\text{carbon dioxide}}{O = C = O} \qquad \underset{\text{hydrogen cyanide}}{H - C \equiv N}$$

Introduction to Allotropy

It was suggested in chapter 4 that *diamond* is a pure form of carbon. This has been verified by direct combustion of diamond in oxygen. The reaction produces carbon dioxide only, and in addition, the mass of carbon dioxide formed is the mass to be expected if diamond *is* a pure form of carbon. Similar experiments with *graphite* lead to a similar conclusion.

Diamond and graphite are evidently the same element, and the different physical properties are apparently due to different crystal structures.

● *Elements which exist in two or more different physical forms are said to exhibit allotropy.*

ALLOTROPY OF CARBON

Diamond

Diamond occurs naturally in South Africa, Brazil, India and Australia, but the South African source is the most important. South Africa exports more than a ton of diamonds annually. Diamonds can also be made synthetically, by subjecting carbon-containing compounds or graphite to very high temperatures and pressures. Diamonds made by this method are quite tiny, but are indistinguishable from naturally occurring diamonds.

A diamond crystal consists of a three dimensional array of carbon atoms, with each carbon atom being bonded tetrahedrally to four other carbon atoms—see fig. 23.1. Crystals of diamond may therefore be regarded as giant molecules.

It has also been seen that the structure of diamond as represented in the model is consistent with its properties.

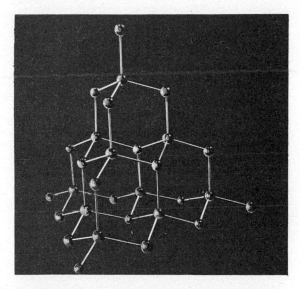

Fig. 23.1. A model representing the structure of diamond.

A giant molecule: the 'Golconda d'Or', a flawless diamond weighing 95.40 carats, and measuring 27.5 × 26.0 mm.

PROPERTIES OF DIAMOND

Diamond is one of the hardest substances known. It sublimes at temperatures above 3500 °C and is a non-conductor of electricity.

One of the properties of diamond which makes it a highly prized gemstone is its property of diffracting (or bending) light to a very marked extent. The cutting of faces on diamond is designed to use this property to give the gem the greatest possible amount of 'fire'. Only a small proportion of the diamonds mined in South Africa are sufficiently pure and clear to be used as gems. The rest are used for industrial purposes.

Diamond is the densest form of carbon and its relatively high density (3.5 g cm^{-3}) reflects the compact nature of the substance.

Diamond is extremely resistant to chemical attack, but it will burn in oxygen if heated to 800 °C.

INDUSTRIAL DIAMONDS

The industrial uses of diamond depend on its great hardness. Black diamonds are used in rock drills and chips of diamond are used in glass cutters. Industrial diamonds are also used as abrasives for grinding very hard materials.

Graphite

Graphite occurs naturally in considerable quantities, principally in Ceylon, Siberia and Korea. It is also manufactured in large quantities in electric furnaces. Coke is packed into the furnace together with a little iron(III) oxide as catalyst, and then covered with a mixture of sand and coke. The mixture is heated at a temperature of about 3000 °C for about 24 hours. Impurities in the coke vapourize, and the product is practically pure graphite.

PHYSICAL PROPERTIES OF GRAPHITE

Graphite is a crystalline solid. It has a soft greasy feeling and leaves black streaks if it is rubbed on paper. Examination of these black streaks shows that they consist of a thin layer of minute crystals. Graphite is less dense than diamond, and has a density of 2.2 g cm^{-3}. It has a metallic lustre and is a fairly good conductor of electricity.

The structure of graphite has been determined by X-ray crystallography in the same way as diamond. The structure is shown in fig. 23.2.

The structure shows that graphite consists of layers. The layers are parallel to one another and there is a relatively wide separation between adjacent layers.

At least three of the four valency electrons of each carbon atom are used in forming bonds with the three adjacent carbon atoms in the layer. The fourth valency electron of each carbon atom is not located by bond formation. The electrical conductivity of graphite can be explained on the basis of the mobility

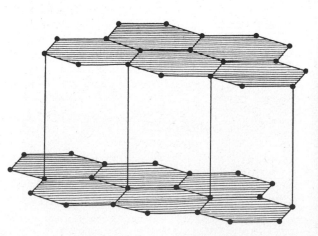
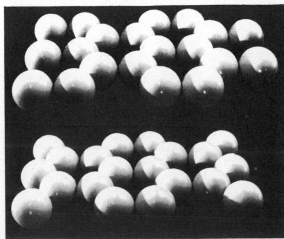

Fig. 23.2. Diagram and model representing the structure of graphite.

of these electrons. The separate layers are too widely spaced to be bound together by covalent bonds and are probably held by weak van der Waal's forces, which explains why graphite is so soft and flaky.

CHEMICAL PROPERTIES OF GRAPHITE

Graphite is somewhat more reactive than diamond. Thus it burns more readily and is attacked by nitric acid and sulphuric acid.

USES OF GRAPHITE

Graphite is used in making 'lead' pencils. The graphite is powdered and mixed with clay, the hardness of the 'lead' depending on the proportion of clay.

Graphite is a highly refractory material with a melting point of about 3500 °C. It is widely used for making crucibles to hold molten metals.

The electrical conductivity of graphite makes it suitable as the electrode material in electric furnaces and arc lights. In addition, graphite is not attacked by chlorine and it is therefore used for electrodes during electrolysis, if chlorine is evolved at the electrode.

Graphite is also a good lubricant, because adjacent layers slide easily over one another. It is particularly useful for lubricating moving parts which are subjected to high temperatures, where mineral oils would burn away.

Other Forms of Carbon

Several further forms of carbon exist. They include coke, wood charcoal, bone charcoal, carbon black and gas carbon. These forms of carbon were at one time regarded as amorphous (non-crystalline) forms, but X-ray investigations have shown that they have the same structure as graphite, although the crystals are exceedingly small.

These forms of carbon are obtained by heating carbon containing substances out of contact with the air so that they cannot burn. Volatile substances are formed and driven off leaving a residue which is mostly carbon. This process is called destructive distillation.

The main residue obtained from destructive distillation of coal is 'coke', large quantities of which are used in the production of iron. Destructive distillation of wood yields 'charcoal' and this has an open porous structure.

ACTIVATED CHARCOAL

Wood charcoal has the special property of being able to adsorb certain substances. One form of wood charcoal which is particularly suitable for this purpose is the highly porous charcoal formed by destructive distillation of coconut shells. This is called coconut charcoal.

The adsorptive capacity of wood charcoal can be increased to a very great extent by steam treatment, and the product is then called activated charcoal. In the activation process, wood charcoal is packed in a steel tube and heated fairly strongly while superheated steam is passed through the tube. This treatment removes adsorbed substances from the porous charcoal. These adsorbed substances are mainly hydrocarbons of high molecular mass (tarry substances) which were formed during the destructive distillation process and were retained by the charcoal. The removal of these makes the whole surface, throughout the porous honeycomb structure of the charcoal, an active adsorber.

Experiment: Adsorption from solution by activated charcoal

This can be conveniently demonstrated by using activated charcoal to adsorb a coloured substance such as methylene blue. About 200 cm³ of fairly deep-blue coloured solution is prepared (e.g. by adding a few cm³ of an approximately 0.5% solution of methylene blue to 200 cm³ of water in a beaker). A small portion of this solution is reserved, and to the remainder, a small amount of activated charcoal is added. The black liquid is stirred for a short time and then filtered to remove the charcoal. The filtrate is a clear and colourless liquid. Some of the reserved sample is now filtered through a clean filter paper, and the filtrate is found to be still blue.

It can be concluded that the decolourizing effect was due to the activated charcoal.

Activated charcoal can also adsorb relatively large volumes of many gases. Given below are the volumes of some gases adsorbed by 1 g of activated charcoal at 15 °C.:

SO_2	380 cm³	CO	9 cm³
Cl_2	235 cm³	O_2	8 cm³
H_2S	180 cm³	N_2	8 cm³

As a general rule, the most readily adsorbed gases are those which are also easily liquefiable or very soluble in water. This property is used to advantage in gas masks designed to protect the wearer against the toxic effects of such gases.

CARBON BLACK

Carbon black or lamp black is produced by burning liquid or gaseous petroleum products in a limited supply of air.

It is a black, greasy, finely divided powder. Considerable quantities are used to toughen rubber for tyres, thereby increasing their resistance to wear. Carbon black is also used in making black paint and printer's ink.

CARBON DIOXIDE

Carbon dioxide can be prepared commercially as follows:

1. BY HEATING LIMESTONE

Limestone is heated in a special kiln to decompose it to quicklime and carbon dioxide (see next chapter):

$$CaCO_3 \rightarrow CaO + CO_2 \text{ (g)}$$

2. BY BURNING CARBON OR CARBON COMPOUNDS

Most fuels produce carbon dioxide on combustion. Such fuels include coke, wood and petroleum products. The gas from such sources contains much nitrogen and other substances. The carbon dioxide must be purified before use.

3. BY FERMENTATION PROCESSES

Carbon dioxide is an important by-product in the production of ethanol by the fermentation of sugar. The fermentation process is brought about by mixing sugar with yeast. The sugar usually employed is sucrose (cane sugar or beet sugar). The yeast produces an enzyme (organic catalyst) which converts the sucrose into simpler sugars, and a further enzyme which brings about the fermentation:

$$C_6H_{12}O_6 \rightarrow 2C_2H_5OH + 2CO_2\text{(g)}$$
simple sugar ethanol

Laboratory Preparation and Collection

Carbon dioxide is usually prepared in the laboratory by the reaction of dilute hydrochloric acid and marble chips. Collection over water will remove the

Commercial production of carbon dioxide.

Behind the road tanker are two sets of towers. Those on the right are used to store petroleum products. Combustion of these yields carbon dioxide, which is absorbed in the towers on the left.

The absorbing agent is potassium carbonate solution. This is converted to potassium hydrogen carbonate solution, which on subsequent heating releases pure carbon dioxide.

Fig. 23.3. Laboratory preparation and collection of carbon dioxide: dilute hydrochloric acid is added to marble chips (calcium carbonate).

acid spray from the gas. A suitable apparatus is shown in fig. 23.3.

The CO_3^{2-} ion is a base. It accepts protons from the H_3O^+ ions in the hydrochloric acid. The reaction can be represented:

$$CO_3^{2-} + 2H^+ \rightarrow H_2CO_3$$
$$H_2CO_3 \rightarrow H_2O + CO_2 \text{ (g)}$$

A Kipp's apparatus is more convenient if an intermittent supply of carbon dioxide is required—see chapter 20.

PROPERTIES OF CARBON DIOXIDE
Physical Properties

It is evident during the collection of carbon dioxide that it is a colourless, odourless gas.

If carbon dioxide gas is poured from a gas jar on to a lighted candle, the candle goes out. This suggests that the gas is denser than air. In fact the gas is about $1\frac{1}{2}$ times as dense as air (the molecular mass of CO_2 is about 44).

Carbon dioxide is not very soluble in water, 100 cm³ of which dissolves about 88 cm³ of the gas at 20 °C and 760 mmHg.

Carbon dioxide is easily liquefied, and cylinders of the liquid can be purchased. It is also easily solidified. The solid sublimes at −78.5 °C under atmospheric pressure and it melts at −56.7 °C under 5 atmosphere pressure.

Chemical Properties
STABILITY OF CARBON DIOXIDE

It will be seen in the next chapter that carbon dioxide is a product of high temperature decomposition of limestone in a lime kiln. It is therefore stable at high temperatures.

COMBUSTION IN CARBON DIOXIDE

If a burning wax taper is introduced into carbon dioxide, the taper goes out and the gas does not ignite. Thus the gas does not burn nor does it support ordinary combustion. However, the gas contains oxygen and this suggests that strongly burning substances might be able to use this oxygen for combustion.

If a strip of burning magnesium is lowered into a gas jar of carbon dioxide, the magnesium continues to burn brightly. The residue after combustion consists of a mixture of black specks and white powder. The black substance is evidently carbon and the white powder is magnesium oxide.

$$2Mg + CO_2 \rightarrow 2MgO + C$$

ACIDIC NATURE OF CARBON DIOXIDE

If water is added to a gas jar of carbon dioxide, and then litmus is added, the litmus turns purplish-red. Evidently the solution of carbon dioxide in water is a weak acid. The acid formed is carbonic acid. It is possible to show that only a very small proportion (about 1%) of the dissolved carbon dioxide actually reacts with the water:

$$CO_2 + H_2O \rightleftharpoons H_2CO_3$$

The carbonic acid formed is a weak acid, that is, it is only slightly ionized.

The acidic nature of carbon dioxide is also shown by reacting it with alkalis. If a test tube full of carbon dioxide is inverted into a beaker of caustic soda solution, the gas is completely absorbed by the alkali:

$$CO_2 + 2OH^- \rightarrow CO_3^{2-} + H_2O$$

Because there is almost certainly an excess of alkali, the equation has been written to show the formation of the carbonate and not the hydrogen carbonate.

If carbon dioxide is bubbled into lime water, the lime water at first turns milky. The milkiness is due to a white precipitate which settles fairly readily. The precipitate can be identified as calcium carbonate.

The reaction is similar to the reaction of carbon dioxide with any other alkali, except that the salt formed is insoluble:

$$CO_2 + 2OH^- \rightarrow CO_3{}^{2-} + H_2O$$
$$Ca^{2+} + CO_3{}^{2-} \rightarrow CaCO_3 \text{ (s)}$$

These reactions can be written in the one equation:

$$Ca^{2+} + 2OH^- + CO_2 \rightarrow CaCO_3 \text{ (s)} + H_2O$$

Prolonged passage of carbon dioxide into this suspension of calcium carbonate causes the liquid to become clear and colourless. Evidently the salt dissolves in excess of carbon dioxide and water, to form the acid salt, calcium hydrogen carbonate, which is soluble in water:

$$CaCO_3 + H_2O + CO_2 \rightarrow Ca^{2+} + 2HCO_3{}^-$$

TEST FOR CARBON DIOXIDE

The effect of the carbon dioxide in turning lime water milky is a convenient test for the gas. Ordinarily, insufficient carbon dioxide is obtained in a test to dissolve the precipitate of calcium carbonate. Thus if a gas is colourless and odourless and turns lime water milky, the gas is carbon dioxide.

Uses of Carbon Dioxide

Carbon dioxide gas is widely used in making soda water and other carbonated drinks. The gas is dissolved in the water under a pressure of several atmospheres. When the bottle is opened, the gas escapes because the pressure is reduced.

Liquid carbon dioxide is used in some fire extinguishers, and is suitable for extinguishing burning oil or petrol. If the nozzle of the cylinder is opened, carbon dioxide gas pours out and blankets the fire.

Extinguishing an oil fire with a carbon dioxide extinguisher.

Dry ice.
Compressed blocks of solid carbon dioxide pass over an automatic weighing machine prior to packaging.

Solid carbon dioxide is called dry ice. It is used for refrigerating ice cream and is packed around the ice cream canisters. Dry ice has two advantages over ordinary ice. It is much colder, and it changes from the solid state directly to the gaseous state, without forming a liquid.

Other Reactions in which Carbon Dioxide is produced

1. *The soda-acid type of fire extinguisher* produces carbon dioxide by the reaction of sulphuric acid with a solution of sodium hydrogen carbonate. A diagram of such an extinguisher is shown in fig. 23.4.

When the extinguisher is inverted, the loosely fitting stopper drops out of the bottle of acid, and the acid spills out. The reaction between the acid and the sodium hydrogen carbonate produces very rapid evolution of carbon dioxide:

$$HCO_3{}^- + H^+ \rightarrow H_2CO_3$$
$$H_2CO_3 \rightarrow H_2O + CO_2 \text{ (g)}$$

The principal value of the carbon dioxide is to build up sufficient gas pressure to force a stream of water from the extinguisher. The liquid issuing from the extinguisher froths because of escaping carbon dioxide, and some of this gas might reach the seat of the fire.

2. *Baking powder* is used in cake mixtures to make the cake rise. Baking powder is a dry mixture of sodium hydrogen carbonate (baking soda) and an acidic substance. This acidic substance can be cream of tartar (sodium hydrogen tartrate, $NaHC_4H_4O_6$) or calcium dihydrogen orthophosphate, $Ca(H_2PO_4)_2$. Carbon dioxide is evolved when the mixture is moistened, and evolution of the gas becomes more rapid on heating. The formation of gas bubbles in the cake mixture causes it to rise.

3. *The rising of bread dough*, prior to baking is also due to the evolution of carbon dioxide. Bread dough is mixed with yeast which reacts with carbohydrates present, to form carbon dioxide.

4. *Effervescent powders* are very similar to baking powder. They contain sodium hydrogen carbonate and citric acid which react to produce carbon dioxide. The gas is evolved fairly rapidly when the powder is placed in water, and this causes the liquid to 'fizz' or effervesce.

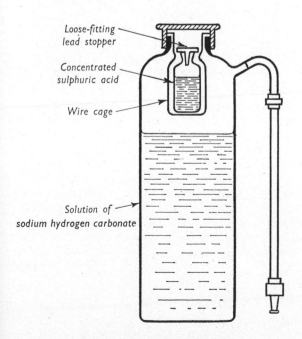

Fig. 23.4. Soda-acid fire extinguisher.

Loose-fitting lead stopper

Concentrated sulphuric acid

Wire cage

Solution of sodium hydrogen carbonate

CARBONIC ACID, CARBONATES AND HYDROGEN CARBONATES

Carbonic Acid

PREPARATION:

Carbon dioxide is passed into water:

$$CO_2 + H_2O \rightleftharpoons H_2CO_3$$

Only a very small part of the dissolved carbon dioxide actually combines with the water (about 1%).

PROPERTIES:

Carbonic acid is an *unstable acid*. If the solution is heated, carbon dioxide is evolved and the acid decomposes:

$$H_2CO_3 \rightarrow H_2O + CO_2$$

Thus the acid cannot be obtained in the dry state.

It has been seen already that carbonic acid is a *weak acid*. The evidence for this lies in its poor electrolytic conductivity and is supported by the litmus test. Carbonic acid is therefore only slightly ionized:

$$H_2CO_3 + H_2O \rightleftharpoons H_3O^+ + HCO_3 \qquad \text{(slight)}$$

The hydrogen carbonate ion is capable of acting as an acid, although it is a much weaker acid than carbonic acid:

$$HCO_3^- + H_2O \rightleftharpoons H_3O^+ + CO_3^{2-} \qquad \text{(very slight)}$$

Thus carbonic acid is a *diprotic acid* and yields two distinct series of salts. The normal salts are called carbonates, e.g., sodium carbonate, Na_2CO_3. The acid salts are called hydrogen carbonates, e.g., sodium hydrogen carbonate, $NaHCO_3$.

Carbonates

All carbonates are white or cream coloured except where the cation is coloured, e.g., $MnCO_3$ is pink. All react with acids in the same way as calcium carbonate, described earlier:

$$CO_3^{2-} + 2H^+ \rightarrow H_2CO_3$$
$$H_2CO_3 \rightarrow H_2O + CO_2$$

The only carbonates which are readily soluble in water are sodium carbonate, potassium carbonate and ammonium carbonate. Carbonates of calcium, magnesium and zinc are well known carbonates which are insoluble in water. Some metals do not appear to form carbonates, e.g., aluminium.

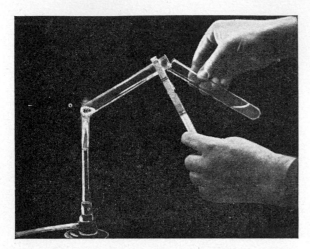

Fig. 23.5. Method for testing the thermal stability of carbonates.

STABILITY OF CARBONATES

The stability of carbonates can be tested by heating them in a hard glass test tube or in an ignition tube. It is possible to test for the evolution of carbon dioxide as shown in fig. 23.5.

If sodium carbonate is heated in a hard glass test tube as described above, the lime water may turn milky. This observation should be interpreted cautiously, because if sodium carbonate is heated in a platinum vessel, it can be melted without the evolution of carbon dioxide. This suggests that the carbon dioxide produced in the glass test tube is not formed by thermal decomposition, but by reaction with silica in the glass. Confusing results can also be obtained by using old samples of sodium carbonate which have been exposed to the air. Such samples can contain considerable amounts of sodium hydrogen carbonate.

Apart from this difficulty with sodium carbonate, the effect of heat on carbonates can be observed as described above and can be summarized:

● *Sodium and potassium carbonates are stable at the temperature of the Bunsen flame. Carbonates of metals below sodium on the activity series decompose into the oxide of the metal and carbon dioxide.*

The effect of heat on ammonium carbonate is treated under ammonium salts.

Hydrogen Carbonates

Not many hydrogen carbonates exist. Sodium and potassium hydrogen carbonates exist as solids. Calcium, magnesium and iron(II) hydrogen carbonates exist in solution only.

These hydrogen carbonates can be formed from the corresponding carbonates by reaction with water and carbon dioxide:

$$CO_3{}^{2-} + H_2O + CO_2 \rightarrow 2\,HCO_3{}^-$$

All the hydrogen carbonates are soluble in water and all react with acids to form carbon dioxide:

$$HCO_3{}^- + H^+ \rightarrow H_2CO_3$$
$$H_2CO_3 \rightarrow H_2O + CO_2$$

The $HCO_3{}^-$ ion is amphiprotic but in this reaction it functions as a base.

STABILITY OF HYDROGEN CARBONATES

If sodium hydrogen carbonate is heated as described earlier for the carbonates, it is observed that drops of a colourless liquid form on the inside of the tube and the lime water turns milky. After allowing the residue to cool, it is found that the addition of acid causes effervescence and the gas evolved will again turn lime water milky.

Sodium hydrogen carbonate decomposes on heating to give carbon dioxide, water and sodium carbonate, and the equation for the reaction is:

$$2NaHCO_3 \rightarrow Na_2CO_3 + H_2O + CO_2$$

This decomposition of sodium hydrogen carbonate will take place even at the temperature of boiling water.

A solution of calcium hydrogen carbonate can be prepared by bubbling carbon dioxide into lime water until the precipitate which first forms is dissolved. If this solution is heated, a white precipitate again forms and the lime water test indicates the evolution of carbon dioxide. The calcium hydrogen carbonate solution evidently decomposes like sodium hydrogen carbonate:

$$Ca^{2+} + 2HCO_3{}^- \rightarrow CaCO_3(s) + H_2O + CO_2$$

This reaction also occurs if the solution is allowed to evaporate at room temperature.

● *In general, all bicarbonates decompose very readily, forming the corresponding carbonate, carbon dioxide and water.*

STUDY QUESTIONS

1. Carbon is in Group IV of the periodic table.
   (a) Draw diagrams to illustrate the bonding of carbon in methane, carbon dioxide, diamond, and graphite.
   (b) Compare and contrast the physical properties of diamond and graphite.
   (c) Explain what is meant by the term 'allotropy', by reference to the allotropes of carbon.
   (d) Cooking odours can be removed from air by activated charcoal. Explain the process involved.

2. Write equations for the reactions which occur when carbon dioxide is absorbed by:
   (a) excess caustic soda solution;
   (b) excess caustic potash solution.
   Name the salt produced in each case.

3. Write equations for the effect of heat on:
   (a) magnesium carbonate;
   (b) zinc carbonate;
   (c) potassium hydrogen carbonate;
   (d) magnesium hydrogen carbonate in solution.

4. Write equations for the reactions which occur when carbon dioxide is bubbled into lime water for a long time, and the resulting liquid is boiled.

5. Calculate the volume at s.t.p. of carbon dioxide formed, by dissolving 10.0 gramme of pure calcium carbonate in excess dilute hydrochloric acid.
   $(Ca = 40.1, \ C = 12.0, \ O = 16.0.)$

6. When carbon dioxide is bubbled into lime water, a precipitate of calcium carbonate forms.
   (a) What simple procedure could be adopted to support this statement?
   (b) What procedures are necessary to prove the statement?

7. When magnesium burns in carbon dioxide, the products are magnesium oxide and carbon.
   Devise tests to support the statement about the identity of the products.

8. Write equations for the reactions which occur when the following pairs of substances are mixed:
   (a) calcium carbonate and dilute hydrochloric acid;
   (b) sodium carbonate and dilute sulphuric acid;
   (c) zinc carbonate and dilute nitric acid.
   Name the salt produced in each reaction.

9. A sample of sodium hydroxide solution is divided into two equal parts.
   (a) The first sample is saturated with carbon dioxide.
   (b) The second sample is added to the solution produced in (a).
   Write equations for all the reactions which occur.

10. Calculate the mass of calcium carbonate precipitated when one dm^3 of carbon dioxide at s.t.p. is completely absorbed by excess calcium hydroxide solution.
    $(Ca = 40.1, \ C = 12.0, \ O = 16.0.)$

11. Describe how to prepare a solution of magnesium hydrogen carbonate starting from a suspension of magnesium carbonate in water. Write an ionic equation for the reaction and rewrite the equation using empirical formulae.

12. Explain why sodium carbonate becomes contaminated with sodium hydrogen carbonate on exposure to the atmosphere.

13. Represent the thermal decomposition of calcium hydrogen carbonate by an ionic equation. Rewrite the equation using empirical formulae.

14. Write equations to illustrate the reactions of the hydrogen carbonate ion acting as an acid and as a base.

Answers to Numerical Problems

5. 2.24 dm^3.
10. 4.46 g.

24 : The Industrial and Natural Derivaties of Limestone

Calcium carbonate is an important compound of carbon because it is available in large quantities from mineral deposits and it can be converted into a number of very useful products which will be discussed in this chapter. In addition, chemical processes due to the effects of water and the atmosphere on calcium carbonate form products which cause great inconvenience and expense, and these will also be discussed.

Occurrence of Calcium Carbonate

Calcium carbonate is the next most abundant mineral in the Earth's crust after the silicates, and is found mainly in the form of limestone and marble, although there are smaller deposits of chalk. In addition calcium carbonate can be obtained from sea shells and coral. A very pure form of calcium carbonate is found as clear crystals called Iceland spar.

INDUSTRIAL PRODUCTS FROM LIMESTONE
Quicklime (Calcium Oxide)

Calcium oxide is formed when the carbonate is heated strongly. The reaction is *reversible*:

$$CaCO_3 \rightleftharpoons CaO + CO_2 (g)$$

The reaction can be driven in the forward direction by using high temperatures and allowing the carbon dioxide to escape

1. *In the laboratory* this can be carried out by heating a piece of marble very strongly in the flame of a Méker burner. To do this the marble may be supported on a loop of stout iron wire (fig. 24.1). The marble becomes incandescent as the reaction proceeds. After about ten minutes, the burner may be turned out, and the residue in the wire loop allowed to cool.

The cold residue 'hisses' when drops of water are added and crumbles to a white powder. With excess

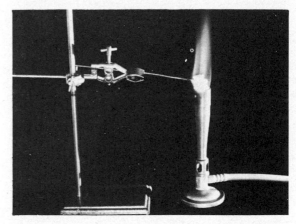

Fig. 24.1. Laboratory decomposition of marble using a Méker burner.

The limestone quarry at Lilydale, Victoria.

The limestone is blasted from the face of the quarry and carried up the hoist (top right). The building next to the hoist is the crushing plant, from which a conveyor takes the limestone to the top of the kiln (top centre). A close-up view of the kiln will be seen on the facing page. This kiln is 21 m tall, with an inside diameter of 3 m, and holds 150 000 kg of limestone.

water, a solution is formed which is alkaline to litmus, whereas the original marble is insoluble in water and does not affect litmus. These tests indicate that the residue formed by the thermal decomposition of calcium carbonate is calcium oxide.

2. *On an industrial scale*, the required conditions of high temperature and removal of carbon dioxide are obtained in a *lime kiln*. If the limestone is available as fairly large lumps, a vertical kiln is used. The limestone is heated either by burning coke in the kiln or by using gaseous fuels (see fig. 24.2). Where finely powdered calcium carbonate must be utilized, a horizontal rotary kiln is used.

In some kilns, the carbon dioxide from the kiln is led off and used in further manufacturing processes, but in others, the gases are allowed to escape to the atmosphere. The quicklime is obtained as hard white lumps containing over 90% calcium oxide.

USES OF QUICKLIME

Quicklime is used for the production of slaked lime (calcium hydroxide) and calcium carbide, both of which are important industrial chemicals. In the laboratory it can be used for drying ammonia gas.

Calcium Carbide. This is prepared by heating quicklime and coke in an electric furnace, to temperatures as high as 2000 °C:

$$CaO + 3C \rightarrow CaC_2 + CO$$

Acetylene is an inflammable gas prepared from calcium carbide by adding water:

$$CaC_2 + 2H_2O \rightarrow C_2H_2(g) + Ca(OH)_2$$

Acetylene is an important chemical used for the manufacture of acetic acid, plastics and synthetic rubber, and in oxyacetylene flames for cutting and welding metals.

Calcium cyanamide is prepared by heating calcium carbide with nitrogen:

$$CaC_2 + N_2 \rightarrow CaCN_2 + C$$

Calcium cyanamide is used in fertilizers because it slowly liberates its nitrogen as ammonia, in the presence of water:

$$CaCN_2 + 3H_2O \rightarrow 2NH_3 + CaCO_3$$

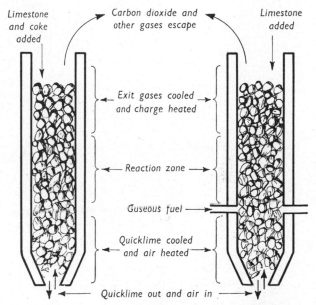

Fig. 24.2. Diagram and photograph of a vertical lime kiln. The photograph corresponds with the diagram using coke as the fuel.

Slaked Lime (Calcium Hydroxide)

When water is added to quicklime a considerable amount of heat is evolved:

$$CaO + H_2O \rightarrow Ca(OH)_2$$

The product is a white powder, calcium hydroxide.

Calcium hydroxide is not very soluble in water but it forms an alkaline solution which is called *lime water*. It is more often used as a thin slurry called *milk of lime*. Calcium hydroxide is widely used as an industrial alkali, because it is cheap. It is also used in the production of caustic soda and bleaching powder, and is a constituent of soda-lime and mortar.

Mortar is a mixture of slaked lime, sand and water and is prepared as a thick paste. When it is used to set bricks, it first sets by drying out. Then over a long period of time it gradually hardens due to absorption of carbon dioxide from the atmosphere:

$$Ca(OH)_2 + CO_2 \rightarrow CaCO_3 + H_2O$$

CEMENT AND CONCRETE

Cement is made by heating limestone with sand and silicates such as clay. It is a mixture of calcium silicates and aluminates with other substances. When cement is mixed with water, a complex series of reactions occurs, including reaction of the silicates with water and the formation of calcium hydroxide. The composition of cement can be varied to give various setting characteristics. Cement is mainly used to make concrete.

Concrete is made by mixing cement with stone screenings, sand and water. *Reinforced concrete* is made by pouring concrete around steel rods or bars.

NATURAL DERIVATIVES FROM LIMESTONE: HARDNESS IN WATER

Limestone or minerals containing calcium carbonate are not soluble in pure water but dissolve slowly in the presence of carbon dioxide. Rain water and the water in streams and rivers contain a small amount of carbonic acid due to dissolved carbon dioxide:

$$CO_2 + H_2O \rightarrow H_2CO_3$$

This carbonic acid reacts with carbonates and converts them to hydrogen carbonates which are soluble in water:

$$CO_3^{2-} + H_2CO_3 \rightarrow 2HCO_3^-$$

As a result of these reactions, natural waters in areas containing limestone are contaminated with dissolved calcium hydrogen carbonate. They also usually contain magnesium hydrogen carbonate and may contain traces of iron(II) hydrogen carbonate.

Other salts such as the chlorides or sulphates of calcium or magnesium may also become dissolved in natural waters due to the weathering of various rocks and minerals. Such natural waters do not readily form a lather when shaken with soap. Instead, a precipitate (or curd) is formed, and the water is said to be *hard water*.

● *Hardness in natural waters is caused by the presence of calcium, magnesium or iron(II) salts in solution.*

Before this property of hardness in water can be discussed, it is necessary to understand something of the nature of soaps.

Soap

Soaps are manufactured from caustic soda and organic fats and oils such as tallow, palm oil and olive oil. Soap is a mixture of sodium salts of long-chain organic acids derived from fats and oils, and the structure of sodium stearate, which is typical of these sodium salts, is shown in the diagram (fig. 24.3).

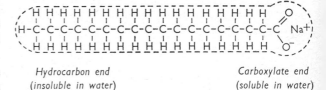

Hydrocarbon end   Carboxylate end
(insoluble in water)   (soluble in water)

Fig. 24.3 Sodium stearate, a typical soap.

The detergent action of soap is essentially due to the fact that it makes water able to 'wet' oils or greases, and this property is due to the different solubilities of the two ends of the soap molecule. The hydrocarbon end is readily soluble in oily or greasy substances but is insoluble in water. The carboxylate end and the sodium ion are readily soluble in water but are insoluble in oils or greases.

When soap is used with water to clean a greasy surface, the hydrocarbon ends of the soap molecules become attached to the grease, and the carboxylate end remains in the water. Agitation or rubbing enables

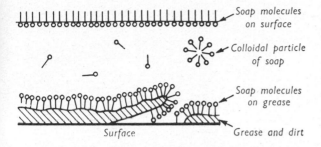

Fig. 24.4. Detergent (cleansing) action of soap.

the grease and entrapped dirt to be removed (see fig. 24.4).

Effect of Hard Water on Soap

If a solution of soap is added to distilled water and shaken, a lather readily forms. However, with hard water a curdy precipitate forms instead of a lather. This precipitate contains calcium and magnesium stearates formed by the calcium and magnesium ions in the hard water. This can be represented by an equation using St^- to represent the stearate ion, $C_{17}H_{35}COO^-$:

$$Ca^{2+} + 2St^- \rightarrow CaSt_2 \text{ (s)}$$
$$Mg^{2+} + 2St^- \rightarrow MgSt_2 \text{ (s)}$$

Once the soap is precipitated, it can no longer make water 'wet' a greasy surface.

If a solution of calcium hydrogen carbonate is prepared in the laboratory by saturating lime water with carbon dioxide, it can be used to observe the formation of the curdy precipitate with soap. When soap solution is shaken with this solution, the curdy precipitate of insoluble calcium stearate forms and slowly floats to the surface of the solution. However, because hydrogen carbonates are unstable to heat, the solution can be 'softened' by being boiled. Thus, if a solution of calcium hydrogen carbonate is boiled, a white precipitate of calcium carbonate forms, and if the soap test is now applied, the resulting liquid will readily lather with the soap, despite the presence of the precipitate.

If a solution of magnesium sulphate is tested in the same way it can be observed that a curdy precipitate forms regardless of whether the solution is tested before or after boiling. This difference in behaviour leads to a classification of hardness into two types.

Types of Hardness

1. *Temporary hardness is hardness which can be removed by boiling.*

Temporary hardness is caused by the presence of calcium, magnesium or iron(II) hydrogen carbonates in solution.

2. *Permanent hardness is hardness which cannot be removed by boiling.*

Permanent hardness is caused by calcium and magnesium chlorides and sulphates in solution.

It should be noted that the presence of any compound capable of forming a precipitate with soap will cause hardness in water, but the above salts are the only ones which commonly occur in natural waters.

Although permanent hardness cannot be removed by boiling, it can be removed by other methods.

Removal of Hardness

The removal of hardness involves the removal of the calcium, magnesium or iron(II) ions from solution. This is usually brought about by precipitating the ions as insoluble salts.

REMOVAL OF TEMPORARY HARDNESS

(i) *By boiling.* The hydrogen carbonate ions are completely converted to carbonate ions at 100 °C. Calcium and magnesium ions are thus *precipitated from solution* as insoluble carbonates.

$$Ca^{2+} + 2HCO_3^- \rightarrow CaCO_3 \text{ (s)} + H_2O + CO_2$$
$$Mg^{2+} + 2HCO_3^- \rightarrow MgCO_3 \text{ (s)} + H_2O + CO_2$$

(ii) *By adding slaked lime.* If just sufficient slaked lime is added to change the hydrogen carbonate ions to carbonate ions the calcium and magnesium ions are *removed by precipitation.*

$$HCO_3^- + OH^- \rightarrow CO_3^{2-} + H_2O$$
$$Ca^{2+} + CO_3^{2-} \rightarrow CaCO_3 \text{ (s)}$$

The reaction between slaked lime and calcium hydrogen carbonate solution can also be represented as:

$$Ca^{2+} + 2HCO_3^- + Ca^{2+} + + 2OH^-$$
$$\rightarrow 2CaCO_3 \text{ (s)} + 2H_2O$$

The magnesium ions are removed in a similar way.

In using this method, it is important not to use excess calcium hydroxide as this would make the water hard.

GENERAL METHODS FOR REMOVING HARDNESS

Several methods are available for removing calcium, magnesium or iron(II) ions from solution *irrespective of the anions present*. These methods may be used for removing both temporary and permanent hardness.

(i) *By using excess soap*. If sufficient soap is used, all of the hardness is removed and the excess soap is available to form a lather. Because soap is more expensive than the other substances which will be discussed below, this method is usually avoided. Not only is it an expensive method, particularly on a commercial scale, but the curdy precipitate forms as a scum on the water.

(ii) *By adding washing soda*. Washing soda or sodium carbonate takes its name from its use in removing hardness. If excess sodium carbonate is added, the calcium and magnesium ions are precipitated, e.g.:

$$Ca^{2+} + CO_3^{2-} \rightarrow CaCO_3 \text{ (s)}$$

(iii) *By adding sodium phosphate*. Trisodium orthophosphate can be used in the same way as sodium carbonate. The reaction precipitates insoluble calcium or magnesium phosphate.

(iv) *By ion-exchange methods*. A number of materials are available under various names such as zeolite or permutit. Zeolite is a type of natural mineral which exists in various forms, one of which is called natrolite and has the empirical formula $Na_2Al_2Si_3O_{10}.2H_2O$. This can be used to soften water because the sodium ions can be replaced by other cations such as calcium ions. Using the formula Na_2Ze to represent a portion of the zeolite containing two sodium ions (i.e. Ze^{2-} represents the unit $Al_2Si_3O_{10}.2H_2O^{2-}$), the equation for the removal of calcium ions can be written as:

$$Ca^{2+} + Na_2Ze \rightleftharpoons 2Na^+ + CaZe$$

This is called an ion-exchange reaction and it is reversible. The direction of the reaction depends on the concentration of the ions in the solution and will proceed in the direction shown when hard water is run over the zeolite. If the zeolite is treated with a fairly concentrated solution containing sodium ions (e.g. Na^+Cl^- because it is cheap), the reaction is reversed:

$$2Na^+ + CaZe \rightleftharpoons Ca^{2+} + Na_2Ze$$

The zeolite is usually placed in a tube and the hard water is allowed to run slowly through the tube.

When the zeolite becomes 'spent', sodium chloride solution is used to regenerate it, and the zeolite can then be used to soften more hard water.

A number of synthetic resins which contain the sodium sulphonate group, $-SO_3Na$, are made as water softeners and they react in the same way as zeolites.

Summing Up, Hardness in Water

Hardness in water can be detected by adding soap to the water. The soap forms a curdy scum on the water and does not lather readily. Any solute containing a cation which forms an insoluble stearate

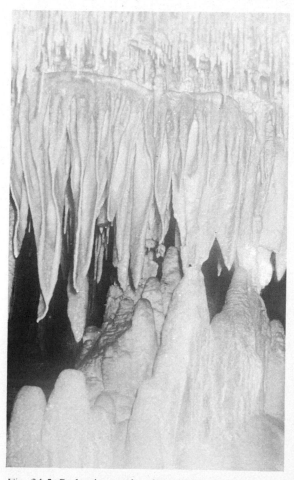

Fig. 24.5. Stalactites and stalagmites in the limestones caves at Buchan, Victoria.

(e.g. Ca^{2+}, Mg^{2+}, Fe^{2+}) causes hardness. These solutes become dissolved in the water due to the effects of the atmosphere and water on limestone and other minerals. Any process (generally precipitation) which will remove such cations from solution will be a softening process.

Temporary hardness is defined as hardness in water which can be removed by boiling. It is caused by the cations magnesium, calcium or iron(II), present in solution as hydrogen carbonates.

Permanent hardness is defined as hardness in water which cannot be removed by boiling. It is caused by the cations calcium or magnesium, present in solution as chlorides or sulphates.

DISADVANTAGES OF HARD WATER

The effect of hardness on soap poses serious problems in commercial laundries and in textile industries. Hardness in water also creates difficulties in steam boilers. Deposits of calcium carbonate form on the inside of the heating pipes and cut down the efficiency of the boiler. Calcium sulphate, which is not very soluble, also forms an insulating layer. Such layers are called *boiler scale*.

Limestone Caves

The effect of water and carbon dioxide on limestone has already been described.

$$CaCO_3 + H_2O + CO_2 \rightarrow Ca^{2+} + 2HCO_3^-$$

Over a long period of time this reaction can lead to the formation of large caverns in the limestone strata. Once these caves, or a series of caves, are formed, stalactites and stalagmites begin to form. As a drop of water containing calcium hydrogen carbonate reaches a cave it may evaporate in currents of air and form a small speck of calcium carbonate:

$$Ca^{2+} + 2HCO_3^- \rightarrow CaCO_3 \text{ (s)} + H_2O + CO_2$$

Over immense periods of time the fantastic shapes seen in limestone caves are built up. The colouring is largely due to the presence of traces of iron salts in the limestone (fig. 24.5).

STUDY QUESTIONS

1. One of the major impurities in limestone is magnesium carbonate. What are the major impurities of quicklime likely to be?

2. Write the equation for the reaction which occurs when milk of lime is used to neutralize hydrochloric acid.

3. Mortar made by mixing slaked lime and sand with water, often becomes weathered out of the space between bricks after many years. What chemical reaction may contribute to the weathering of lime mortar?

4. Describe and explain the reaction between soap (e.g. sodium stearate) and magnesium chloride solution.

5. Explain why a deposit forms on the inside of kettles used to boil hard water. Write the equations for any reactions involved.

6. A commercial method for the removal of both types of hardness from water is known as the lime-soda process. Excess slaked lime is first added to the water and then washing soda is added. Explain how both types of hardness are removed and write equations for any reactions which take place.

7. Explain how a synthetic resin containing $-SO_3^-$ Na^+ units can be used to soften hard water. Write an equation to represent the removal of hardness and also the regeneration of the resin.

8. Explain how stalactites and stalagmites form in limestone regions.

9. Calculate the volume of carbon dioxide expressed at s.t.p., produced by complete decomposition of 1 kilogramme of pure limestone.
(C=40, C=12, O=16.)

Answer to Numerical Problem
9. 224 dm³.

25 : Carbon Monoxide, Fuels and their Combustion

CARBON MONOXIDE

Preparation of Carbon Monoxide

Carbon monoxide can be prepared in the laboratory by a number of methods, but because of the extremely poisonous nature of the gas, all methods require elaborate safety precautions. As little as one part of carbon monoxide per 100 parts of air is fatal within a few minutes.

METHOD (a)

One method of preparation is to pass oxygen slowly through a long copper or hard glass tube packed with pieces of carbon. The tube is heated from the outside to make the carbon red hot. The oxygen first combines with the carbon forming carbon dioxide:

$$C + O_2 \rightarrow CO_2$$

The carbon dioxide is reduced by the hot carbon to carbon monoxide:

$$CO_2 + C \rightarrow 2CO$$

The reaction is not complete, but any unchanged carbon dioxide can be absorbed by soda-lime and the carbon monoxide can be collected over water.

A variation of the above method is to use carbon dioxide instead of oxygen, in which case only the second reaction takes place.

METHOD (b)

Carbon monoxide may be prepared by heating organic acids with concentrated sulphuric acid. The sulphuric acid dehydrates the organic acid, e.g., with formic acid:

$$HCOOH + H_2SO_4 \rightarrow CO + H_3O^+ + HSO_4^-$$

If oxalic acid is used, a mixture of carbon monoxide and carbon dioxide is formed:

$$(COOH)_2 + H_2SO_4$$
$$\rightarrow CO + CO^2 + H_3O^+ + HSO_4^-$$

The reaction with formic acid is quite convenient to use in the laboratory although it must be remembered that carbon monoxide is colourless and odourless but very poisonous.

Some concentrated sulphuric acid is placed in a flask and warmed gently. Formic acid is added to the warm sulphuric acid drop by drop so that the carbon monoxide is evolved slowly (see fig. 25.1).

Fig. 25.1. Laboratory preparation and collection of carbon monoxide: formic acid is added to concentrated sulphuric acid.

The gas is collected over water and when the gas jar is nearly full, the addition of formic acid is stopped and the gas jar is replaced by a gas jar full of water, taking care that no gas is allowed to escape. The gas jar of carbon monoxide should be fitted with a cover slip while still under water.

If a lighted taper is plunged well into the gas, the taper goes out but the gas itself burns with a blue flame. The product of combustion turns lime water milky whereas the original gas does not affect lime water. Thus carbon dioxide is the product of combustion of carbon monoxide.

The properties of carbon monoxide are listed below.

Physical Properties of Carbon Monoxide

1. A colourless, odourless gas.
2. Has about the same density as air. (The molecular mass of CO is about 28.)
3. Only very slightly soluble in water (100 cm³ of water dissolves about 3 cm³ of carbon monoxide at 20 °C and 760 mmHg).
4. It is difficult to liquefy (b.p. = −190 °C).

Chemical Properties of Carbon Monoxide

1. It is stable to heat.
2. It does not support ordinary combustion.

3. Carbon monoxide burns with a blue flame in air or oxygen. Mixtures of carbon monoxide and air may explode violently when ignited:

$$2CO + O_2 \rightarrow 2CO_2$$

4. Its solution in water is neutral and it is classified as a neutral oxide. However, it does have some acidic character because it reacts with hot concentrated caustic soda solution under pressure, forming sodium formate:

$$CO + Na^+ + OH^- \rightarrow HCOO^- Na^+$$

5. The reducing property of carbon monoxide is shown by its reaction with oxygen. It is used to reduce metallic oxides industrially and in the laboratory:

$$CuO + CO \rightarrow Cu + CO_2$$
$$Fe_2O_3 + 3CO \rightarrow 2Fe + 3CO_2$$

6. Carbon monoxide has the unusual ability of forming covalent bonds with metals. Thus it forms a volatile compound called *tetracarbonylnickel*, $Ni(CO)_4$.

The property of forming covalent bonds explains the poisonous nature of carbon monoxide. It forms a very stable compound with the haemoglobin of the blood, which is then unable to absorb oxygen.

Test for Carbon Monoxide

Carbon monoxide burns with a blue flame to form carbon dioxide as the only product.

FUELS

Many fuels contain carbon monoxide or form carbon monoxide on combustion in limited supplies of air. The more important fuels will be discussed.

Hydrocarbon Fuels

The main source of hydrocarbon fuels is petroleum from oil wells. Petroleum is a term used to describe all of the carbon-containing substances obtained from an oil well. The main products are natural gas and oil.

The natural gas is mainly methane, CH_4, but also contains some ethane, C_2H_6, and very small amounts of other gases. The main use of natural gas is as an industrial and domestic fuel. It is compressed after being piped from the wells and the small amounts of

heavier hydrocarbons are condensed for use as liquid fuels.

The oil contains a wide range of hydrocarbons and these are separated by fractional distillation (chapter 7) and then further treated and purified to obtain liquid fuels such as:

(i) *petrol or gasoline*, a mixture of hydrocarbons with 5 to 11 carbon atoms per molecule;

(ii) *kerosene*, a mixture containing hydrocarbons with 12 to 16 carbon atoms per molecule;

(iii) *gas oil and diesel oil*, a mixture containing hydrocarbons with 13 to 18 carbon atoms per molecule;

(iv) *fuel oil*, a mixture of hydrocarbons with more than 16 carbon atoms per molecule.

The Combustion of Hydrocarbon Fuels

When methane, CH_4, or ethane, C_2H_6, is burnt in efficient burners, carbon dioxide and steam are the only products:

$$CH_4 + 2O_2 \rightarrow CO_2 + 2H_2O$$
$$2C_2H_6 + 7O_2 \rightarrow 4CO_2 + 6H_2O$$

These gases form violently explosive mixtures with air, and methane has been responsible for many terrible explosions in coal mines. Coal miners call methane 'fire damp'.

The oil rig *Glomar III* seen from a supply helicopter.

Storage of liquefied petroleum gas.
The two spherical vessels are known as Horton Spheres, and contain the liquefied petroleum gas under pressure.

Smog in Melbourne.
Smog contains entrapped pollutants which arise from combustion of fuels, mainly by motor vehicles and industrial furnaces.

Propane, C_3H_8, is a constituent of liquid petroleum gas (see below) and its combustion is similar to methane and ethane:

$$C_3H_8 + 5O_2 \rightarrow 3CO_2 + 4H_2O$$

Butane, C_4H_{10}, is stored as a liquid under pressure. It is the major constituent of 'liquid petroleum gas' (L.P. gas). When the pressure is released, the L.P. gas vaporizes and the vapour can be burnt in air:

$$2C_4H_{10} + 13O_2 \rightarrow 8CO_2 + 10H_2O$$

Higher hydrocarbons are used in internal combustion engines under the name of *petrol* or *gasoline*. *Octane*, C_8H_{18}, is a typical constituent of petrol. When it is burnt in excess air, carbon dioxide and water are formed:

$$2C_8H_{18} + 25O_2 \rightarrow 16CO_2 + 18H_2O$$

However, inside the cylinder of an engine, this reaction does not go to completion and the exhaust gases from automobiles contain appreciable quantities of carbon monoxide. Because of this a petrol engine should never be allowed to operate for any length of time in a closed space such as a garage.

Kerosene is a mixture of hydrocarbons with 12 to 16 carbon atoms per molecule. It is used as fuel for heating and as the fuel for jet engines. Its combustion is similar to the other hydrocarbons in excess air and it can also form poisonous carbon monoxide if not burnt efficiently.

Reformed Petroleum Gas

As a result of the distillation and catalytic cracking processes carried on in an oil refinery, a wide range of hydrocarbon products is available. Some of these are used as fuels but those which are unsuitable may be catalytically reformed to make domestic gas.

A catalytic reforming plant is represented in the diagram fig. 25.2. The petroleum products are referred to as *feedstock* and may vary in composition, depending on the operation of the oil refinery. They can be reformed into a suitable domestic gas by adjusting the operation of the plant to suit the feedstock.

The plant is operated in a cycle of four steps:

(i) *Heat*. Feedstock and air are introduced and burnt to heat the catalyst which is a layer of inert material impregnated with nickel. The hot exit gases pass through the waste heat boiler and then up the chimney.

(ii) *Steam purge.* The feedstock and air are shut off and steam is blown through to sweep out the combustion products.

(iii) *Gas-make.* Feedstock, steam and air are admitted in the required proportions and pass downwards through the catalyst layer. Some of the hydrocarbon feedstock is consumed in the production of hydrogen, e.g.

$$C_3H_8 + 3H_2O \rightarrow 3CO + 7H_2$$

Other hydrocarbon molecules are cracked (i.e. break into smaller molecules) and react with the hydrogen, e.g.

$$C_3H_8 + H_2 \rightarrow CH_4 + C_2H_6$$

These reactions are endothermic and the catalyst cools down. The product gas passes through the waste heat boiler, the washbox and the water tower before passing into the gas main.

(iv) *Steam purge.* The feedstock and air are shut off and the system is purged with steam before repeating the cycle.

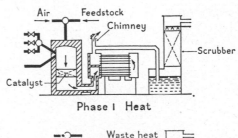

Phase 1 Heat

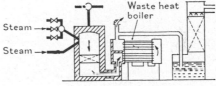

Phase 2 Purge

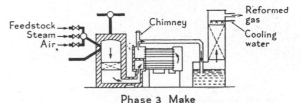

Phase 3 Make

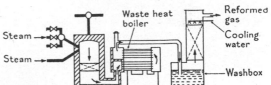

Phase 4 Purge

Fig. 25.2. A catalytic reforming plant.

Above: diagrams showing the four stages in the cycle.

In the photograph at left air, steam and petroleum products are carried through pipes (left) to the large cylindrical catalyst chamber. The other main items shown are (left to right) the gas scrubber box, the waste heat boiler and the stack.

Rocket Fuels

Rocket engines develop their thrust from a jet of hot gases which the engine forces from a nozzle at the rear of the rocket. Most rocket engines use the combustion of a fuel with a suitable oxidizer, to produce the hot gases. Both the fuel and the oxidizer are carried in the rocket and so a rocket engine differs from other kinds of engines in that it does not require a supply of oxygen from the atmosphere. Rocket fuels may be either liquids or solids.

LIQUID FUELS

Various types of liquid fuelled rockets are used in the Apollo system for Moon Landing. The Saturn launch vehicle, which comprises the first two stages at lift-off burns a mixture of kerosene and liquid oxygen. This mixture is pumped to each engine at the rate of $15\,000$ kg^{-1} and is ignited to produce a thrust of 6.7×10^6 N. Five of these engines burn simultaneously during the first stage and one burns during the second stage.

The next two stages of the Apollo system burn a mixture of liquid oxygen and liquid hydrogen. Each engine develops a thrust of 11×10^5 N. Five engines burn during the third stage and one burns during the fourth stage. This single-engined fourth stage is stopped and restarted in space to adjust the Earth parking orbit.

The remaining rocket engines in the command module and the Lunar landing vehicle and also the fifty small attitude control rockets use hydrazine, H_2NNH_2, and dimethylhydrazine, $(CH_3)_2NNH_2$, as fuel and dinitrogen tetroxide, N_2O_4, as oxidizer. When these two liquids are mixed they ignite spontaneously and such a fuel system is called *hypergolic* (self igniting). These engines are particularly critical to the success of the whole manoeuvre. The engine in the command module must be started and stopped several times to adjust the Lunar orbit and to return the astronauts to Earth. The engine in the Lunar descent stage can be throttled to deliver any thrust between 4900 to 44 000 N. This is essential to achieve a fully controlled Lunar landing.

SOLID FUELS

Solid rocket propellants are simpler to use than the liquid fuels described above but have not been developed as rapidly. Three main types of solid propellants are at present in use:

The Apollo 11 liquid-fuelled rocket blasts off for the moon, July 11th, 1969.

Rockets of this type can be very precisely controlled, but cannot be prepared long in advance or kept in immediate readiness for any length of time. As a result they are ideal for space research but of little military value.

(i) *Cordite type.* A mixture of nitrocellulose and nitroglycerine.

(ii) *Composite type.* Granules of potassium or ammonium perchlorate in a combustible binder such as asphalt, rubber or plastic.

(iii) *Pressed charge.* Ammonium nitrate as the oxidizer and gas producing agent, with some combustible binder and catalysts.

Solid propellants have been used mainly for small rockets such as the small emergency escape rockets and the retrorockets of the Apollo space craft. The advantage of a solid fuel is that it can be loaded into the rocket well in advance of the time of firing and thus the rocket is always ready for use. For this reason, solid fuels have been used for the ballistic Polaris missiles fired from nuclear submarines. The disadvantage of a solid fuel is that once ignited, the rocket fires until all the fuel is consumed. This is suitable for military rockets where the maximum speed is used, but is not so suitable for a satellite launch vehicle where fine control is needed.

GAS BURNERS

Because inefficient burning of hydrocarbons produces carbon monoxide, it is important to burn domestic gas efficiently. The essential features of the design of a gas burner are due to R. W. Bunsen. The component parts of a typical laboratory *Bunsen burner* are shown in fig. 25.3 and a typical design used for domestic gas burners is also shown (fig. 25.4).

If the air hole of a Bunsen burner is opened and the gas is lit, it burns with a blue cone surmounted by an almost invisible flame. If a wire gauze is held in the flame below the level of the top of the blue cone, a circular ring of red-hot metal with a relatively cool centre can be observed. It is concluded that the gas in the blue cone is relatively cool and that the flame is therefore hollow. The hottest part of the flame is just above the top of the blue cone. If the air hole is opened

Launching a Polaris rocket.
The solid propellant in this rocket can be prepared long in advance and kept in constant readiness.

Fig. 25.3. (left) The component parts of a Bunsen burner.
Fig. 25.5. (right) A Méker burner.

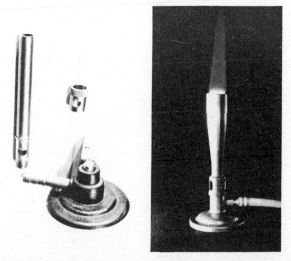

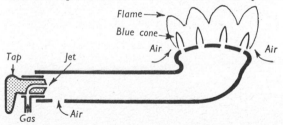

Fig. 25.4. (left) A domestic gas burner.

fully and the gas supply is slowly turned down, the flame becomes unstable and is likely to 'strike back' down the tube and burn at the jet. This is potentially very dangerous as the whole burner may become hot enough to melt the rubber tubing and cause a serious fire.

The flame of a Bunsen burner is almost colourless and may not be noticed when reaching for reagent bottles or other apparatus. Therefore, when a burner is not in actual use but is allowed to remain alight, the air hole should be closed so that a yellow smoky luminous flame is seen. The gas may then be turned down without risk of 'striking back'. It is safer to maintain only a small yellow flame and it is also more economical because of the lower consumption of gas.

A development of the Bunsen burner called the *Méker burner* is shown in fig. 25.5. This has several air holes to admit a large amount of air. Such an arrangement is likely to make the flame 'strike back' but this is prevented by a heavy metal or ceramic grid at the top of the tube. The grid conducts away the heat of the flame and prevents ignition of the gas-air mixture in the tube. Because of the greater proportion of air, combustion is more rapid and the flame consists of very short cones and is much hotter than the ordinary Bunsen burner flame.

It must not be assumed from these comments that the Bunsen burner flame is not very hot. In fact it

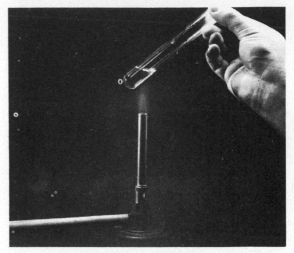

Fig. 25.6. Heating a liquid in a test tube over a low Bunsen flame.

is too efficient for the familiar laboratory process of heating a small amount of liquid in a test tube. For this purpose the air hole should be partly closed. The flame should be almost invisible but not yellow or smoky and the gas supply should be adjusted so that only a small flame is obtained (fig. 25.6).

In using gas as a source of heat it should always be borne in mind that the gas is poisonous.

STUDY QUESTIONS

1. Suggest a suitable arrangement of laboratory apparatus for the preparation and collection of a sample of carbon monoxide from the reaction of oxalic acid (a solid) with concentrated sulphuric acid.

2. Iron(II, III) oxide is reducible by carbon monoxide to iron. Write the equation for the reaction.

3. Calculate the approximate densities of each of the following gases relative to air at the same temperature and pressure:
   (a) methane, CH_4;   (c) propane, C_3H_8.
   (b) acetylene, C_2H_2;
               ($C = 12$, $H = 1$.)

4. Hydrazine and dinitrogen tetroxide ignite spontaneously to form nitrogen and steam. With dimethylhydrazine, carbon dioxide is also formed. Write equations for these reactions.

5. The combustion of diesel fuel oil in a poorly maintained diesel engine produces a smoky exhaust. What pollutants are discharged into the atmosphere and why are they formed?

6. Natural gas is odourless but it is deliberately contaminated with other gases to give it an odour. Give an explanation for this procedure.

7. The formula of pentane is C_5H_{12}. Write the equation for its complete combustion in air.

8. Calculate the volume of oxygen needed for complete combustion of 1 dm³ of:
   (a) carbon monoxide; (b) methane.
(All volumes at the same temperature and pressure.)

Answers to Numerical Problems

3. (a) about ½ as dense;   8. (a) ½ dm³;
   (b) about the same;   (b) 2 dm³.
   (c) about 1½ times as dense.

26 : Aspects of Organic Chemistry

All living things, plants and animals alike, contain carbon compounds which are vital to their life. In addition, natural fuels such as coal and petroleum contain carbon. The number of carbon compounds which are known is enormous and they are called 'Organic' compounds. To understand why there is such a large number of organic compounds, it is helpful to consider two features of the way in which carbon atoms form bonds.

A carbon atom has four valency electrons in its outer shell and all of these must be shared in covalent bonds when the atom is in a molecule. The two features of this bonding which will be considered are:
(i) the various ways in which a carbon atom uses its electrons when bonded to *different atoms*;
(ii) the various ways in which a carbon atom uses its electrons when bonded to *another carbon atom*.

Carbon Atom Bonded to Different Atoms

Some examples will be used to illustrate the bonding of a carbon atom to atoms of hydrogen, oxygen and nitrogen. These are chosen because an H-atom can share one electron, an O-atom can share two electrons and an N-atom can share three electrons.

$$H \cdot \quad : \overset{..}{\underset{.}{O}} \cdot \quad \cdot \overset{..}{N} \cdot$$

CARBON AND HYDROGEN.

If a carbon atom is bonded to hydrogen atoms, each hydrogen atom will use one of the electrons from the carbon atom to form electron pairs. Thus four hydrogen atoms are required.

$$\begin{array}{c} H \\ | \\ H - C - H \\ | \\ H \end{array} \quad \text{or } CH_4 \text{ methane}$$

This molecule is tetrahedral because this is the shape which allows the four regions of electrons to be as far apart as possible.

CARBON AND OXYGEN.

If a carbon atom is bonded to oxygen atoms, each oxygen atom will use two electrons from the carbon atom to form a double bond. Thus two oxygen atoms are required.

$$O = C = O \quad \text{or } CO_2 \text{ carbon dioxide.}$$

This molecule is linear because this is the shape which allows the two regions of electrons to be as far apart as possible.

CARBON, OXYGEN AND HYDROGEN.

In the above examples, carbon has formed four bonds, oxygen has formed two bonds and hydrogen has formed one bond. This statement is true irrespective of whether single or double bonds are formed. For example, if a carbon atom is bonded to an oxygen atom, a double bond is formed. This satisfies the bonding requirements of the oxygen atom but the carbon atom must still form two more bonds. These could be formed with two hydrogen atoms. Thus:

$$\begin{array}{c} H \\ \diagdown \\ C = O \\ \diagup \\ H \end{array} \quad \text{or } CH_2O \text{ formaldehyde}$$

This molecule is triangular because this is the shape which allows the three regions of electrons to be as far apart as possible.

Another arrangement of atoms of these same elements which satisfies the bonding requirements can be drawn. If only a single bond is formed between the carbon and the oxygen, then the carbon atom could bond to three hydrogen atoms. The oxygen must also bond to a hydrogen atom to complete its two bonds.

$$\begin{array}{c} H \\ | \\ H - C - O - H \\ | \\ H \end{array} \quad \begin{array}{l} \text{or } CH_3\text{-OH} \text{ methanol} \\ \text{(also called methyl alcohol)} \end{array}$$

This molecule has a shape which cannot be simply described in words, but the four atoms around the carbon are arranged tetrahedrally, because this is the shape which allows the four regions of electrons to be as far apart as possible.

CARBON, NITROGEN AND HYDROGEN.

If a carbon atom is bonded to a nitrogen atom, three pairs of electrons are shared. This leaves one further electron on the carbon atom to be shared. Therefore the carbon atom cannot bond to another nitrogen atom but it can bond to a hydrogen atom.

$$H - C \equiv N \quad \text{or} \quad HCN \quad \text{hydrogen cyanide}$$

This molecule is linear because it has two regions of electrons around the carbon atom.

Using the examples of compounds containing oxygen as a guide, other possible arrangements can be drawn:

$$\begin{matrix} H \\ \diagdown \\ \quad C = N \\ \diagup \quad \diagdown \\ H \qquad\; H \end{matrix} \qquad \text{and} \qquad \begin{matrix} H \\ | \\ C \\ \end{matrix}$$

Summary.

From these examples it can be seen that different atoms may form single, double or triple bonds and that these can be formed in any combination. The only restriction is that the total number of bonds must not exceed one for hydrogen, two for oxygen, three for nitrogen and four for carbon.

Carbon Bonded to Carbon

The examples used here will be for molecules containing two carbon atoms and the only other atoms which will be used are hydrogen atoms. Since carbon can form single, double or triple bonds, three molecules are possible.

$$H - C \equiv C - H \qquad \text{or} \quad C_2H_2 \quad \text{acetylene}$$

$$\begin{matrix} H \\ \diagdown \\ \quad C = C \\ \diagup \quad \diagdown \\ H \qquad\quad H \end{matrix} \qquad \text{or} \quad C_2H_4 \quad \text{ethylene}$$

$$\begin{matrix} H \qquad\qquad H \\ \diagdown \qquad\quad\; \diagup \\ \quad C - C \\ \diagup \qquad\; \diagdown \\ H \qquad\qquad H \end{matrix} \qquad \text{or} \; C_2H_6 \; \text{ethane}$$

Hydrocarbon Compounds

The very large number of compounds which contain *only carbon and hydrogen* is due to the different ways in which carbon atoms can be bonded together. Any particular carbon atom must form four bonds, no matter how large and complicated the rest of the molecule may be. At least one of these four bonds must be to another carbon atom in the molecule, except in the case of the hydrocarbon methane. Therefore there must be no more than three hydrogen atoms bonded to any one carbon atom. The possible parts of a hydrocarbon are therefore:

The dots represent lone electrons on the carbon which are available for sharing with another atom.

It is possible to draw the formulae of all of the millions of different hydrocarbon compounds by fitting these pieces together in various combinations.

EXAMPLE:

How many combinations are possible for molecules containing four carbon atoms? The ·CH₃ group forms one bond so it could be used at each end of the chain of carbon atoms. (N.B. the ·CH₃ group is so common that it is given a special name—the methyl group). The :CH₂ group forms two bonds so it could be used to build up the chain. Thus:

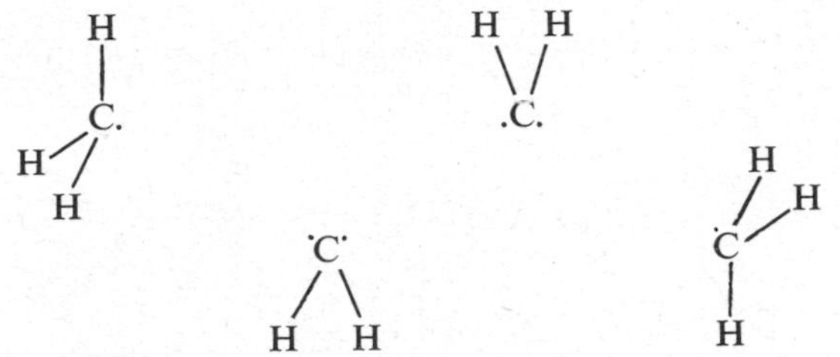

Further examples of molecules containing four carbon atoms can be obtained by including a double bond between two of the carbon atoms. There are four possible structures.

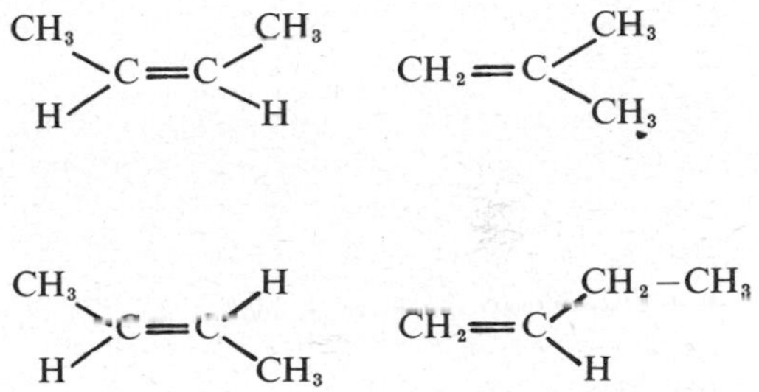

By combining two ·CH₃ groups and two :CH₂ groups the molecule can be drawn.

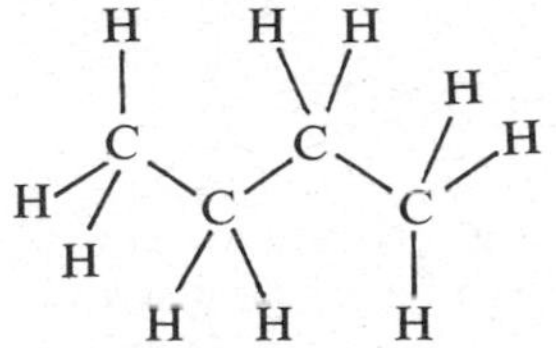

or $CH_3-CH_2-CH_2-CH_3$ or C_4H_{10}

If a :CH group is used, it can be bonded to three ·CH₃ groups. The full diagrams of the hydrocarbon groups and the derived molecule will not be given. Instead the more compact form of the formula will be used.

$$CH_3\!-\!CH\!-\!CH_3$$
$$\mid$$
$$CH_3$$

or C_4H_{10}

NOTE. These molecules both have the same molecular formula. They are both molecules of butane but they have different structural formulae. They are called the two *isomers* of butane. The ending -ane in the name of a hydrocarbon means that only single bonds are present in the molecule.

Isomers

● *Compounds which have identical molecular formulae but different structural formulae are called Isomers.*

These compounds all have the same molecular formula C_4H_8 and they are the four possible isomers of butene. The ending -ene in the name of a hydrocarbon means that a double bond is present in the molecule.

If a molecule containing four carbon atoms has a triple bond then there are two possible structures.

$$CH_3-C \equiv C-CH_3 \qquad CH \equiv C-CH_2-CH_3$$

These are the two possible isomers of butyne, C_4H_6. The ending -yne means a triple bond is present.

There are other possible structures for molecules of hydrocarbons with four carbon atoms. For example:

$$CH_2 = CH - CH = CH_2 \text{ or } C_4H_6 \text{ butadiene}$$

This compound has the same molecular formula as the isomers of butyne.

In addition to the 'chain' type compounds described above, 'cyclic' molecules can also be formed with four carbon atoms. For example, if four :CH₂ groups are joined to form a chain, there are two electrons which have not been paired off—one at each end of the chain.

$$·CH_2 - CH_2 - CH_2 - CH_2·$$

The two electrons can be paired with each other. This can be done by bringing the two ends together:

$$CH_2 - CH_2$$
$$\mid \qquad \mid$$
$$CH_2 - CH_2$$

or C_4H_8 cyclobutane

Another cyclic structure is possible if one ·CH_3 side group is included in the formula:

$$\begin{array}{c} CH_2 \\ | \quad \diagdown \\ \quad \quad CH-CH_3 \\ | \quad \diagup \\ CH_2 \end{array} \quad \text{or} \quad C_4H_8 \quad \text{methylcyclopropane}$$

These two compounds are isomers and they also have the same formula as the isomers of butene.

Some of the structures of molecules containing four carbon atoms have been described. It will be appreciated that if longer chains and rings are considered there will be an enormous number of possible organic compounds. Even using carbon and hydrogen as the *only* elements, the number of possibilities is almost unlimited.

Simplified Formulae

Structural formulae of hydrocarbons are conventionally simplified by not explicitly writing the carbon and hydrogen atoms. Cyclobutane and its isomer (given above) are represented as follows:

A carbon atom is understood to be at each corner or end. There are as many hydrogen atoms on each carbon as are required to complete the four bonds per carbon atom.

Cyclohexane, C_6H_{12}, has a molecule which is hexagonal when viewed from the front. However, the molecule is not planar. It is described as 'chair' shaped as can be shown by drawing a perspective view.

front view

perspective view

cyclohexane, C_6H_{12}

Some Unusual Compounds

The following compounds were prepared only in the last five years and their discoverers have given them descriptive names.

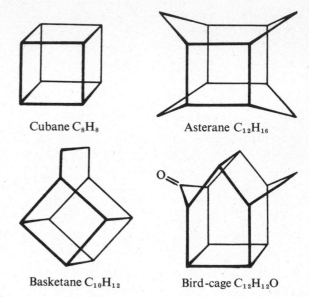

Cubane C_8H_8 Asterane $C_{12}H_{16}$

Basketane $C_{10}H_{12}$ Bird-cage $C_{12}H_{12}O$

Benzene Compounds

The formula of benzene is C_6H_6 and the molecule is composed of six :CH groups arranged in the form of a hexagon. If these groups are drawn in this arrangement, it will be seen that an electron on each carbon atom is not paired off in the usual C−C covalent bond.

The six electrons are spread uniformly over the whole cyclic molecule. They do not bond any particular pair of carbon atoms together, but instead, they contribute to the bonding of all of the carbon atoms in the ring. This is a special type of bonding and is represented by drawing a closed circle inside the hexagon. As for the cyclic hydrocarbons discussed above, a simplified form of the structural formula is used. It is particularly important to note that the presence or absence of a circle drawn inside the hexagon gives a different meaning to the formula. Thus:

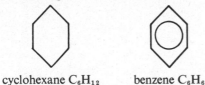

cyclohexane C_6H_{12} benzene C_6H_6

In both of these symbols, a carbon atom is understood to be at each corner. In cyclohexane, each carbon atom has two electrons used in bonds to the adjacent carbon atoms and each is also bonded to two hydrogen atoms to complete the sharing of its four electrons. Hence its formula is C_6H_{12}. In benzene, each carbon has three electrons used in bonds to the adjacent carbon atoms in the ring, and each is bonded to only one hydrogen atom to complete the sharing of its four electrons. Hence its formula is C_6H_6.

Compounds which have this benzene-type bonding are called 'aromatic' compounds. Compounds without this type of bonding are called 'aliphatic' compounds.

If one or two $\cdot CH_3$ groups are introduced in place of hydrogen atoms in the benzene structure, the structures of toluene and xylene are obtained.

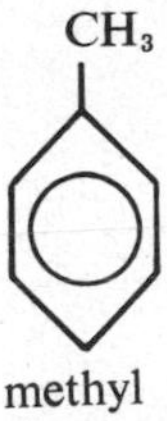

methyl benzene (also called toluene)

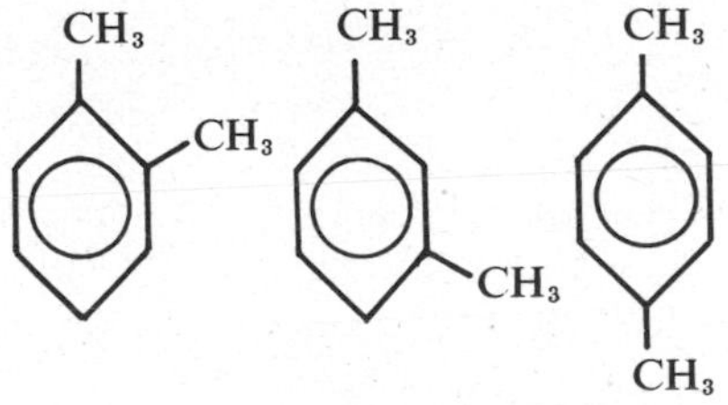

isomers of dimethylbenzene (also called xylene)

Notice that the convention is to show the group bonded to one of the ring carbon atoms only if it is not hydrogen. If a corner has a $\cdot CH_3$ group attached, it means that there is no hydrogen atom in that position, but instead there is a $\cdot CH_3$ group. If a corner has nothing shown attached to it, it means that there is a hydrogen atom in that position.

Benzene type compounds may have more than one six-membered ring. For example, naphthalene has the molecular formula $C_{10}H_8$ and its formula can be

built up from eight $\cdot C - H$ groups and two carbon atoms:

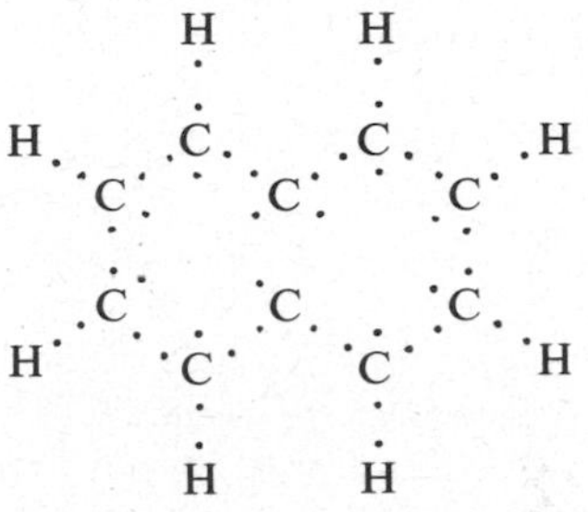

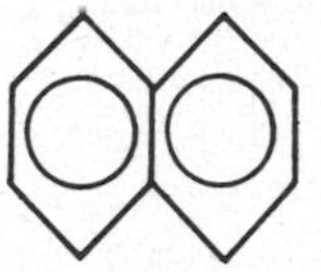

naphthalene $C_{10}H_8$

The main use of naphthalene is for the manufacture of plasticizers (see later). However, it is also a well-known chemical in the household in the form of moth balls. The odour of moth balls is due to the naphthalene. Another well-known chemical which has a similar odour is paradichlorobenzene.

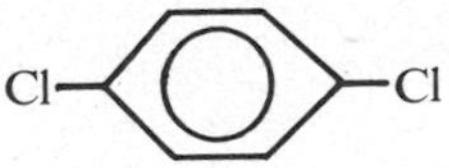

Cl —   — Cl

This compound is widely used as a deodorant in toilets, sometimes as the pure compound and sometimes scented with artificial odours.

The prefix para- in the name of this compound indicates the relative positions of the two groups attached to the benzene ring. The three possible arrangements for two groups are:

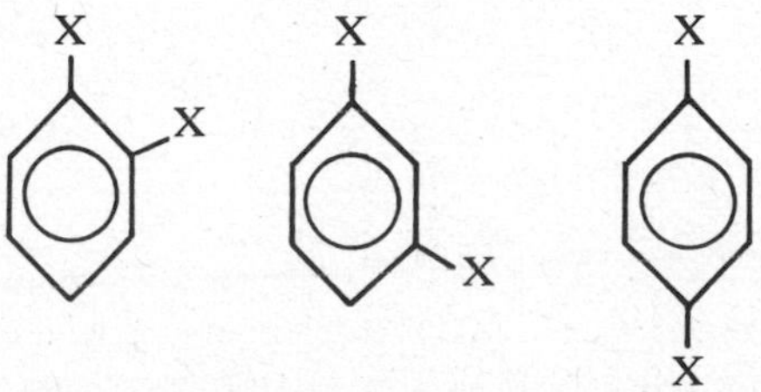

ortho-   meta-   para-

FUNCTIONAL GROUPS IN ORGANIC COMPOUNDS

In addition to the vast number of ways in which carbon and hydrogen atoms can be arranged, still further variety is obtained in organic molecules by the introduction of other atoms such as chlorine, oxygen, nitrogen, etc., or groups of these atoms. Some of these groups have already been described, e.g. the $\cdot NH_2$ group (the amine group) and the $\cdot OH$ group (the hydroxy group). A particular functional group has certain properties which are not greatly altered by the size and shape of the rest of the molecule. The properties of organic compounds are usually the properties of these functional groups.

Some important functional groups containing oxygen will be described. The hydrocarbon part of the molecule will be represented by the single letter R. When R is written in a formula, it represents any one of many possible groups of hydrogen and carbon atoms such as:

$\cdot CH_3$ called the methyl group

$\cdot CH_2 - CH_3$ called the ethyl group

     called the phenyl group

Ethers R-O-R and Alcohols R-O-H

Well known examples of these classes of compounds are:

$CH_3 - CH_2 - O - CH_2 - CH_3$ diethyl ether ('anaesthetic' ether)

$CH_3 - CH_2 - OH$ ethanol (also called ethyl alcohol. This is ordinary 'alcohol')

Notice that both of these compounds have a hydrocarbon group bonded to the oxygen. They differ only in that an alcohol has a hydrogen atom bonded to the oxygen atom but an ether does not have such a hydrogen atom. While this may appear to be only a slight difference, the properties of these compounds are very different.

Aldehydes R-CHO and Ketones R-CO-R

The formula of an aldehyde is often written as $R - CHO$. The R group in this formula could be a hydrogen atom as well as a hydrocarbon group. Thus H-CHO is formaldehyde. However, the R groups in a ketone $R - CO - R$ must not be hydrogen atoms. If one or both of these R groups was replaced by an H-atom, the compound would be an aldehyde and not a ketone. Again, although these formulae are only slightly different, the properties are very different. Well-known examples of these classes of compounds are:

$CH_3 - C \begin{smallmatrix} O \\ H \end{smallmatrix}$ acetaldehyde (a pungent smelling liquid)

$\begin{smallmatrix} CH_3 \\ CH_3 \end{smallmatrix} C = O$ acetone (a pleasant smelling liquid)

Carboxylic Acids R-COOH and Esters R-CO-O-R

Once more these formulae are quite similar to each other and might appear to be merely combinations of ketone, alcohol and ether groups. However, they are quite different in properties from each other and from these other types of compounds. Well-known examples of these classes of compounds are:

$CH_3 - C \begin{smallmatrix} O \\ OH \end{smallmatrix}$ acetic acid (a sharp smelling liquid)

$CH_3 - C \begin{smallmatrix} O \\ O - CH_2 - CH_3 \end{smallmatrix}$ ethyl acetate (a fruity smelling liquid)

There are many other kinds of functional groups but these will not be covered here. The examples given are sufficient to illustrate that the properties of a compound depend on the functional groups present in the molecule. In addition to the many ways in which carbon and hydrogen atoms can be arranged, there are also many positions in which the functional groups can be placed in the molecules and many different combinations of functional groups can be achieved. These are the reasons why there is such a large number of compounds of carbon.

USEFUL ORGANIC COMPOUNDS

The preceding discussion has shown how the bonding requirements of the four electrons on a carbon atom can be satisfied in four ways.

(i) four single bonds;

(ii) a double bond and two single bonds;

(iii) a triple bond and one single bond;

(iv) three electrons shared in a benzene ring.

With these arrangements, a vast number of compounds can be made, many of which are very useful. So that the complicated structural formulae can be written down, carbon atoms and the hydrogen atoms attached to them are often omitted.

Some compounds which are widely used will now be described to give examples of organic compounds and give practice with interpreting organic chemical formulae. The examples are not intended to give a comprehensive coverage and they are not organized in any particular sequence. The intention is to describe compounds with everyday uses but not to provide material to be memorized for examinations.

Insecticides

If benzene, C_6H_6, is reacted with chlorine, an aliphatic cyclic compound is obtained which has the formula $C_6H_6Cl_6$.

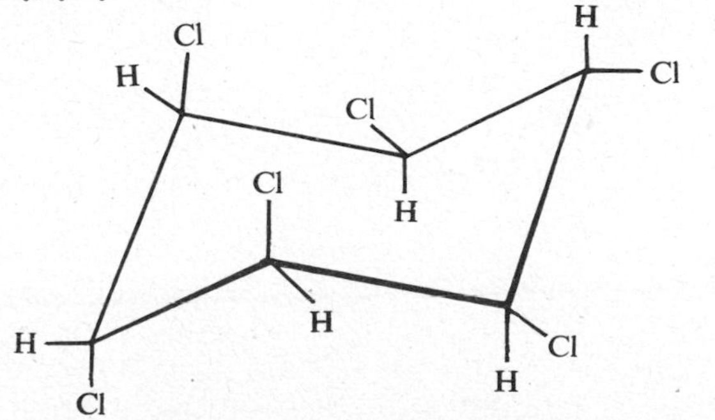

γ-hexachlorocyclohexane

This formula is similar to cyclohexane C_6H_{12} but has six chlorine atoms in place of six of the hydrogen atoms. Hence it is called hexachlorocyclohexane. The molecule is 'chair' shaped and one particular isomer, called the gamma isomer, is a very active insecticide.

This isomer, sold as 'Lindane', 'Gammexane' or 'Hexachlor' is used to control the boll weevil. A solution, called 'Gamma Wash' is used to kill insect pests on animals.

An insecticide which has been very widely used is DDT, an abbreviation for DichloroDiphenylTrichloro ethane.

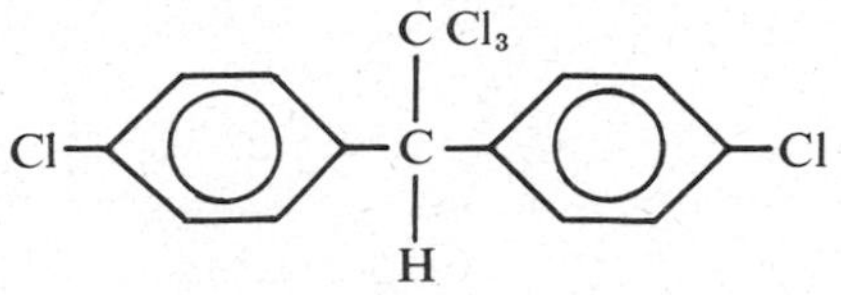

dichlorodiphenyltrichloroethane, **DDT**

This compound has been used in such vast quantities that it has become a very serious pollutant. It is not destroyed by bacteria or enzymes and so is accumulated in food and hence in our bodies.

Two other chloro compounds which are also very powerful insecticides are 'Dieldrin' and 'Chlordane'.

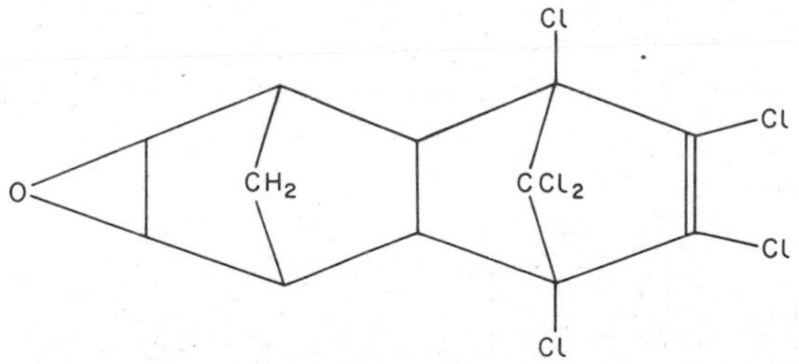

Dieldrin

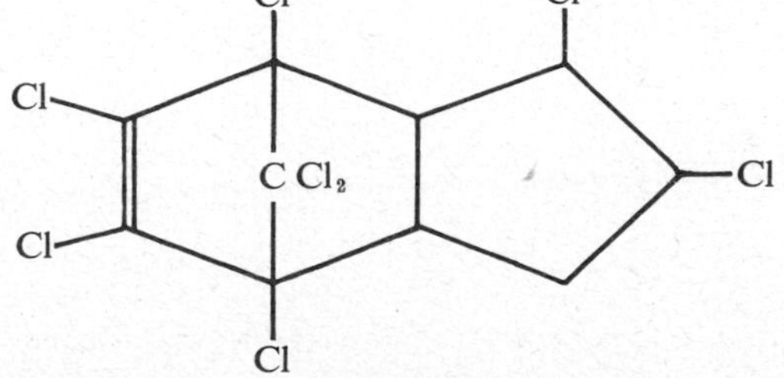

Chlordane

Drugs

Aspirin, a widely used 'pain-killer', is made from salicylic acid by reaction with acetic acid and acetic anhydride.

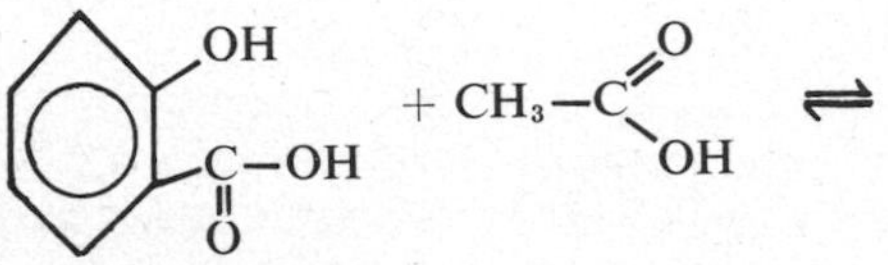

salicylic acid · · · · · acetic acid

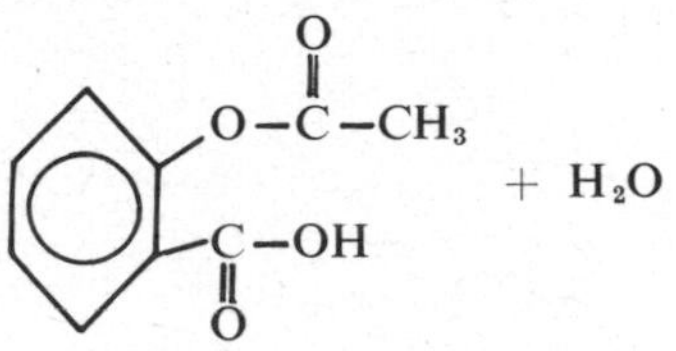

acetyl salicylate (aspirin)

This partially complete reaction can be made to go to completion by removing the water. The water is removed by reaction with acetic anhydride.

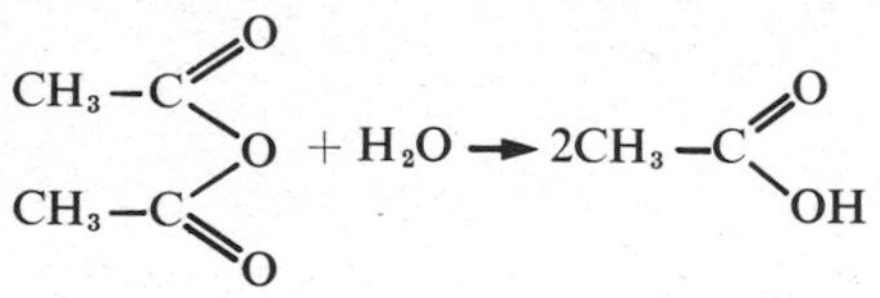

acetic anhydride · · · · · acetic acid

Salicylic acid can also be reacted with methanol to form 'oil of wintergreen' which is used as a pharmaceutical and as a flavouring or as an odourant.

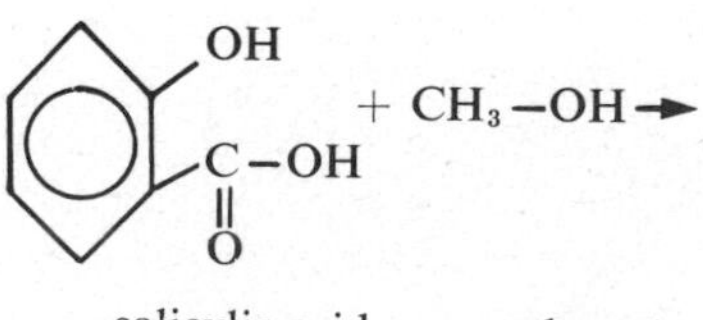

salicylic acid · · · · methanol

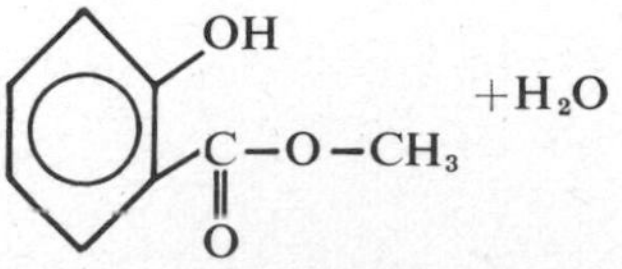

methyl salicylate ('Oil of Wintergreen')

These reactions both involve an $-OH$ group (alcohol group) reacting with a $-CO-OH$ group (carboxylic acid group). Such a reaction is called *esterification* and the products are called *esters*. Esters contain the $-CO-O-$ group (ester group).

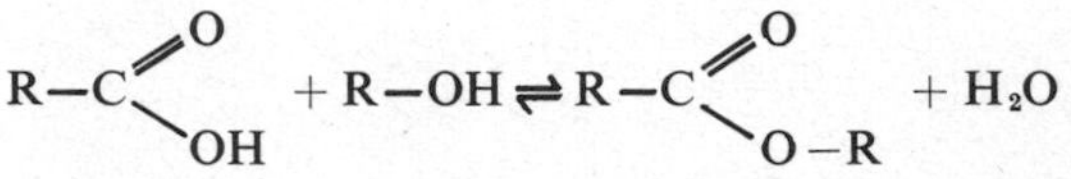

acid · · · · alcohol · · · · ester · · · · water

Sweeteners

Sucrose, $C_{12}H_{22}O_{11}$, has a structure containing two rings. One is six membered and the other is five membered. The six membered ring is chair shaped, similar to cyclohexane, but one corner is occupied by an oxygen atom.

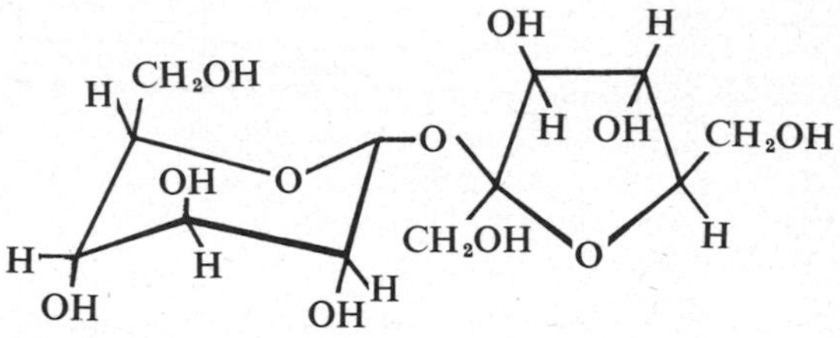

sucrose

Artificial sweeteners are widely used in place of sucrose. Saccharins and cyclamates are both salts containing sodium or potassium cations. Saccharin tends to leave a bitter after-taste and the excessive use of cyclamates is dangerous.

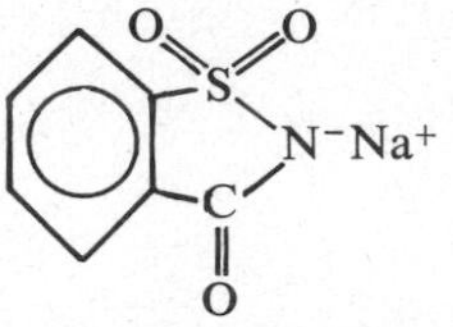

saccharin

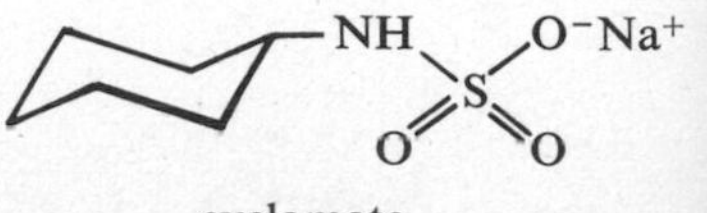

cyclamate

Many compounds are being examined to find a completely satisfactory artificial sweetener. One example of such a compound is the methyl ester of apartyl phenylalanine :

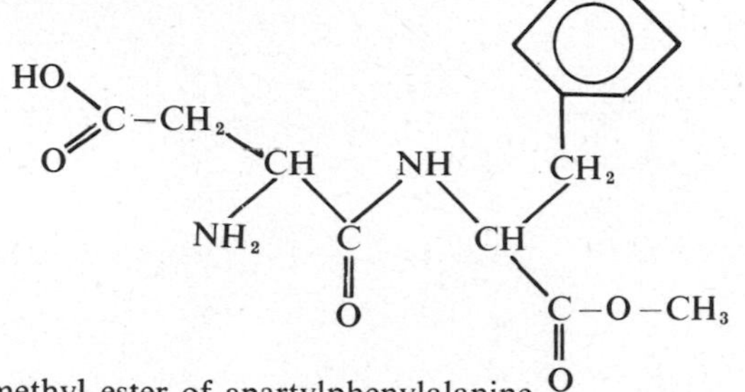

methyl ester of apartylphenylalanine

Detergents

Compounds containing double bonds can join their molecules together to form bigger molecules. For example two molecules of propylene, $CH_3-CH=CH_2$ can react by transferring a hydrogen atom :

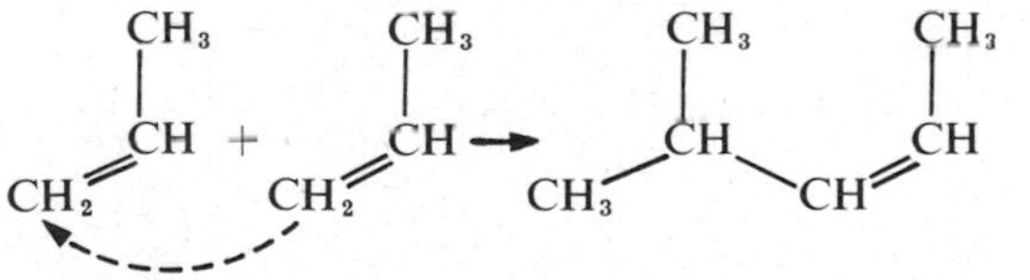

The product can then add on another propylene molecule and the process can continue. A product with twelve carbon atoms can be made :

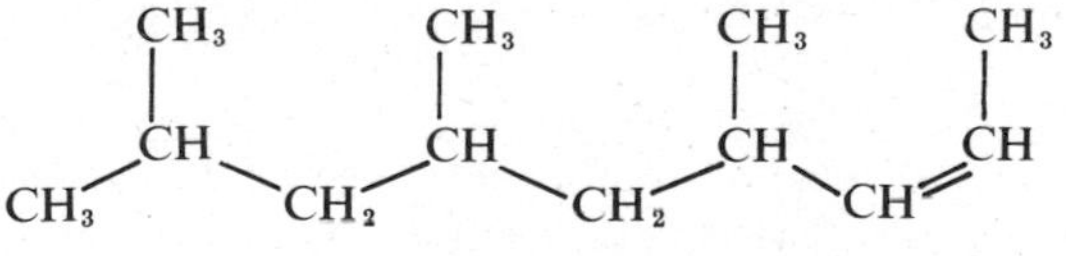

or $C_{12}H_{24}$

The double bond in this molecule can be used to attach the chain to benzene. This also occurs by transfer of a hydrogen atom so the formula of benzene will be drawn with one H-atom written in :

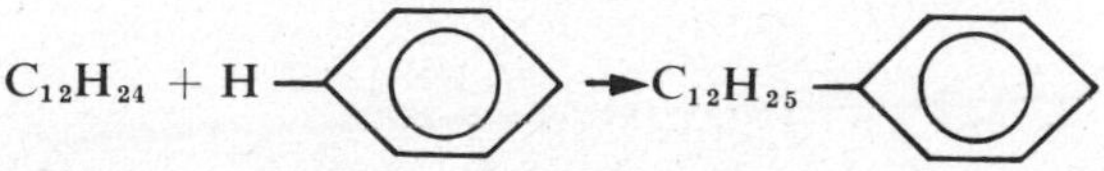

Reaction with sulphuric acid introduces the sulphonic acid group $\cdot SO_3H$ by a process which replaces a hydrogen as shown by the general equation :

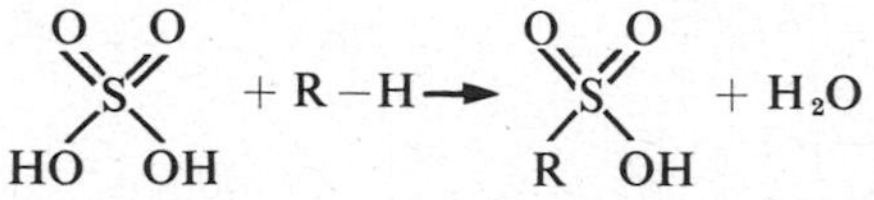

The $\cdot SO_3H$ group is introduced into the position opposite to the $\cdot C_{12}H_{25}$ chain :

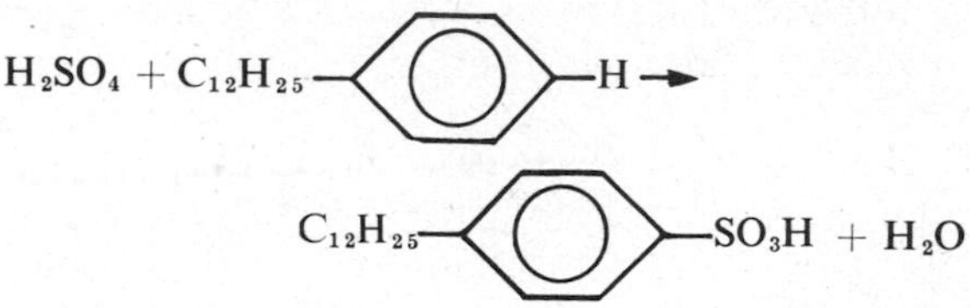

Neutralization of this acid with caustic soda gives the sodium salt which is widely used as a household detergent.

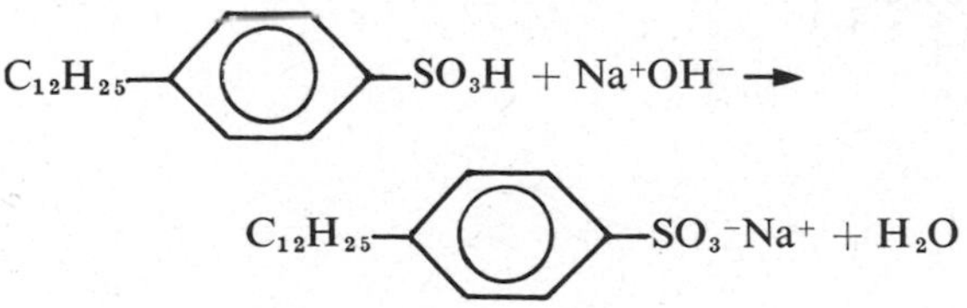

These detergents are a major cause of water pollution because the bacteria in sewage plants cannot completely break down these organic compounds. The bacteria begin to destroy the molecule from the end of the side chain but bacterial attack on the branched chain of carbon atoms is slow. For this reason, straight chain $C_{15}H_{30}$ separated from petroleum oil is now being used in place of the branched molecule made from propylene. Sewage bacteria can rapidly degrade detergent molecules with straight chain hydrocarbon groups.

Polypropylene and Similar Polymers

The process of forming long chains from propylene, which was described above, can be made to produce very long molecules which may contain thousands of carbon atoms in the chain.

Propylene is $CH_2=CH-CH_3$. Any molecule represented by $CH_2=CH-X$ can be polymerized in a similar way.

$$\underset{CH_2=CH}{\overset{\overset{\displaystyle X}{|}}{}} \xrightarrow{\text{polymerized}} \underset{[-CH_2-CH-]}{\overset{\overset{\displaystyle X}{|}}{}}$$

By using compounds with different groups represented by X, many different kinds of polymers can be obtained.

(i) If X is ·H, the polymer is polyethylene or 'Polythene': $[-CH_2-CH_2-]$

(ii) If X is ·CH₃, the polymer is polypropylene:

$$\underset{[-CH_2-CH-]}{\overset{\overset{\displaystyle CH_3}{|}}{}}$$

Polyethylene and polypropylene can be melted and extruded to form fibre, rod or sheet. A tube can be extruded by blowing air into the inside of the tube while it is being extruded. This prevents the tube from collapsing while the plastic sets. If a large diameter, thin-walled tube is blown and then later flattened out, it can be sealed at intervals with a hot roller. When the sealed lengths are cut off, they form plastic bags.

Polyethylene is a clear plastic which is chemically unreactive and impervious to air and water. It is therefore widely used to package and store a wide variety of materials including foods, acids and other corrosive liquids. Polypropylene is stronger than polyethylene but it is more expensive.

(iii) If X is ·C₆H₅, the polymer is polystyrene:

$$[-CH_2-CH-]$$

Polystyrene has good electrical insulating properties. It can be formed as a clear plastic. It can also be formed as a foam by mixing it with a substance which releases a gas when it is heated, e.g.

$$CH_3-\underset{\underset{\displaystyle CH_3}{|}}{\overset{\overset{\displaystyle CH_3}{|}}{C}}-N=N-\underset{\underset{\displaystyle CH_3}{|}}{\overset{\overset{\displaystyle CH_3}{|}}{C}}-CH_3 \qquad \begin{array}{l}\text{Compound which}\\\text{releases } N_2 \text{ gas when}\\\text{heated.}\end{array}$$

Polystyrene foam is widely used in packaging and in the manufacture of insulating food containers. It is also used to make lightweight articles such as surfboards, etc.

(iv) If X is ·CN, the polymer is polyacrylonitrile:

$$\underset{[-CH_2-CH-]}{\overset{\overset{\displaystyle CN}{|}}{}}$$

This polymer turns black without melting at about 200 °C so it cannot be melt-extruded. It is dissolved in solvents and formed into a fibre by evaporating off the solvent. The fibre is called 'Orlon' or 'Acrilan'.

(v) If X is ·OOC —CH₃, the polymer is polyvinylacetate:

$$\begin{array}{c}O=C-CH_3\\|\\O\\|\\[-CH_2-CH-]\end{array}$$

This polymer, called PVA, is used in water based 'plastic' paints and in adhesives.

(vi) If X is ·Cl, the polymer is polyvinylchloride:

$$\underset{[-CH_2-CH-]}{\overset{\overset{\displaystyle Cl}{|}}{}}$$

Polyvinylchloride, called PVC, has excellent electrical

insulating properties and is relatively flame resistant compared to other plastics. It is a fairly brittle plastic but it is used to make chemically inert pipes, plumbing fittings, sinks, etc.

By adding a plasticizer (see below) to the PVC, it becomes quite flexible and can be used to make insulating covers for electrical wires, artificial leather and cloth. It can also be made into clear flexible sheets for covering car seats, chairs, etc.

Plasticisers

These are materials which, when added to a plastic, make it less brittle. The most widely used plasticiser is dioctylphthalate; DOP. The first step in its manufacture is the oxidation of xylene or naphthalene.

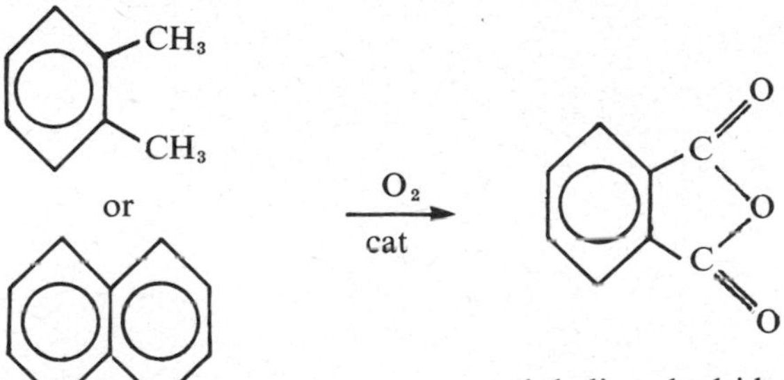

phthalic anhydride

The product is reacted with an isomer of octyl alcohol:

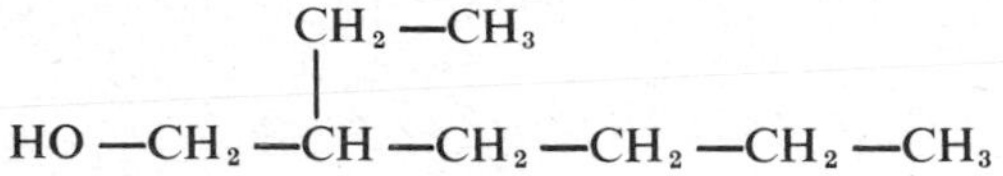

The following ester is formed:

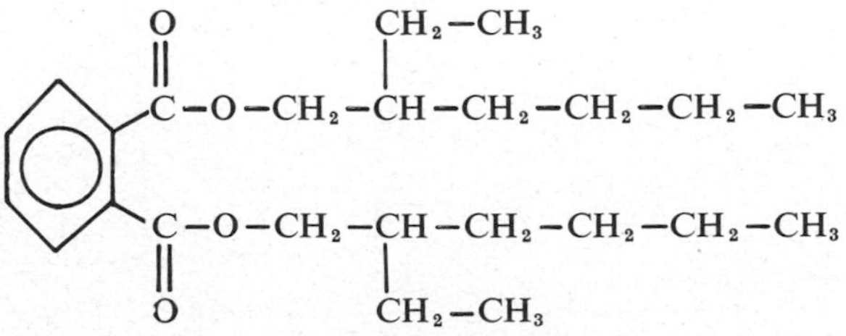

di-octyl phthalate, DOP

Polyesters

If an acid with two carboxylic acid groups is reacted with an alcohol containing two alcohol groups, an ester is formed which can react with further groups and so form a polymer.

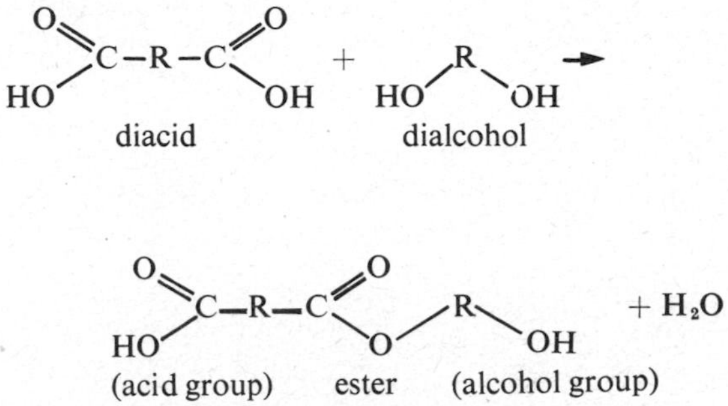

diacid   dialcohol

(acid group)   ester   (alcohol group)

The ester has functional groups at each end of its molecule. The acid group end can react with more of the dialcohol. The alcohol group end of the ester can react with more of the diacid.

In practice, the formation of a long chain polymer is a process which requires very careful choice of starting materials. The acid should have 3 or 4 carbon atoms between its functional groups and the alcohol should have only two carbons. The conditions and sequence of reactions must also be very carefully controlled. These requirements can be met by the following compounds:

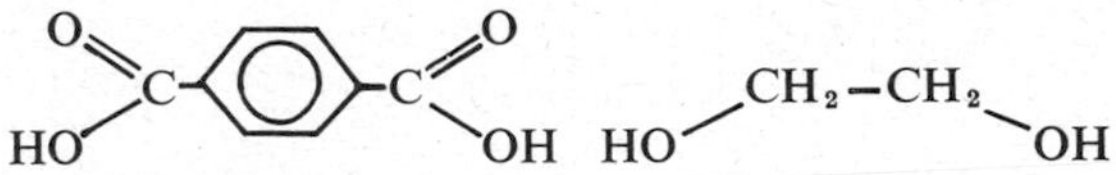

terephthalic acid   ethylene glycol

The polymer formed from these melts at about 270 °C and can be melt spun to form a fibre called 'Dacron', 'Terylene' or 'Teron'. It can also be melt extruded as a film called 'Mylar' or 'Cronar'.

Polyester plastics can also be made from ordinary phthalic acid and glycerol. They are called alkyd resins or 'Glyptal' resins.

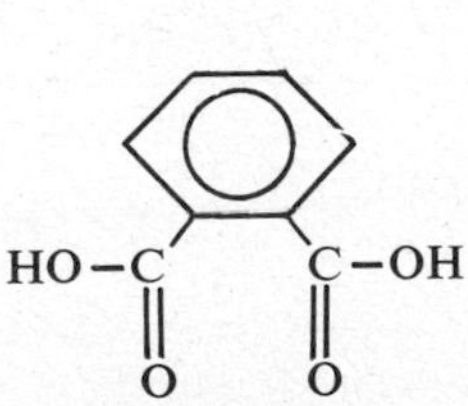

phthalic acid 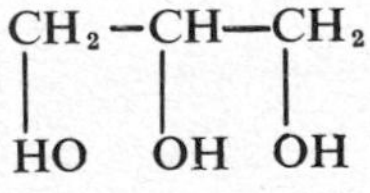   glycerol

These react at 150 °C to form a linear polymer, because under these conditions, the centre ·OH group of the glycerol does not react. The polymer is soluble in solvents and is used in lacquers and paints. If more phthalic acid is used and the temperature raised to 200 °C, the centre ·OH groups of the glycerol react and the product is a tough insoluble solid used for making moulded objects.

Nylon

There are several different kinds of nylons but they all have :N–H groups spaced along the chain. The most common nylon is made from hexamethylene-diamine and adipic acid:

Summary

It will be appreciated that there are so many organic compounds that it is impossible for any one person to study all of them. Even after several years of study, only a fairly small portion of this vast subject could be covered. However, the principles of bonding and of functional groups introduced in this chapter can be applied to the study of all organic compounds.

The substances which have been described here have been chosen because they are in common every-day use. They have been presented to give familiarity in interpreting organic formulae and provide a first approach to the significance of structure, isomerism and functional groups in organic chemistry.

hexamethylenediamine

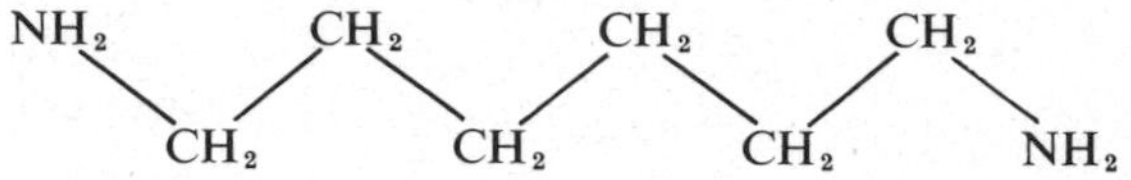

adipic acid

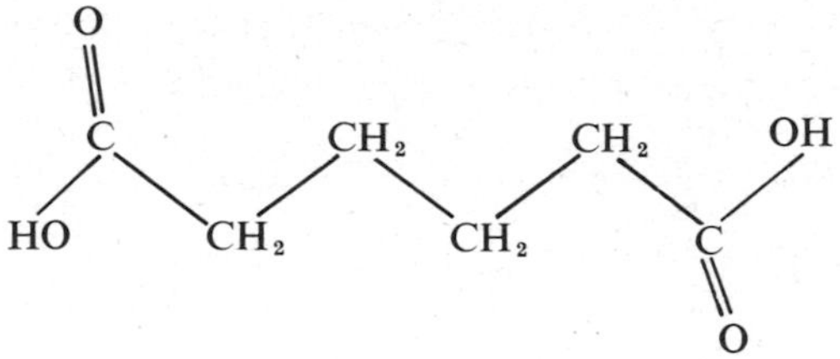

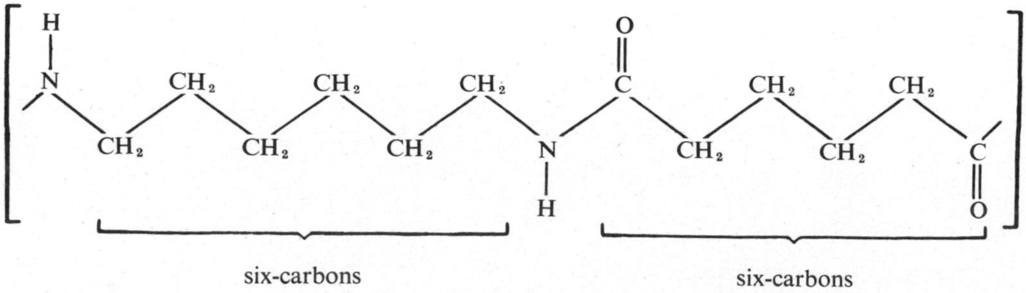

six-carbons   six-carbons

This nylon is called six-six-nylon or 66 nylon because there are six carbons between the :NH groups in each of the two parts of the repeating unit in the chain.

STUDY QUESTIONS

1. Draw diagrams to represent the shapes of the molecules:
   (a) $HO(CH_2)_2OH$;
   (b) CH_3CN;
   (c) $HCOOH$;
   (d) cyclopentane C_5H_{10}.

2. Write formulae for the isomers of:
   (a) C_3H_6;
   (b) $C_2H_2Cl_2$.

3. The molecular formula of a compound is C_3H_8O. Write formulae for all possible structures and label each structure as an alcohol or an ether.

4. The molecular formula of a compound is C_3H_6O. Write formulae for all possible structures and label each structure as an aldehyde or a ketone.

5. The molecular formula of a compound is $C_3H_6O_2$. Write formulae for all possible structures and label each structure as a carboxylic acid or an ester.

6. Cyclohexane, C_6H_{12}, has 'chair' shaped molecules. At room temperature, a few molecules in each thousand are excited to an unstable 'boat' shape. Draw diagrams to represent these two different shapes of cyclohexane molecules.

7. How many structures are possible for:
   (a) cyclohexadiene, C_6H_8;
   (b) cyclopentadiene, C_5H_6?

8. What are the structures of ortho-, meta-, and para-xylene?

9. There are only two isomers of anthracene, $C_{14}H_{10}$. Draw their structures.

10. Glycine, $C_2H_5NO_2$ is a carboxylic acid and it is also an amine. It reacts with itself to form $C_4H_8N_2O_3$. Write formulae for these compounds.

11. 'Aspirin' dissolves in caustic soda solution to form a sodium salt. Write the equation for this reaction.

12. The powerful insecticide obtained from pyrethrin flowers consist of a mixture of esters. Hydrolysis of one of these esters yields the following two compounds:

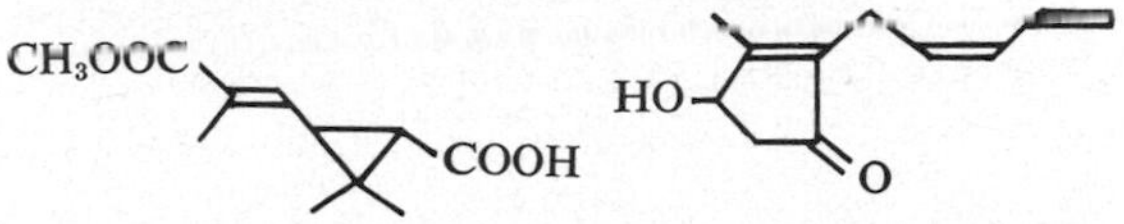

What are the functional groups in these compounds? One of these compounds is called pyrethric acid and the other is called pyrethrolone—name each compound. What are the molecular formulae of these compounds? What is the structural formula of the ester formed from them?

13. What serious air pollution problem results from burning articles made of PVC plastic.

14. A new plastic has a high resistance to 'pilling', i.e. formation of small balls of fluff on fabrics made from the fibre. The synthetic fibre is of a polyethylene terephthalate type. What is the formula of the repeating unit in this polymer?

15. Nylon runners are used in fitting sliding doors to cupboards. What advantages do these have over oiled metal runners or races of ball bearings?

27 : Silicon and Silica

Occurrence of Silicon

Silicon is the second most abundant element in the Earth's crust and comprises about one-quarter of the Earth's crust by mass. The element does not occur in the uncombined state.

Silicon occurs widely as the oxide, silicon dioxide (called silica), and as silicates. Almost all rocks (except limestone) contain silicates. Thus granite is a mixture of silicon dioxide (quartz), with felspar and mica, both of which are silicates.

The weathering of rocks such as granite frees the particles of quartz, which ultimately accumulate as sand. Chemical weathering of the silicates leads to the formation of clay which contains aluminium silicate, $Al_2O_3.2SiO_2,2H_2O$, together with varying amounts of iron oxide and other substances.

Silica itself occurs naturally as quartz (rock crystal), sand, and various amorphous forms such as flint and opal. An amorphous form of silicon dioxide is kieselguhr. This is a sedimentary deposit of the skeletons of minute sea creatures.

The Element Silicon

The element silicon can be produced by reduction of sand with coke in an electric furnace. It is a crystalline non-metal and is used in the making of silicon steel and silicones. It is also used in the manufacture of silicon transistors.

Valency of Silicon

Silicon is in the same group of the periodic table as carbon. It has four electrons in its outer shell and like carbon it has a covalency of four. Apart from this, there is little similarity between the two elements.

SILICON DIOXIDE OR SILICA

The structure of silicon dioxide was discussed in chapter 4. It is similar to diamond.

Physical Properties

Silica, SiO_2, has a very high melting point (about 1600 °C) and an oxy-hydrogen flame or an electric furnace is needed to soften it. It is used to make high quality chemical apparatus such as test tubes and dishes. However, such apparatus is expensive and glass is more commonly used. One important advantage of silica apparatus is that it can be heated until

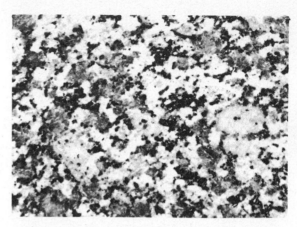

A piece of granite showing a flat, polished surface. Granite contains crystals of quartz (colourless), felspars (white or pink), and mica (black).

A crystal of quartz showing its hexagonal symmetry.

it is red hot, and then plunged into cold water, without risk of cracking. This property of being able to resist thermal shock is due to its very low coefficient of expansion.

Silica is practically insoluble in water.

Chemical Properties

Silica is a fairly unreactive substance as would be expected from its occurrence in nature. It does not react with water or with ordinary acids, but it will react with hydrofluoric acid:

$$SiO_2 + 4HF \rightarrow SiF_4 + 2H_2O$$

Silicon tetrafluoride is a gas, and escapes during the reaction.

Silicon dioxide reacts with fused caustic alkalis, or more slowly with boiling alkali solutions, to form a salt and water:

$$SiO_2 + 2OH^- \rightarrow SiO_3^{2-} + H_2O$$

The product is sodium or potassium silicate.
Silica also reacts with basic oxides if strongly heated, e.g.:

$$SiO_2 + CaO \rightarrow CaSiO_3$$
$$\text{calcium silicate}$$

Silicon dioxide is therefore an *acidic oxide*.

Sodium and calcium silicates can also be formed by reacting silica with sodium or calcium carbonate at high temperatures:

$$Na_2CO_3 + SiO_2 \rightarrow Na_2SiO_3 + CO_2$$
$$CaCO_3 + SiO_2 \rightarrow CaSiO_3 + CO_2$$

These reactions are the basis of the manufacture of glass, and they proceed because SiO_2 displaces the more volatile CO_2.

Silicon dioxide can be reduced by carbon at the temperature of an electric furnace:

$$SiO_2 + 2C \rightarrow Si + 2CO$$

If more carbon is present, *silicon carbide (carborundum)* is produced:

$$SiO_2 + 3C \rightarrow SiC + 2CO$$

Carborundum is a very hard material and it is made into grinding wheels, hones and various other abrasive products.

Uses of Silicon Dioxide

Pure silica (quartz) is used for making chemical apparatus. Quartz crystals are also used to control the frequency in radio transmission.

Sand is very widely used for making carborundum, mortar, concrete, sodium silicate and glass.

SODIUM SILICATE (water glass)

Sodium silicate is produced commercially by heating a mixture of sodium carbonate and sand very strongly:

$$Na_2CO_3 + SiO_2 \rightarrow Na_2SiO_3 + CO_2$$

The molten product is cooled, broken up, and dissolved in water. A thick solution is produced which is called water glass.

Sodium silicate is used in soaps and washing powders. It is also used in treating wood to make it fire resistant and as a glue in making cardboard cartons.

GLASS

Ordinary glass or soda-lime glass is composed of sodium silicate and calcium silicate. Sodium silicate is fairly easy to melt but is soluble in water. Calcium silicate is difficult to melt but is insoluble in water. A mixture of the two gives suitable properties.

Glass is produced commercially as a molten substance and it is cooled in such a way that it does not crystallize. It may therefore be regarded as a super-cooled liquid consisting of a solution of the two silicates in each other.

Production of Glass

The substances used to make glass are sand, anhydrous sodium carbonate and limestone. These substances are charged into a furnace, together with a certain amount of broken glass. The charge is heated to about 1 400 °C by the combustion of fuel oil in preheated air. A glass furnace is similar to the open hearth furnace used for the production of steel (see chapter 18).

The reactions which occur in the furnace are:

$$Na_2CO_3 + SiO_2 \rightarrow Na_2SiO_3 + CO_2$$
$$CaCO_3 + SiO_2 \rightarrow CaSiO_3 + CO_2$$

Carbon dioxide gas is evolved quietly during the reaction, and by the time the mixture has reached the working end of the furnace, all bubbles of gas have escaped.

Most glass products are machine made. Sheet glass is made by pressing a horizontal rod into the molten glass and then lifting the rod vertically. As the rod is lifted, glass clings to it, forming a sheet. The sheet is then drawn along by rollers. Plate glass is similar to ordinary sheet glass but is usually thicker for the sake of greater strength. Both sides of the plate are ground flat so that the sheet is free from distortions. Bottles and flasks are usually mechanically blown into shape, from small portions of molten glass.

The glass products are cooled slowly to avoid setting up strains which would make the glass liable to crack easily. This slow cooling is called *annealing*.

Special Types of Glass

1. *Hard glass* is obtained by using potassium carbonate instead of sodium carbonate in the charge. This glass is composed of potassium silicate and calcium silicate.

metallic oxides in small amounts. The colour is due in some cases to the formation of coloured silicates. Thus a blue glass can be produced by adding cobalt oxide to a soda-lime glass and the colour is due to cobalt silicate. Similarly chromium oxide makes the glass green due to a silicate of chromium. On the other hand, the element selenium is used to make a red glass but the colour is due to a finely dispersed form of the element itself.

Ceramics

Pottery, porcelain, bricks and tiles are made from silicate minerals, mainly clay. The clay is mixed to a plastic mass with water and then shaped. After drying, the objects are heated ('fired') at a high temperature in a kiln, when complex chemical changes take place. Ceramic bricks and tiles are usually coloured because of the presence of iron compounds. Pure white porcelain is made from white clay.

SILICONES

Silicones were first prepared in 1904 but were not developed as industrial products until World War II.

Silicones are compounds consisting of very large molecules containing alternate silicon and oxygen atoms, with the silicon atoms also bonded to hydrocarbon groups. One example of a silicone molecule is a long chain molecule built up by repetition of the unit:

$$\cdots \; \underset{\underset{CH_3}{|}}{\overset{\overset{CH_3}{|}}{Si}} - O - \underset{\underset{CH_3}{|}}{\overset{\overset{CH_3}{|}}{Si}} - O \; \cdots$$

The $-CH_3$ group is called the methyl group and is the simplest member of a whole series of hydrocarbon groups. More complex hydrocarbon groups are related to the higher hydrocarbons, e.g., $-C_2H_5$, ethyl; $-C_3H_7$ propyl; $-C_4H_9$, butyl; etc.

By controlling the length of the silicon-oxygen chains and by using various hydrocarbon groups, it is possible to produce silicones which range from thin oils to elastic solids.

Silicones are good electrical insulators and are water repellent. In addition, they retain these properties under extreme conditions of heat and cold. Silicones find widespread uses as lubricants, electrical insulators and in waxes and polishes.

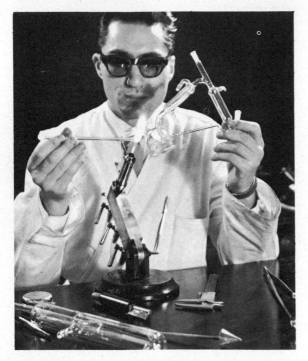

A glass-worker is shown constructing a piece of laboratory glass ware for fractional distillation of a very small quantity of liquid.

2. *Flint glass* consists mainly of potassium silicate and lead silicate, produced by heating sand, potassium carbonate and red led (Pb_3O_4). It is used in making cut glass articles.

3. *Heat resistant glass*, marketed under such names as 'pyrex', is made from sand, sodium carbonate, aluminium oxide and borax. This glass consists of borosilicates and has good resistance to breakage through mechanical and thermal shock. It is widely used in laboratory apparatus.

4. *Safety glass* can be produced in two ways. One way is to cement a sheet of clear plastic between two sheets of glass. Another method is to deliberately develop strain in the glass, by rapidly and uniformly cooling the outer surface of a glass sheet, which has just been softened by heating. Such a glass does not produce sharp splinters when shattered. It is used in automobiles.

5. *Coloured glass* is produced by adding various

STUDY QUESTIONS

1. Write equations for the reaction of silicon dioxide with:
   (a) anhydrous potassium carbonate;
   (b) fused caustic potash;
   (c) lead(II) oxide (PbO).

2. The structure of silicon carbide is similar to diamond. Sketch a small portion of the structure.

3. Silica vessels are very useful in a laboratory.
   (a) What properties make such vessels very useful as equipment?
   (b) Under what circumstances should use of silica ware be avoided?

4. Compare and contrast carbon dioxide and silicon dioxide with respect to:
   (a) structure;
   (b) physical properties;
   (c) chemical properties.

5. Calculate the mass of carborundum which is theoretically obtainable from 3 kg of pure sand, by reaction with coke in an electric furnace.
   ($Si = 28$, $O = 16$, $C = 12$.)

6. Indicate the probable electronic arrangement in molecules of the gas silicon tetrafluoride and the probable shape of the molecules.

7. If sodium silicate solution is treated with concentrated hydrochloric acid, a gelatinous precipitate of silicic acid, H_2SiO_3, forms. Write equations for:
   (a) the formation of the acid;
   (b) its dehydration on heating;
   (c) its reaction with sodium hydroxide solution.

8. If a crystal of cobalt(II) chloride is placed in a fairly dilute solution of sodium silicate, long blue threads grow from the crystal into the solution. What explanation can be offered for this fact?

9. Pottery is often 'glazed' by covering it with a thin layer of a coloured glass. What advantages does glazed pottery have over unglazed pottery? How are glazes coloured?

10. What explanation can be offered for:
   (a) the water repellent property of silicones;
   (b) their lack of electrical conductivity?

Answer to Numerical Problem
   5. 2 kg.

28 : Nitrogen and its Compounds

Valency of Nitrogen

Nitrogen is in group V of the periodic table. It has five electrons in its outer shell and therefore has vacancies for three more electrons. By sharing three electrons in covalent bonds, a nitrogen atom can be surrounded by four pairs of electrons, as in the molecules:

nitrogen   nitrosyl chloride   ammonia

Electrovalent compounds containing N^{3-} ions are uncommon. The only one which is readily formed is magnesium nitride, Mg_3N_2. The bonding in this compound is mainly ionic: $(Mg^{2+})_3(N^{3-})_2$.

Occurrence of Nitrogen

Approximately four-fifths of the atmosphere consists of nitrogen in the uncombined state. Compounds of nitrogen do not occur in the earth's solid crust to any considerable extent, apart from the deposits of sodium nitrate in Chile. However, compounds of nitrogen are found in all living things in the form of complex molecules of proteins containing the elements nitrogen, carbon, hydrogen and oxygen. Some proteins also contain the elements phosphorus and sulphur.

Proteins are polymers of amino-carboxylic acids. There are many different kinds of amino-acid groups in natural proteins but only the simplest will be given to illustrate the structure. The simplest amino-acid is *glycine*, NH_2-CH_2-COOH. This can polymerize, the first step in the formation of a chain being the formation of a dimer.

glycine   glycine

dimer of glycine

The chain can continue to build up. In general, proteins are polymers of the type

For glycine, X represents a hydrogen atom, but for other amino acids, X represents a group of atoms such as methyl, $\cdot CH_3$, etc.

Preparation of Nitrogen

Nitrogen is used industrially in very large quantities in the production of ammonia. The nitrogen is sometimes obtained by liquefaction of air, but more often it is obtained by removing oxygen from the air by chemical methods, principally using coke. This is described in detail at the end of the chapter under the Haber Process.

Laboratory Preparation and Collection of Nitrogen

Pure nitrogen can be obtained in the laboratory by gently heating a mixture of concentrated solutions of ammonium chloride and sodium nitrite—see fig. 28.1.

$$NH_4^+ + NO_2^- \rightarrow N_2 + 2H_2O$$

Fig. 28.1. Laboratory preparation and collection of nitrogen: a solution containing a mixture of ammonium chloride and sodium nitrite is heated.

Physical Properties of Nitrogen

Nitrogen is a colourless, odourless gas, which is slightly less dense than air. The molecular mass of nitrogen is about 28. The gas is only slightly soluble in water, 100 cm³ of water dissolving about 2 cm³ of the gas under room conditions. The boiling point of nitrogen is -195.8 °C at atmospheric pressure.

Chemical Properties of Nitrogen

Nitrogen is stable to heat, and it neither burns nor supports combustion. Its solution in water is neutral to litmus.

Although nitrogen is a relatively inactive gas, it does form some compounds by direct combination with other elements. Lightning flashes in the atmosphere cause small but important amounts of nitrogen monoxide to form:

$$N_2 + O_2 \rightleftharpoons 2NO$$

The nitrogen monoxide is ultimately converted to nitrates. These nitrates can be absorbed by the roots of plants and used to build up plant proteins. Nitrogen monoxide can be formed also by passing air through an electric arc. In this reaction the nitrogen is oxidized.

Nitrogen gas will also react with hydrogen gas under suitable conditions to form ammonia:

$$N_2 + 3H_2 \rightleftharpoons 2NH_3$$

In this reaction, the nitrogen is reduced.

At high temperatures, nitrogen combines directly with some metals, such as magnesium and aluminium, forming nitrides:

$$3Mg + N_2 \rightarrow Mg_3N_2$$
$$2Al + N_2 \rightarrow 2AlN$$

Boron nitride, BN, a compound of boron and nitrogen, may prove to be a very important compound, because, when formed under extreme pressure and at high temperature, it is as hard as diamond. The hardness can be attributed to two facts:
(1) A boron atom has 3 valency electrons and a nitrogen atom has 5, making overall the same number of valency electrons as two carbon atoms.
(2) When formed under these drastic conditions, boron nitride has the same structure as diamond.

The Nitrogen Cycle

All plants require compounds of nitrogen in order to build up proteins for the growth of new tissue.

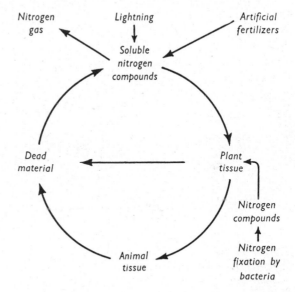

Fig. 28.2. The nitrogen cycle.

The supply of nitrogen in nature is supplemented by the manufacture of artificial fertilizers. The utilization of nitrogen by living things is a cyclic process, in which various bacteria play an important part, particularly in nitrogen fixation and decay. See fig. 28.2.

NITRIC ACID, HNO₃

Laboratory Preparation of Nitric Acid

Nitric acid is prepared by heating about equal masses of sodium or potassium nitrate with concentrated sulphuric acid at a temperature of about 130 °C. The temperature is kept low, so that the nitric acid will not decompose very much.

No cork or rubber should be used in the apparatus because the acid vapour reacts with them very rapidly, producing nitrogen dioxide. An all-glass apparatus is used and a suitable arrangement is shown in fig. 28.3.

The nitrate ion is a very weak base. It does not accept protons from H_3O^+ ions. Thus dilute sulphuric acid cannot be used. However, it does accept protons from the very strong proton donor H_2SO_4. This is why concentrated sulphuric acid must be used. The equation for the reaction is:

$$H_2SO_4 + NO_3^- \rightarrow HNO_3 + HSO_4^-$$

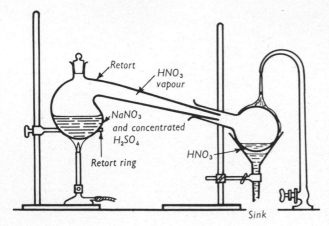

Fig. 28.3. Preparation of nitric acid.

The acid obtained has a slightly yellowish colour due to dissolved nitrogen dioxide. This gas is produced by decomposition of some of the nitric acid:

$$4HNO_3 \rightarrow 2H_2O + 4NO_2 + O_2$$

The nitrogen dioxide may be removed by bubbling air through the mixture, but it is a fairly slow process.

The residue in the retort is molten sodium hydrogen sulphate and it sets to a hard cake on cooling.

At rather higher temperatures, the sodium hydrogen sulphate formed in the above reaction would react with further sodium nitrate, according to the equation:

$$HSO_4^- + NO_3^- \rightarrow HNO_3 + SO_4^{2-}$$

However, this is avoided because the higher temperatures required would cause very substantial decomposition of the nitric acid vapour.

The acid produced by the method described above is a solution containing more than 90% HNO_3. It fumes strongly in air and is called *fuming nitric acid*. This acid is particularly corrosive in its reaction with skin or clothes and great care must be exercised in working with it.

The fuming liquid can be identified as nitric acid by pouring it on copper. The reaction is surprisingly slow, but if a little water is added to the liquid, a very vigorous effervescence occurs with the production of nitrogen dioxide. Apparently the water causes ionization of the fuming nitric acid and the copper metal reduces the nitrate ions more rapidly than the un-ionized acid.

$$Cu + 2NO_3^- + 4H^+ \rightarrow Cu^{2+} + 2NO_2 + 2H_2O$$

Commercial Production of Nitric Acid

Nitric acid is produced commercially by two methods. One of these methods is simply a large scale adaptation of the laboratory method described above. The other method, which uses ammonia as a starting substance, is described at the end of this chapter.

Properties of Nitric Acid

PHYSICAL PROPERTIES

Nitric acid is a colourless volatile liquid.

The boiling point of an aqueous solution of nitric acid depends on both the concentration of the solution and the atmospheric pressure. The graph shown in fig. 28.4 shows the variation of boiling point with concentration, at a pressure of one atmosphere.

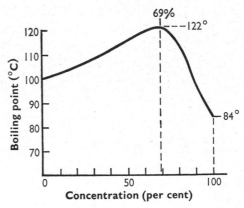

Fig. 28.4. Boiling point curve of aqueous solutions of nitric acid at one atmosphere pressure.

If a solution of nitric acid is boiled, the concentration of the residue gradually approaches a concentration of 69%. This happens regardless of whether a dilute or a very concentrated solution is boiled. The acid of 69% concentration is called a constant boiling or *azeotropic* mixture and this mixture does not change in concentration as it is boiled away.

CHEMICAL PROPERTIES

A dilute solution of nitric acid is a very good electrolytic conductor. Nitric acid is therefore a *strong acid*:

$$HNO_3 + H_2O \rightarrow H_3O^+ + NO_3^-$$

Production of nitro-glycerine.

The Biazzi plant at the Deer Park factory, Melbourne, carries out production of nitroglycerine by continuous blending of nitric acid and glycerine. When installed in 1956 the plant was the world's largest.

It is a *monoprotic acid*, forming only one series of salts, called nitrates.

Nitric acid slowly turns yellow if exposed to sunlight and the acid also turns yellow if boiled. The acid decomposes as follows:

$$4HNO_3 \rightarrow 2H_2O + 4NO_2 + O_2$$

Nitric acid is a *powerful oxidizing agent*. Its oxidation of metals has been discussed in chapter 16, and further oxidizing reactions will be discussed as they arise in subsequent chapters.

Uses of Nitric Acid

Nitric acid is used in the production of nitrates, some of which are useful fertilizers.

A very important industrial use of nitric acid depends on its reaction with certain organic compounds to form *nitro-compounds*. Thus with benzene, C_6H_6, it forms nitrobenzene, $C_6H_5NO_2$, which is an important intermediate in the production of certain dyes. Other nitro-compounds are used in explosives,

for a variety of industrial and military purposes, e.g., nitrocellulose (gun cotton), trinitrotoluene (T.N.T.), nitroglycerine (used in dynamite).

Explosives

A chemical explosive is a substance or a mixture of substances which can suddenly produce a large amount of heat and a large volume of gaseous products, leading to the sudden development of a high pressure. Nitroglycerine, for example, decomposes explosively according to the equation:

$$4C_3H_5(NO_3)_3 \rightarrow 12CO_2 + 10H_2O + 6N_2 + O_2$$

Because of the high temperature of the reaction, the volume of the gaseous products is about 10 000 times greater than the volume of the nitroglycerine itself.

Nitroglycerine is a constituent of many industrial explosives, but the pure substance (a liquid) is so sensitive to shock that it is usually mixed with other substances to moderate its sensitivity.

Dynamite is a general term for explosives based on nitroglycerine. The nitroglycerine is often mixed with sodium or potassium nitrate and a small amount of very fine sawdust (wood meal). The percentage of nitroglycerine is varied from 20% to 75%. For example, gelignite is a dynamite which consists of 60% nitroglycerine, 5% nitrocellulose, 27% potassium nitrate and 8% wood meal. Various dynamites are widely used for industrial blasting as in quarries, mines, demolition and oil search.

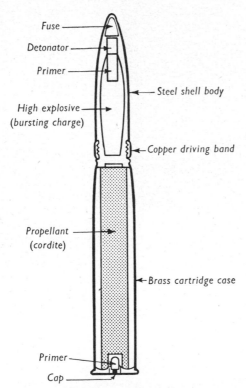

Fuse
Detonator
Primer
Steel shell body
High explosive (bursting charge)
Copper driving band
Propellant (cordite)
Brass cartridge case
Primer
Cap

Section through a 9.4 cm high explosive shell.

Cordite is an explosive which contains 65% nitrocellulose, 30% nitroglycerine and 5% mineral jelly (vaseline). It is used as a propellant for rifle bullets and shells because it undergoes extremely rapid combustion, but does not detonate.

Trinitrotoluene (T.N.T.) is used for filling military shells and bombs. It is able to withstand the shock of the propellant but it explodes with great violence when detonated (see below). It is commonly mixed with ammonium nitrate (up to 80%), which is itself an explosive and is much cheaper than T.N.T.

Most industrial and military explosives need to be initiated by a primer (or detonator). The primer consists of a much more sensitive explosive such as mercury fulminate, $Hg(ONC)_2$ or lead azide, $Pb(N_3)_2$. The primer can be set off by impact or by a fuse or electrically.

● *Under no circumstances should anyone but a skilled operator attempt to make any explosive substance, for the omission of a single precaution could lead to serious accidents.*

NITROUS ACID, HNO$_2$

Nitrous acid is a weak acid (largely un-ionized) and therefore the NO_2^- ion is a moderately strong base. It will accept protons from H_3O^+ ions.

$$H_3O^+ + NO_2^- \rightarrow HNO_2 + H_2O$$

On the addition of dilute sulphuric acid to sodium nitrite, it is observed that the liquid turns blue and effervesces. The blue substance is probably nitrous acid or its anhydride, dinitrogen trioxide, N_2O_3. The gas evolved during the effervescence is a colourless gas which turns brown as it mixes with the air. The colourless gas is nitrogen monoxide formed by decomposition of the nitrous acid:

$$3HNO_2 \rightarrow H_3O^+ + NO_3^- + 2NO$$

NITRATES

Solubility of Nitrates

All the common nitrates are soluble in water.

Thermal Stability of Nitrates

SODIUM NITRATE

If sodium nitrate is heated fairly strongly in an ignition tube, a yellow liquid forms. If this liquid is heated very strongly, there is an effervescence of a colourless gas which re-ignites a glowing string. If the residue is cooled, it sets to a white solid. Addition of dilute sulphuric acid to the residue produces a blue liquid and effervescence of a gas which turns brown in air.

The residue is evidently sodium nitrite, $NaNO_2$, and because this contains less oxygen than sodium nitrate, $NaNO_3$, the gas evolved was oxygen, rather than dinitrogen monoxide (N_2O) which also re-ignites a glowing string. The liquid first formed was molten sodium nitrate. The reaction can be represented by the equation:

$$2NaNO_3 \rightarrow 2NaNO_2 + O_2$$

POTASSIUM NITRATE

The effect of heat on potassium nitrate is similar:

$$2KNO_3 \rightarrow 2KNO_2 + O_2$$

ZINC NITRATE

If hydrated zinc nitrate, $Zn(NO_3)_2,6H_2O$, is heated gently in an ignition tube, a clear liquid is produced. On stronger heating, clouds of white fumes are evolved. With continued heating a yellow solid appears in the tube and a dense brown gas is evolved. A glowing string placed in the mouth of the ignition tube bursts into flame. The yellow residue in the tube turns white on cooling and is soluble in hot dilute hydrochloric acid without effervescence, producing a colourless solution.

The clear liquid formed on gentle heating is evidently zinc nitrate dissolved in its water of crystallization, and the white fumes evolved later are steam. The brown gas is nitrogen dioxide and the glowing string test suggests the presence of oxygen. The residue is zinc oxide. The reactions which occur can be represented by the equations:

$$Zn(NO_3)_2,6H_2O \rightarrow Zn(NO_3)_2 + 6H_2O$$
$$2Zn(NO_3)_2 \rightarrow 2ZnO + 4NO_2 + O_2$$

NITRATES OF MAGNESIUM, COPPER AND LEAD

The effect of heat on magnesium nitrate, $Mg(NO_3)_2,6H_2O$, and on copper(II) nitrate, $Cu(NO_3)_2,3H_2O$, is similar to the effect of heat on zinc nitrate. The decomposition of lead nitrate, $Pb(NO_3)_2$, is also similar except that this salt does not contain water of crystallization.

NITRATES OF SILVER AND MERCURY

The effect of heat on the nitrates of silver and mercury is to produce the metal, although silver nitrate produces silver nitrite on moderate heating and must be heated very strongly to produce the metal:

$$2AgNO_3 \rightarrow 2Ag + 2NO_2 + O_2$$

SUMMARY OF THE EFFECT OF HEAT ON METALLIC NITRATES

Nitrates of metals high on the activity series decompose into the corresponding nitrites and oxygen. Nitrates of metals from magnesium to copper on the activity series decompose into a metal oxide, nitrogen dioxide and oxygen. Nitrates of the metals mercury and silver decompose on *strong* heating to form the metal, nitrogen dioxide and oxygen.

AMMONIUM NITRATE

If ammonium nitrate is heated it first forms a clear liquid. On stronger heating, an effervescence occurs and clouds of white fumes are evolved. A glowing string held at the mouth of the tube may burst into flame. When only a small amount of solid is left it explodes if the heating is continued.

Ammonium nitrate decomposes at moderate temperatures into water and dinitrogen monoxide:

$$NH_4NO_3 \rightarrow 2H_2O + N_2O$$

At higher temperatures ammonium nitrate decomposes explosively. Thus, if ammonium nitrate is used to prepare dinitrogen monoxide, the ammonium nitrate must not be heated strongly. Moreover, the heating should be stopped when about two-thirds of the solid has decomposed.

NITROGEN DIOXIDE, NO₂

Preparation

Nitrogen dioxide can be prepared by the reaction of concentrated nitric acid with copper. It is collected by upward displacement of air (see fig. 28.5).

The equation for the reaction is:

$$Cu \rightarrow Cu^{2+} + 2e^-$$
$$[NO_3^- + 2H^+ + e^- \rightarrow NO_2 + H_2O] \times 2$$
$$\overline{Cu + 2NO_3^- + 4H^+ \rightarrow Cu^{2+} + 2NO_2 + 2H_2O}$$

Physical Properties of Nitrogen Dioxide

Nitrogen dioxide is a red-brown gas with a pungent

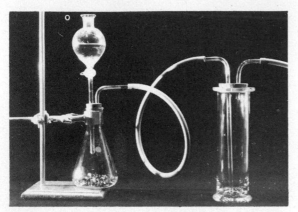

Fig. 28.5. Preparation and collection of nitrogen dioxide: concentrated nitric acid is added to copper.

penetrating odour. It is *very poisonous* and attacks the tissues of the throat and lungs.

It can be seen during the collection of the gas that nitrogen dioxide is much denser than air.

It is readily soluble in water and, therefore, cannot be collected over water.

Nitrogen dioxide is easily liquefied and solidified. Its boiling point is 22 °C and its freezing point is −10 °C. The reaction of copper with concentrated nitric acid, produces a mixture of nitrogen dioxide and a little nitrogen monoxide. Pure liquid nitrogen dioxide is therefore best prepared from lead nitrate.

LIQUEFYING THE BROWN GAS

Lead nitrate does not contain water of crystallization but it may be damp, and so the crystals should be crushed to a powder and then heated gently to drive off any water present. The lead nitrate is then

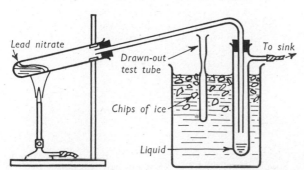

Fig. 28.6. Preparation of liquid nitrogen dioxide.

heated as shown in fig. 28.6. The reaction which occurs is:

$$2Pb(NO_3)_2 \rightarrow 2PbO + 4NO_2 + O_2$$

The liquid nitrogen dioxide may be poured into the cold drawn-out test tube shown, and the tube can be quickly sealed at the constriction. The nitrogen dioxide remains in the liquid state as it warms up to room temperature.

Chemical Properties

STABILITY OF NITROGEN DIOXIDE

It will be observed during the liquefaction of nitrogen dioxide that the colour of the gas fades as it is cooled, and that it condenses to a pale yellow liquid. On further cooling it forms an almost colourless solid.

The converse can be observed if the gas is heated.

Fig. 28.7. A flask fitted with a cork and a glass tube drawn to a fine jet which can be used to show the effect of heat on a mixture of N_2O_4 and NO_2.

Thus, if the flask shown in fig. 28.7 is filled with nitrogen dioxide and heated, the colour of the gas deepens to a dark brown. The colour becomes paler if the gas is allowed to cool.

Determinations of the density of the gas at different temperatures show that at 154 °C, the molecular mass of the gas corresponds to the formula NO_2, but as the gas is cooled it contains increasing proportions of N_2O_4. At −10 °C, the substance is a solid, and it consists entirely of N_2O_4—see fig. 28.8.

The forward change is known as *dissociation* and the reverse reaction is known as *association* or *polymerization*.

If nitrogen dioxide is heated *strongly*, the dark brown colour which first forms becomes paler, and a colourless gas is produced. The gas turns brown again

$$N_2O_4 \xrightleftharpoons[\substack{\text{cool} \\ \text{(complete at } -10\,°C)}]{\substack{\text{heat} \\ \text{(complete at } 154\,°C)}} 2NO_2$$

dinitrogen tetroxide   nitrogen dioxide
(colourless)   (dark brown)

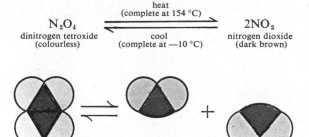

Fig. 28.8. Dissociation of dinitrogen tetroxide.

on cooling. At temperatures above 160 °C approximately, nitrogen dioxide decomposes into nitrogen monoxide and oxygen:

$$2NO_2 \underset{\text{cool}}{\overset{\text{heat}}{\rightleftharpoons}} 2NO + O_2$$

COMBUSTION IN NITROGEN DIOXIDE

If a feebly burning taper is introduced into a gas jar of nitrogen dioxide, the gas does not ignite and the taper goes out. However, a strongly burning taper continues to burn brightly in the gas. The product of reduction of the nitrogen dioxide is a colourless gas which turns brown in air, nitrogen monoxide. But strongly burning magnesium reduces nitrogen dioxide to nitrogen:

$$4Mg + 2NO_2 \rightarrow 4MgO + N_2$$

Combustion in nitrogen dioxide indicates that it is a strong oxidizer.

ACIDIC PROPERTIES OF NITROGEN DIOXIDE

A solution of nitrogen dioxide in water is acidic to litmus. A mixture of nitric acid and nitrous acid is formed:

$$2NO_2 + H_2O \rightarrow H^+ + NO_3^- + HNO_2$$

Nitrogen dioxide is called a *mixed acid anhydride* because it forms a mixture of acids with water. The reaction with caustic alkalis produces a mixture of nitrate and nitrite ions:

$$2NO_2 + 2OH^- \rightarrow NO_3^- + NO_2^- \, H_2O$$

TEST FOR NITROGEN DIOXIDE

Nitrogen dioxide is a pungent red-brown gas which turns litmus red and has a sweet smell when very dilute.

AMMONIA, NH₃

Ammonia is obtained commercially as a useful by-product in the destructive distillation of coal. It is also produced on a very large scale from nitrogen and hydrogen.

Laboratory Preparation and Collection of Ammonia

Ammonium chloride, NH_4Cl, slaked lime, $Ca(OH)_2$, and water are mixed to a paste and heated gently. The NH_4^+ ion is an acid and it donates a proton to the base OH^-:

$$NH_4^+ + OH^- \rightleftharpoons NH_3 \, (g) + H_2O$$

The heating removes the ammonia, and the reaction proceeds to the right.

The ammonia gas is dried with quicklime and collected by downward displacement of air—see fig. 28.9.

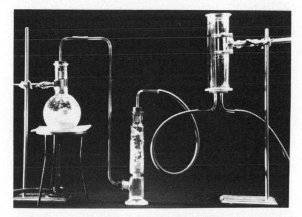

Fig. 28.9. Laboratory preparation and collection of ammonia: a paste of ammonium chloride, calcium hydroxide and water is heated and the gas is dried over calcium oxide.

Physical Properties of Ammonia

Ammonia is a colourless gas which fumes in moist air. It has a pungent choking odour and is poisonous. The gas is about half as dense as air (the molecular mass of NH_3 is about 17).

If a jar of ammonia is inverted into water and the cover is removed from the jar, the water surges violently into the jar. Ammonia is exceedingly soluble in water, and this fact can also be demonstrated by

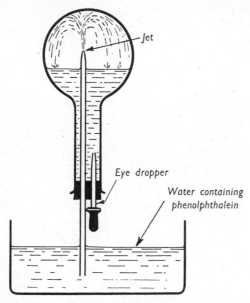

Fig. 28.10. The fountain experiment.

the so-called *fountain experiment*—see fig. 28.10. In this experiment the flask is filled with ammonia gas and the end of the glass tubing is placed in water to which an indicator such as phenolphthalein may be added. A few drops of water are then squirted into the flask using the eye dropper, in order to start the fountain. As the ammonia dissolves, the pressure inside the flask is reduced and more water enters the flask. The phenolphthalein turns red, showing that the solution is alkaline. Under ordinary room conditions, one volume of water dissolves about 700 volumes of ammonia.

Ammonia is easily liquefied, its boiling point being −33 °C at one atmosphere pressure.

Chemical Properties

STABILITY

Ammonia does not decompose at room temperatures, but at moderate temperatures it forms nitrogen and hydrogen:

$$2NH_3 \rightarrow N_2 + 3H_2$$

COMBUSTION

If a burning taper is thrust into a jar of ammonia the taper goes out and the gas does not burn. However,

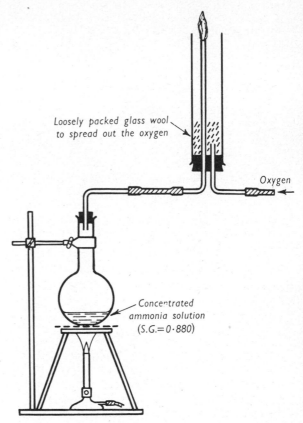

Fig. 28.11. Combustion of ammonia in oxygen.

if the taper is slowly lowered into ammonia a brief flash of brown flame can usually be seen. This suggests that the ammonia-air mixture at the top of the jar burns briefly. Sustained combustion of ammonia in a richer supply of oxygen can be illustrated by using the apparatus shown in fig. 28.11.

The source of ammonia is a saturated aqueous solution which has a relative density of 0.880 and is often referred to as 0.880 ammonia. Very gentle heating of this solution causes evolution of ammonia. When a litmus test shows that ammonia is escaping from the jet, the ammonia can be lit and it burns in the oxygen with a yellow-brown flame:

$$4NH_3 + 3O_2 \rightarrow 2N_2 + 6H_2O$$

NATURE OF SOLUTION

A solution of ammonia in water is alkaline to litmus.

The solution is a poor conductor of electricity and hence contains only a relatively few ions—see fig. 28.12.

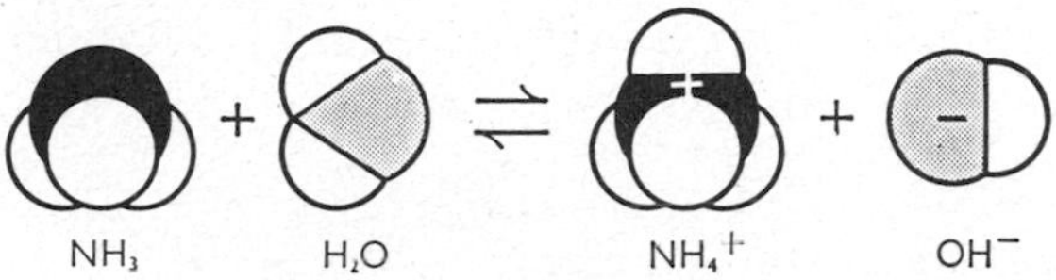

$$NH_3 + H_2O \rightleftharpoons NH_4^+ + OH^-$$

Fig. 28.12. Reaction of ammonia and water.

This equilibrium can be shifted to the right by removing the OH^- ions from the solution (by the addition of an acid) or the equilibrium can be shifted to the left by removing NH_3 from the system (by boiling). Other reactions to be discussed later can also shift the equilibrium to the right or to the left, and the solution can therefore function either as an alkali or as a solution of NH_3 molecules.

REACTION WITH METALS

If ammonia is passed over hot magnesium, the magnesium displaces the hydrogen and forms magnesium nitride:

$$3Mg + 2NH_3 \rightarrow Mg_3N_2 + 3H_2$$

Hot sodium metal, however, forms sodamide:

$$2Na + 2NH_3 \rightarrow 2NaNH_2 + H_2$$

OXIDATION OF AMMONIA

If ammonia gas is passed over hot copper(II) oxide, the copper(II) oxide is reduced to copper and the ammonia is oxidized to produce water and nitrogen:

$$3CuO + 2NH_3 \rightarrow 3Cu + 3H_2O + N_2$$

Strong oxidizers will oxidize ammonia in *aqueous solutions*. Thus reaction of chlorine gas and ammonia solution also produces nitrogen:

$$3Cl_2 + 2NH_3 \rightarrow N_2 + 6H^+ + 6Cl^-$$

This is a simplified equation in that it shows hydrogen ions being produced. A better representation of the reaction would show the formation of ammonium ions:

$$3Cl_2 + 8NH_3 \rightarrow N_2 + 6NH_4^+ + 6Cl^-$$

If chlorine is used in excess, an explosive oil, nitrogen trichloride, is formed:

$$3Cl_2 + NH_3 \rightarrow NCl_3 + 3H^+ + 3Cl^-$$
$$or \quad 3Cl_2 + 4NH_3 \rightarrow NCl_3 + 3NH_4^+ + 3Cl^-$$

Some Reactions of Ammonia Solution

If ammonia solution is added to solutions of salts containing cations of metals from magnesium to copper on the activity series, precipitates of the hydroxides form, e.g.:

$$Fe^{3+} + 3OH^- \rightarrow Fe(OH)_3 \ (s)$$
$$Mg^{2+} + 2OH^- \rightarrow Mg(OH)_2 \ (s)$$

In these reactions, hydroxide ions are removed from the equilibrium:

$$NH_3 + H_2O \rightleftharpoons NH_4^+ + OH^-$$

The equilibrium therefore shifts to the right and *the ammonia molecules* react with the water molecules to form more NH_4^+ and OH^- ions. If the OH^- ions are continually removed, the equilibrium continually shifts to the right. In these reactions the ammonia solution functions as a source of hydroxide ions.

However, excess of ammonia solution will dissolve some hydroxides, such as copper(II) hydroxide and zinc hydroxide. The solutions formed can be shown to contain tetramminecopper(II) ions, $Cu(NH_3)_4^{2+}$, and tetramminezinc(II) ions, $Zn(NH_3)_4^{2+}$, respectively. The dissolution of the hydroxides can be represented:

$$Cu(OH)_2 + 4NH_3 \rightarrow Cu(NH_3)_4^{2+} + 2OH^-$$
$$Zn(OH)_2 + 4NH_3 \rightarrow Zn(NH_3)_4^{2+} + 2OH^-$$

In these reactions, the ammonia solution functions as a *solution of ammonia molecules*.

Ammonia solution will also dissolve precipitates of some silver salts such as silver chloride, with the formation of diamminesilver(I) ions, $Ag(NH_3)_2^+$:

$$AgCl + 2NH_3 \rightarrow Ag(NH_3)_2^+ + Cl^-$$

It is not necessary for the metal ions to be present in the form of a precipitate, in order to form complex ammines. Thus if ammonia solution is added to silver nitrate solution, the silver ions are converted to diamminesilver(I) ions without the formation of a precipitate, although very careful addition of the ammonia can result in the intermediate formation of silver oxide.

Tests for Ammonia

Ammonia is a colourless gas which is easily detected by its distinctive smell. It is the *only* common gas which turns litmus blue and it forms white clouds with hydrogen chloride gas:

$$NH_3 + HCl \rightarrow NH_4Cl$$

Uses of Ammonia

Large quantities of ammonia are converted to *ammonium sulphate* for use as a fertilizer. Ammonia is also converted to *nitric acid* on a very large scale.

Ammonia is easily liquefied by pressure alone and liquid ammonia has a large latent heat of vapourization. These two properties make ammonia a suitable *refrigerant* in large scale refrigerating plants and ice works.

An important chemical produced from ammonia is urea, NH_2CONH_2. Urea is produced by first reacting ammonia with carbon dioxide to form ammonium carbamate:

$$2NH_3 + CO_2 \rightarrow NH_2COO^-NH_4^+$$

This is heated under pressure to produce urea:

$$NH_2COO^-NH_4^+ \rightarrow NH_2CONH_2 + H_2O$$

Urea is being used increasingly as a fertilizer and it is an important intermediate for making certain plastics.

Ammonia solution is used in some *cleaning fluids*. Its value lies in its ability to remove grease, e.g. from glass or painted woodwork.

Ammonium Salts

Ammonium salts are commonly prepared by absorbing ammonia gas in the appropriate acid, e.g.:

$$NH_3 + HCl \quad \rightarrow NH_4Cl$$
$$NH_3 + HNO_3 \rightarrow NH_4NO_3$$
$$2NH_3 + H_2SO_4 \rightarrow (NH_4)_2SO_4$$

Ammonium carbonate is usually prepared by heating a dry mixture of ammonium sulphate and calcium carbonate:

$$(NH_4)_2SO_4 + CaCO_3 \rightarrow (NH_4)_2CO_3 + CaSO_4$$

The heating causes the ammonium carbonate to sublime, but the product is not pure. It contains some ammonium hydrogen carbonate, NH_4HCO_3, and ammonium carbamate, NH_2COONH_4.

Ammonium chloride, sulphate, nitrate and carbonate are white crystalline salts and they are all readily soluble in water. They all contain the univalent ammonium ion NH_4^+ and react with alkalis, like all ammonium salts, to evolve ammonia:

$$NH_4^+ + OH^- \rightarrow NH_3\ (g) + H_2O$$

Effect of Heat on Ammonium Salts

AMMONIUM CHLORIDE

If ammonium chloride is heated in an ignition tube and moist litmus paper is held at the mouth of the tube, the litmus paper usually first turns blue and then turns red. At the same time a white smoke is produced and a white deposit forms on the cool part of the tube.

The white solid forming on the cool part of the tube can be shown to be ammonium chloride and the litmus test suggests that some ammonia and hydrogen chloride are evolved. The ammonium chloride decomposes on heating to form ammonia gas and hydrogen chloride gas, and these gases largely recombine on cooling:

$$NH_4Cl \overset{\text{heat}}{\underset{\text{cool}}{\rightleftharpoons}} NH_3 + HCl$$

Thus ammonium chloride sublimes on heating.

AMMONIUM SULPHATE

On moderate heating, ammonium sulphate melts and evolves ammonia. The residue is a mixture of ammonium hydrogen sulphate and unchanged ammonium sulphate:

$$(NH_4)_2SO_4 \rightarrow NH_3\ (g) + NH_4HSO_4$$

On very strong heating, further reaction occurs but the change is a complex one. Sulphur dioxide can be detected among the products.

AMMONIUM CARBONATE

This salt loses ammonia on standing in air and forms the hydrogen carbonate:

$$(NH_4)_2CO_3 \rightarrow NH_3\ (g) + NH_4HCO_3$$

On quite gentle heating, ammonium carbonate decomposes completely:

$$(NH_4)_2CO_3 \rightarrow 2NH_3\ (g) + H_2O\ (g) + CO_2\ (g)$$

AMMONIUM NITRATE and AMMONIUM NITRITE

The thermal decomposition of ammonium nitrate has been discussed earlier in this chapter under 'thermal stability of nitrates':

$$NH_4NO_3 \rightarrow N_2O + 2H_2O$$

The effect of heat on ammonium nitrite was also discussed earlier in this chapter under 'preparation of nitrogen'. For convenience, a solution containing ammonium ions and nitrite ions was used:

$$NH_4^+ + NO_2^- \rightarrow N_2 + 2H_2O$$

Uses of Ammonium Salts

Ammonium sulphate is a widely used fertilizer. It provides plants with a source of soluble nitrogen compounds and, in addition, the plants benefit from the presence of the soluble *sulphur* compound. Both nitrogen and sulphur are used by plants in building up proteins.

Ammonium nitrate would appear to be a very useful fertilizer, but it is very soluble in water and is rapidly washed down into the sub-soil. It is widely used in explosives.

Ammonium chloride, or sal-ammoniac, is used in dry cells and as a flux in soldering. Its use as a flux depends on its reaction with oxide layers on the metal to be soldered. The oxides are converted to volatile chlorides and hence the metal surface is cleaned.

Ammonium carbonate, or sal-volatile, is used in smelling salts because it slowly evolves ammonia at room temperatures.

Hydrazine

Hydrazine, N_2H_4, is an important compound of nitrogen and hydrogen. It is prepared by oxidizing ammonia or urea with sodium hypochlorite, $NaOCl$, in the presence of a little glue, which helps to prevent side reactions.

Hydrazine is a colourless fuming liquid and it is dangerous to handle because it burns readily in air and reacts vigorously with many substances. Hydrazine and some of its derivatives are used as rocket fuels.

Hydrazine   Dimethylhydrazine

Commercial Production of Ammonia— The Haber Process

Most ammonia produced in the world at the present time is synthesized from nitrogen and hydrogen. The source of nitrogen is the air and the source of hydrogen is water. Coke is the other raw material, and it is used to remove the oxygen from the air and from steam.

A mixture of steam and air is blown through red hot coke. The reactions which occur are mainly:

$$C + H_2O \rightarrow CO + H_2$$
$$C + O_2 \rightarrow CO_2$$
$$CO_2 + C \rightarrow 2CO$$

The gas mixture which is produced consists mainly of carbon monoxide, hydrogen and nitrogen (from the air).

In order to remove the carbon monoxide from the gas, it is mixed with steam and passed over an iron catalyst at 500 °C. The carbon monoxide is oxidized to carbon dioxide and more hydrogen is produced:

$$H_2O + CO \rightarrow CO_2 + H_2$$

The carbon dioxide so formed is removed from the other gases by dissolving it in water under pressure.

The gas mixture is now mainly nitrogen and hydrogen in the required proportion. The gas is compressed to a high pressure, about 350 atmosphere being typical, and before going to the catalyst converter the last traces of carbon monoxide and carbon dioxide are absorbed by special absorbers. The compressed gases are passed over a catalyst at a temperature of about 550 °C. The catalyst is iron granules containing small amounts of promoters such as aluminium oxide, and potassium oxide. These increase the efficiency of the catalyst. The reaction occurring in the catalyst chamber is:

$$N_2 + 3H_2 \rightleftharpoons 2NH_3$$

About 20% conversion occurs with each passage of the gases over the catalyst. The ammonia is removed from the gases by cooling as this causes the ammonia to liquefy under the high pressure. The unconverted gases are mixed with more nitrogen and hydrogen, and are recirculated through the catalyst chamber.

After some time, noble gases from the air accumulate in the unconverted gases and the gas mixture is then discarded instead of being recirculated to the catalyst chamber.

Commercial Production of Nitric Acid from Ammonia—The Ostwald Process

In this process, ammonia gas is oxidized by air in the presence of a catalyst to nitrogen monoxide. The nitrogen monoxide is ultimately converted to nitric acid.

Hot ammonia gas and air are passed downwards through a fine wire gauze of platinum maintained at 1000 °C. The ammonia is oxidized:

$$4NH_3 + 5O_2 \rightarrow 4NO + 6H_2O$$

At first the platinum gauze is heated electrically, but the oxidation of ammonia is exothermic and, after a short time, the heat of the reaction is sufficient to maintain the catalyst at 1000 °C.

The hot gases produced consist of nitrogen monoxide, steam and nitrogen from the air. They are cooled and mixed with air, causing oxidation of the nitrogen monoxide to nitrogen dioxide:

$$2NO + O_2 \rightarrow 2NO_2$$

The gases are then passed up towers fed with water or dilute nitric acid:

$$4NO_2 + O_2 + 2H_2O \rightarrow 4HNO_3$$

The nitric acid produced is about 50% HNO_3 and is suitable for some purposes such as production of sodium nitrate. If a more concentrated acid is required, the nitric acid can be mixed with concentrated sulphuric acid and distilled. The sulphuric acid retains much of the water and 95% nitric acid is produced.

The oxidation of ammonia by air using a platinum catalyst can be illustrated by using the apparatus shown in fig. 28.13. A platinum wire, preferably in the form of a spiral, is made red hot in the Bunsen flame, and is then quickly lowered into the ammonia-air mixture. Despite the fact that the platinum cools down while being transferred to the flask, it becomes red hot when placed in the gas mixture. The platinum will remain red-hot for a considerable time and brown fumes appear in the flask. If oxygen is used instead of air the platinum becomes white hot and explodes the ammonia-oxygen mixture.

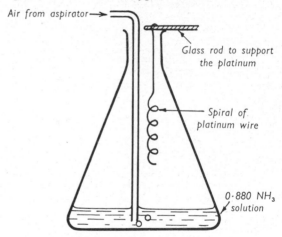

Fig. 28.13. Oxidation of ammonia by air using platinum as a catalyst.

SUMMARY OF VALENCIES ON NITROGEN

The compounds of nitrogen discussed in this chapter show that nitrogen can have an electrovalency of -3, e.g. Mg_3N_2. Most of the compounds of nitrogen are covalent and the usual covalency is 3, e.g. NH_3. However, nitrogen can show a wide range of covalencies as illustrated by its oxides and related compounds.

COVALENCY OF NITROGEN	OXIDE	ACID	SODIUM SALT
1	N_2O (dinitrogen monoxide)	$H_2N_2O_2$ (hyponitrous acid)	
2	NO (nitrogen monoxide)		
3	N_2O_3 (dinitrogen trioxide)	HNO_2 (nitrous acid)	$NaNO_2$ (sodium nitrite)
4	NO_2 (nitrogen dioxide) and N_2O_4 (dinitrogen tetroxide)		
5	N_2O_5 (dinitrogen pentoxide)	HNO_3 (nitric acid)	$NaNO_3$ (sodium nitrate)

STUDY QUESTIONS

1. Write equations for the reactions between the following pairs of substances:
(a) concentrated sulphuric acid and sodium nitrate;
(b) concentrated nitric acid and copper;
(c) dilute nitric acid (35%) and silver.

2. Explain what is meant by saying that nitric acid forms an azeotropic mixture with water.

3. Write equations for the effect of heat on:
(a) sodium nitrate (anhydrous);
(b) magnesium nitrate hexahydrate;
(c) copper(II) nitrate trihydrate;
(d) ammonium nitrate (anhydrous).

4. If copper is treated with concentrated nitric acid and the gas produced is liquefied, a blue-green liquid which contains N_2O_3 is produced.
What explanation can be offered for the production of the blue-green liquid?

5. Write equations to illustrate the reactions which occur if:
(a) nitrogen dioxide is cooled;
(b) burning magnesium is lowered into nitrogen dioxide;
(c) nitrogen dioxide is dissolved in cold water;
(d) nitrogen dioxide is dissolved in caustic soda solution.

6. When lead nitrate is heated, nitrogen dioxide and oxygen are produced as gases.
Devise an experiment by which you could collect the oxygen and established that it is not dinitrogen monoxide.

7. When dilute sulphuric acid is added to sodium nitrite an acid-base reaction occurs. Write an equation to emphasize the acid-base nature of the reaction and label the acids and bases involved.

8. Write equations for the reactions which occur when the following pairs of substances are heated:
(a) caustic soda solution and ammonium sulphate;
(b) slaked lime and ammonium nitrate.

9. Write equations to illustrate the reactions of ammonia gas with:
(a) water;
(b) hot copper (II) oxide;
(c) hydrogen chloride gas;
(d) hot magnesium;
(e) nitric acid;
(f) hot sodium metal.

10. Write equations for the effect of moderate heat on:
(a) ammonium chloride;
(b) ammonium sulphate;
(c) ammonium carbonate;
(d) ammonium nitrate.
What generalizations can be made concerning the effect of heat on ammonium salts?

11. When dinitrogen tetroxide is heated gently, the bond between the two nitrogen atoms is broken, and nitrogen dioxide forms.
(a) What information does this give concerning the strength of the bond between the two nitrogen atoms?
(b) How can you account for the bond re-forming when the substance cools?

12. The following table gives the percentage of ammonia in equilibrium, in a mixture of 3 parts of hydrogen to one part of nitrogen by volume:

Temp. °C	Percentage Partial Pressure of Ammonia at Equilibrium		
	at 1 atm	at 100 atm	at 1000 atm
300	2.18	52.1	92.6
500	0.129	10.4	57.5
700	0.022 3	2.14	12.9
900	0.006 9	0.68	—

Using the data in the table, discuss the factors which will determine the choice of conditions for commercial production of ammonia from hydrogen and nitrogen.

13. A mixture of fuming nitric acid and concentrated sulphuric acid can be used to nitrate organic compounds, that is, it supplies the nitro group ($-NO_2$). It is thought that this is due to the ionization of HNO_3 in H_2SO_4 to form NO_2^+ ions. Write an equation to represent this ionization assuming that the H_2SO_4 reacts to form HSO_4^- ions.

14. Calculate the mass of sulphuric acid required to convert 100 g of potassium nitrate to potassium hydrogen sulphate and nitric acid.
(K = 39.1, N = 14.0, O = 16.0)

15. Calculate the density of ammonia gas at s.t.p. in gram per dm³.
(N = 14.0, H = 1.0.)

16. Air and water can be used as a source of ammonia. Air and ammonia can be used to make nitric acid. Ammonia and nitric acid can be used to make ammonium nitrate and then this salt can be used to produce dinitrogen monoxide. Write chemical equations for all reactions involved in this sequence of chemical changes.

Answers to Numerical Problems

14. 97 g.
15. 0.76 gram per dm³.

29 : Phosphorus and its Compounds

Valency of Phosphorus

Phosphorus is in the same group of the periodic table as nitrogen. Both elements have atoms with five outer shell electrons. However, there is not much similarity between the two elements. Nitrogen molecules are N_2 but molecules of phosphorus are P_4 or P_∞ (structure discussed below). Nitrogen has a covalency of three but phosphorus may have covalencies of three or five, e.g. in PF_3 and PF_5.

phosphorus   phosphorus
trifluoride   pentafluoride

In PF_3, the phosphorus atom shares 3 electrons and has one unshared pair of electrons but in PF_5, the phosphorus atom shares all 5 electrons and then has no unshared electrons.

Occurrence of Phosphorus

Phosphorus is available from mineral deposits of calcium orthophosphate, $Ca_3(PO_4)_2$, and also smaller quantities are available from organic sources such as bone ash. Complex phosphorus compounds are found in all living things, and so phosphorus is continually removed from soils, particularly where commercial crops are grown. Because of this it is necessary to add phosphorus compounds in a form available to plants.

Australian sources of phosphate are Nauru and Ocean Island in the Pacific Ocean and Christmas Island in the Indian Ocean.

Preparation of Phosphorus

A mixture of calcium phosphate, silica and coke is heated in an electric furnace. The overall reaction probably takes place in two stages, forming calcium silicate and phosphorus pentoxide, followed by reduction of the oxide by the coke. This is represented by the equations, using empirical formulae:

$$Ca_3(PO_4)_2 + 3SiO_2 \rightarrow 3CaSiO_3 + P_2O_5$$
$$P_2O_5 + 5C \rightarrow 2P + 5CO$$

Alternatively, one equation can be written to represent the overall change:

$$Ca_3(PO_4)_2 + 3SiO_2 + 5C$$
$$\rightarrow 3CaSiO_3 + 2P + 5CO$$

The calcium silicate is called 'slag' and is tapped from the bottom of the furnace, while phosphorus vapour and carbon monoxide are led off from near the top.

Fig. 29.1. Extraction of phosphate, Nauru.

The phosphate is found as a deposit among pinnacles of coral. It is removed by a mechanical grab, leaving the coral exposed. The denuded coral is, of course, quite useless for agriculture, and rehabilitation of the worked sections would be an extremely expensive proposition. By the time working is completed, virtually the whole island will be denuded in this way.

The phosphorus vapour is condensed under water and the carbon monoxide is used as a fuel. The crude molten phosphorus is filtered, washed under water and then cast into sticks which are stored under water. The solid obtained in this way is pale yellow in colour. It is called white or yellow phosphorus.

This can be converted into a red form by heating it in the absence of air. At about 250 °C in the presence of substances which act as 'initiators' (i.e. they start the reaction), the yellow phosphorus changes to red phosphorus very slowly, the complete change taking about a week. Iodine is often used as an initiator.

Forms of Phosphorus

Phosphorus Vapour: Molecular mass determinations of the vapour indicate that above 800 °C, phosphorus exists as diatomic molecules, P_2. Below 800 °C the molecular mass of the vapour indicates tetratomic molecules, P_4. The P_4 molecule is tetrahedral in shape (fig. 29.2).

White or yellow phosphorus is a fairly soft solid which is slightly soluble in water but readily soluble in organic solvents such as carbon disulphide. Sticks of white phosphorus are usually coated with a thin layer of the red form, due to the effect of light which initiates the change. If white phosphorus is exposed to air it emits a pale glow called 'phosphorescence' (this can only be seen in a darkened room) and eventually bursts into flame.

SPECIAL NOTE: Because it is spontaneously inflammable in air and also because it is poisonous, **white phosphorus must always be stored under water.** *If a piece of white phosphorus is required, a stick should be removed from storage by a pair of tongs.* **Never allow it to come in contact with the skin.** *The stick may be cut under water with a knife and then returned to the storage container.* **It is very dangerous to cut phosphorus in the air** *as the friction of the knife blade may ignite the phosphorus.*

Red phosphorus is a purplish-red amorphous powder which is insoluble in water and in organic solvents. It is much less reactive than white phosphorus and *can be safely stored in contact with the air.* It becomes coated with a viscous film of a syrupy liquid which is

Fig. 29.2. Models representing the structure of a P_4 molecule showing: (a) the atoms as spheres in contact (left); (b) the relative positions of the centres of the atoms (right).

phosphoric acid, formed by oxidation of the phosphorus and subsequent absorption of water from the atmosphere.

Further comparisons of the properties of white and red phosphorus are given in the table fig. 29.3.

PROPERTY	WHITE PHOSPHORUS	RED PHOSPHORUS
Odour	*Like garlic*	*Odourless*
Melting point	44 °C	*About* 600 °C
Density (g cm^{-3})	1.83	2.05—2.39
Solubility in water	*Slightly soluble*	*Insoluble*
Solubility in CS_2	*Soluble*	*Insoluble*
Ignition temperature	*About* 50 °C	*About* 260 °C
In air	*Phosphoresces and may ignite spontaneously*	*Does not phosphoresce*
Summing up	*Very reactive and very poisonous*	*Not so reactive and not poisonous*

Fig. 29.3. Properties of white and red phosphorus.

Structure of Phosphorus Molecules

There is some doubt as to how many allotropes of phosphorus exist, although solid phosphorus has been obtained in at least three different allotropic forms known as white, violet and black phosphorus. X-ray experiments indicate that the white or yellow form described above is composed of tetrahedral P_4 molecules identical to those found in phosphorus vapour. When white phosphorus is changed to red phosphorus, bonds in these P_4 molecules are broken and the phosphorus atoms link together either in a haphazard way or in a regular pattern forming giant molecules. It seems clear that white phosphorus is an allotropic form although red phosphorus probably contains a mixture of allotropes. Black phosphorus is obtained from white phosphorus by the effect of very high pressures.

The effect of iodine as an initiator in converting white phosphorus to red phosphorus is to form bonds with the phosphorus atoms in the tetrahedral molecules, thus breaking bonds between phosphorus atoms. The iodine is chemically combined in the product and therefore is not acting as a catalyst because it is consumed during the reaction.

Uses of Phosphorus

Phosphorus is used mainly to manufacture soluble compounds for use as *fertilizers*. Ammonium phosphate is a particularly useful salt as it supplies both nitrogen and phosphorus to the soil.

Small amounts of phosphorus are used in making special bronze alloys, in the preparation of rat-poisons and in anti-personnel and smoke bombs for military purposes.

Matches were once made from white phosphorus, but because of its poisonous nature, this is now prohibited.

Safety matches do not contain phosphorus. The head contains readily combustible materials such as antimony sulphide, Sb_2S_3, and sulphur, together with an oxidizer. The striking surface on the box contains red phosphorus and ground glass. When the match is struck against the prepared surface, the phosphorus is heated by friction and burns in the oxidizing mixture of the match head. This ignites the match head which then ignites the rest of the match.

Match-making machinery.

Newly cut match sticks are loaded into frames on a continuous belt, and pass through a shallow bath of the tipping material (above, right). After leaving the bath (above, left), the freshly tipped matches pass slowly over rollers (below) so designed that they dry in the atmosphere before being released from the frames and packed into boxes.

Oxides of Phosphorus

Phosphorus forms a number of oxides but only two will be discussed. The empirical formulae of these two oxides are P_2O_5 and P_2O_3 and hence they are called phosphorus pentoxide and phosphorus trioxide respectively. Molecular mass determinations indicate that the molecular formulae of these oxides in the vapour state are P_4O_{10} and P_4O_6 respectively. However, the names derived from the empirical formulae are still commonly used in referring to these oxides. The structure of the molecules of the pentoxide and trioxide are shown in fig. 29.4.

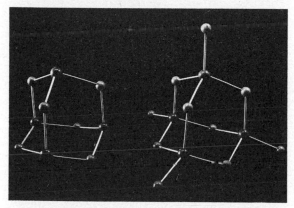

Fig. 29.4. Models representing the structures of the molecules P_4O_6 (left) and P_4O_{10} (right). The dark spheres represent the centres of phosphorus atoms and the white spheres represent the centres of oxygen atoms.

Phosphorus Pentoxide, P

PREPARATION

Phosphorus pentoxide is usually prepared by burning the phosphorus vapour in air as it leaves the electric furnace. The carbon monoxide is also burnt at the same time and the heat obtained from these reactions is used to generate steam to provide energy to operate some of the ancillary equipment. A plentiful supply of air must be used or P_4O_6 may be formed instead of P_4O_{10}:

$$2P_2 + 5O_2 \rightarrow P_4O_{10}$$

PROPERTIES

Phosphorus pentoxide is a white powder which reacts vigorously with water and because of this it is often used as a drying agent for gases.

With cold water, P_4O_{10} forms a white gelatinous solid which can be identified as metaphosphoric acid, HPO_3:

$$2H_2O + P_4O_{10} \rightarrow 4HPO_3$$

With hot water, a colourless solution of ortho-phosphoric acid is formed:

$$6H_2O + P_4O_{10} \rightarrow 4H_3PO_4$$

Orthophosphoric acid is also the final product obtained if the metaphosphoric acid is boiled with water. From these reactions it can be seen that P_4O_{10} is an acidic oxide,

Phosphoric Acids

1. *Orthophosphoric acid*, H_3PO_4, is a weak triprotic acid. It is usually made from the phosphorus vapour produced in the electric furnace, by burning the vapour in air and then absorbing the P_4O_{10} in a spray of water. The mist of orthophosphoric acid formed is precipitated in an electrostatic precipitator. Phosphoric acid is usually sold as a syrupy liquid which is a concentrated solution containing up to 98% H_3PO_4.

H_3PO_4 ionizes in water in three stages, in each of which a proton is donated to an H_2O molecule:

$$H_3PO_4 + H_2O \rightleftharpoons H_3O^+ + H_2PO_4^- \qquad \text{(partial)}$$
$$H_2PO_4^- + H_2O \rightleftharpoons H_3O^+ + HPO_4^{2-} \qquad \text{(slight)}$$
$$HPO_4^{2-} + H_2O \rightleftharpoons H_3O^+ + PO_4^{3-} \qquad \text{(very slight)}$$

It follows that H_3PO_4 is an acid and PO_4^{3-} is a base. The other two anions $H_2PO_4^-$ and HPO_4^{2-} are amphiprotic.

2. *Pyrophosphoric acid.* When orthophosphoric acid is heated to about 250 °C it loses water to form *pyrophosphoric acid*:

$$2H_3PO_4 \rightarrow H_4P_2O_7 + H_2O$$

3. *Metaphosphoric acid.* If the temperature is raised to 600 °C the pyrophosphoric acid is further dehydrated to *metaphosphoric acid*:

$$H_4P_2O_7 \rightarrow 2HPO_3 + H_2O$$

Both of these acids react with hot water to form orthophosphoric acid:

$$H_4P_2O_7 + H_2O \rightarrow 2H_3PO_4$$
$$HPO_3 + H_2O \rightarrow H_3PO_4$$

NOMENCLATURE OF PHOSPHORIC ACIDS

The prefixes ortho-, meta- and pyro- are used in naming these acids according to the following principles:

(i) *ortho-* is used for the most highly hydroxylated acid which an element forms, i.e., the acid contains the greatest number of hydroxyl groups per atom of the element (in this case phosphorus).

(ii) *meta-* is used for the acid formed by the loss of one molecule of water from one molecule of the ortho-acid.

(iii) *pyro-* is used for the acid formed by the loss of one molecule of water from two molecules of the ortho-acid.

The structures of these three acids are shown in the diagram, fig. 29.5. It should be noted that the structure of metaphosphoric acid corresponds to the dehydration of one molecule of the ortho-acid. Pyrophosphoric acid corresponds to the joining together of two molecules of ortho-acid by the elimination of a molecule of water.

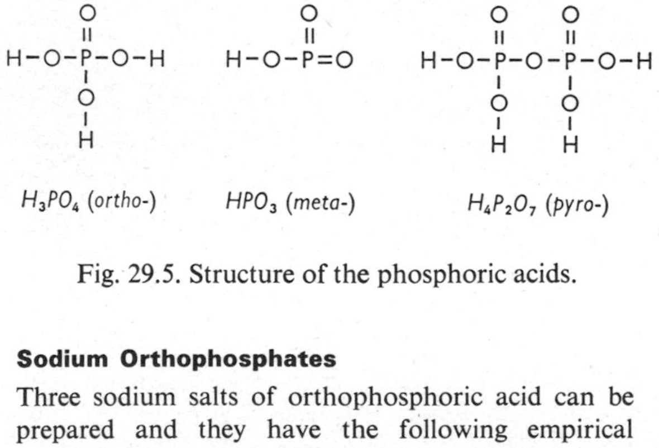

H_3PO_4 (ortho-) HPO_3 (meta-) $H_4P_2O_7$ (pyro-)

Fig. 29.5. Structure of the phosphoric acids.

Sodium Orthophosphates

Three sodium salts of orthophosphoric acid can be prepared and they have the following empirical formulae:

(i) $Na_3PO_4, 2H_2O$, trisodium orthophosphate;

(ii) $Na_2HPO_4, 12H_2O$, disodium hydrogen orthophosphate;

(iii) NaH_2PO_4, H_2O, sodium dihydrogen orthophosphate.

If aqueous solutions of each of these salts are tested in turn with litmus, methyl orange and phenolphthalein, the observations obtained are as shown in the table (fig. 29.6).

SALT SOLUTION	LITMUS	METHYL ORANGE	PHENOL-PHTHALEIN
Na_3PO_4	Blue (Alkaline colour)	Yellow (Alkaline colour)	Red (Alkaline colour)
Na_2HPO_4	Pale blue (Alkaline colour)	Yellow (Alkaline colour)	Pale pink (Neutral colour)
NaH_2PO_4	Red (Acidic colour)	Orange (Neutral colour)	Colourless (Acidic colour)

Fig. 29.6. Effect of sodium orthophosphates on indicators.

PREPARATION OF SODIUM ORTHOPHOSPHATES

It is apparent that the use of methyl orange and phenolphthalein as indicators provides a convenient method for the preparation of these salts.

1. To use *methyl orange*, orthophosphoric acid is added to a solution of caustic soda until the resulting solution turns the indicator orange. This indicates the volumes of acid and alkali needed to form the monosodic salt, NaH_2PO_4.

Once the necessary volumes of each solution are known the three salts can be prepared without the need for using the indicator. The same volume of orthophosphoric acid is used for each of three preparations. The predetermined volume of caustic soda solution is added to the volume of acid, to prepare sodium dihydrogen orthophosphate:

$$H_3PO_4 + OH^- \rightarrow H_2PO_4^- + H_2O$$

If twice the volume of caustic soda solution is used, disodium hydrogen orthophosphate is prepared:

$$H_3PO_4 + 2OH^- \rightarrow HPO_4^{2-} + 2H_2O$$

If three times the volume of caustic soda solution is used, trisodium orthophosphate is prepared:

$$H_3PO_4 + 3OH^- \rightarrow PO_4^{3-} + 3H_2O$$

2. To use *phenolphthalein*, caustic soda solution is added to orthophosphoric acid until the resulting solution turns the indicator pale pink. This indicates the volumes of acid and alkali needed to form the disodic salt, Na_2HPO_4. The three salts can be prepared

by using the same volume of orthophosphoric acid with half the determined volume of caustic soda solution, the volume to the phenolphthalein end point, and $1\frac{1}{2}$ times the determined volume respectively.

The salts can be obtained by evaporation of the three solutions.

REACTIONS WITH SILVER NITRATE SOLUTION

It is found that if excess silver nitrate solution is added to solutions of each of the three sodium orthophosphates, a yellow precipitate of silver orthophosphate, Ag_3PO_4, is formed in each case. If each of the solutions remaining after addition of *excess* silver nitrate is tested with litmus, the observations shown in fig. 29.7 are obtained.

SALT SOLUTION	EFFECT OF THE SOLUTION ON LITMUS:	
	(a) before adding $AgNO_3$	(b) after adding excess $AgNO_3$
Na_3PO_4	*Alkaline*	*Neutral*
Na_2HPO_4	*Slightly alkaline*	*Acidic*
NaH_2PO_4	*Acidic*	*Acidic*

Fig. 29.7. Tests to distinguish the sodium orthophosphates.

These observations can be explained by the following equations:

$$PO_4^{3-} + 3Ag^+ \rightarrow Ag_3PO_4 \qquad \text{(solution neutral)}$$
$$HPO_4^{2-} + 3Ag^+ \rightarrow Ag_3PO_4 + H^+ \text{(solution acidic)}$$
$$H_2PO_4^- + 3Ag^+ \rightarrow Ag_3PO_4 + 2H^+ \text{(solution acidic)}$$

These tests provide a convenient method of distinguishing the three sodium orthophosphates.

Calcium Orthophosphates

The three calcium orthophosphates are:

(i) $Ca_3(PO_4)_2$, tricalcium orthophosphate.

(ii) $CaHPO_4$, calcium monohydrogen orthophosphate.

(iii) $Ca(H_2PO_4)_2$, calcium dihydrogen orthophosphate.

The first two of these salts are insoluble in water but the $Ca(H_2PO_4)_2$ is slightly soluble and is therefore suitable to use as a fertilizer because it can be slowly absorbed in solution by plants.

Superphosphate Fertilizers

To convert mineral rock phosphates to a soluble form, the mineral is treated with an acid to change the phosphate to the dihydrogen phosphate:

$$2H^+ + PO_4^{3-} \rightarrow H_2PO_4^-$$

The stoichiometric equation for the reaction using sulphuric acid is:

$$Ca_3(PO_4)_2 + 2H_2SO_4 \rightarrow Ca(H_2PO_4)_2 + 2CaSO_4$$

The calcium sulphate absorbs the water in the mixture and crystallizes as $CaSO_4,2H_2O$. Superphosphate is therefore a mixture of two salts. It is the most widely used fertilizer in Australia, particularly for growing crops such as wheat.

Production of Superphosphate.
Crushed rock phosphate and sulphuric acid are mixed and the resulting slurry dropped into reaction 'dens'. Twelve hours later the reaction is almost complete, and 'green' superphosphate is extracted from the dens, as shown above.

Superphosphate is usually used with small amounts of nitrogen and potassium compounds which are also essential to productive plant growth. $NaNO_3$, $(NH_4)_2SO_4$, NH_4NO_3 and KCl are the most common sources of these elements.

Triple superphosphate fertilizer, so called because it contains about three times the percentage of available phosphorus in the form of soluble phosphorus compounds, is coming into use where transport costs are relatively high. It is often favoured where fertilizers are spread from crop-dusting aeroplanes, because a much smaller mass is needed to supply the same amount of phosphorus to the soil. Triple superphosphate is made by treating mineral phosphates with orthophosphoric acid:

$$Ca_3(PO_4)_2 + 4H_3PO_4 \rightarrow 3Ca(H_2PO_4)_2$$

The production of phosphate fertilizers is a very important aspect of the chemistry of phosphorus because it accounts for well over half the production of phosphorus compounds.

STUDY QUESTIONS

1. A mixture of calcium phosphate, sand and coke is heated in an electric furnace. The gaseous products are mixed with a large excess of air and burnt. The hot products of combustion are absorbed in a spray of water. Write equations for all the chemical reactions involved in these steps.

2. Describe the precautions necessary in the use of white phosphorus in the laboratory.

3. How can white phosphorus be changed to the red form and how could the reverse change be accomplished?

4. Write the equation for the reaction of phosphorus pentoxide with cold water. Write the equation for the reaction which occurs when the resulting mixture is boiled.

5. Write equations for the ionization of orthophosphoric acid in water. Name all of the entities present in the solution and label each entity in each equation as functioning either as an acid or as a base.
Hence, explain the effect of a solution of each of the three sodium orthophosphates on litmus.

6. The usual laboratory reagent called "sodium phosphate" has the following properties. Its solution is alkaline to litmus, but after addition of excess silver nitrate solution, the solution becomes acidic. Which of the three sodium orthophosphates is sold under the name of "sodium phosphate"? Write the equation for the reaction with silver nitrate solution.

7. Calculate the mass of rock phosphate, $Ca_3(PO_4)_2$, required to produce one kg of phosphorus.
   $(Ca = 40, P = 31, O = 16.)$

8. What volume of oxygen expressed at s.t.p., is required to completely convert 31 g of phosphorus to phosphorus pentoxide?
   $(P = 31.)$

Answers to Numerical Problems

7. 5 kg.
8. 28 dm³.

30 : Oxygen

OXYGEN

It has been estimated that oxygen comprises about 50% by mass of the earth's crust, and is by far the most abundant element. The next most abundant element is silicon, which comprises about 25% of the earth's crust by mass.

Valency of Oxygen

Oxygen is in group VI of the periodic table. There are six electrons in the outer shell of electrons and therefore oxygen usually has a covalency of two, e.g.

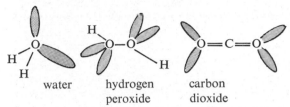

| water | hydrogen peroxide | carbon dioxide |

An oxygen atom may also accept two electrons from metal atoms to form the O^{2-} ion, i.e. it may have an electrovalency of -2, e.g.

$$(Na^+)_2(O^{2-}), (Mg^{2+})(O^{2-}), (Al^{3+})_2(O^{2-})_3$$

The element is a gas consisting of diatomic O_2 molecules.

Commercial Production of Oxygen

The production of oxygen from liquid air has been described in chapter 21 and its production by electrolysis has been described in chapter 13. Electrolysis of an aqueous solution of barium hydroxide produces very pure oxygen at the anode.

Oxygen is supplied commercially as liquid oxygen or as oxygen gas under pressure in cylinders.

The most convenient way of obtaining oxygen in quantity for the laboratory is to use a cylinder of oxygen. Hence the laboratory methods described below can be regarded as a means of obtaining small quantities of the gas. All the familiar laboratory methods of preparing oxygen involve the decomposition of oxygen compounds.

Preparation of Oxygen from Hydrogen Peroxide

Hydrogen peroxide, H_2O_2, is quite stable if *very pure*, but most samples decompose slowly on standing. Light hastens this decomposition and it is therefore usually stored in a brown bottle to protect it from

Bulk liquid oxygen for use in hospitals.

The tanker (above) is pumping liquid oxygen into an evaporator, from which it is piped in gaseous form to points of usage. The incubator (below) provides a fully controlled atmosphere for premature babies, drawing oxygen from the outlet on the wall behind the nurse's head.

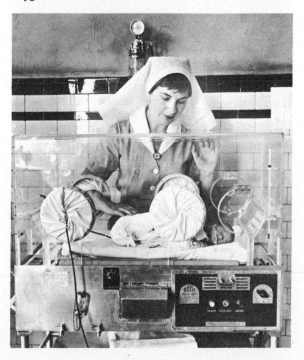

the sunlight. Decomposition proceeds rapidly on heating, producing water and oxygen.

$$2H_2O_2 \rightarrow 2H_2O + O_2$$

However, many substances are able to catalyse the decomposition so that it occurs rapidly at room temperature. Manganese(IV) oxide is one such substance.

● *A catalyst is a substance which alters the rate of a chemical change without itself being consumed.*

Hydrogen peroxide of ordinary household concentration (10 volume or 3%) is suitable for the preparation. The hydrogen peroxide is dropped on a layer of manganese(IV) oxide contained in a suitable vessel, as shown in fig. 30.1. The gas is collected over water.

Fig. 30.1. Preparation and collection of oxygen: hydrogen peroxide solution is dropped on a layer of manganese(IV) oxide powder.

The specification of the concentration of hydrogen peroxide as '10 volume', means that the thermal or catalytic decomposition of 1 volume of the hydrogen peroxide solution will yield 10 volumes of oxygen measured at s.t.p. Hence 100 cm³ of 10 volume hydrogen peroxide will yield 1000 cm³ of oxygen measured at s.t.p. This is sufficient to fill 4 or 5 gas jars at room conditions. The first gas jar will probably contain mainly air displaced from the preparation flask.

Hydrogen peroxide can be purchased as 20 volume or even 100 volume solutions but these higher concentrations ought to be avoided, because they cause painful white blisters on contact with the skin.

If hydrogen peroxide solution is acidified with dilute sulphuric acid and then poured on manganese(IV) oxide, oxygen gas is evolved and the black manganese(IV) oxide dissolves. This is not a catalytic reaction because the manganese(IV) oxide is consumed; it is a redox reaction:

$$MnO_2 + 4H^+ + 2e^- \rightarrow Mn^{2+} + 2H_2O$$
$$H_2O_2 \rightarrow 2H^+ + O_2 + 2e^-$$

$$\therefore MnO_2 + 2H^+ + H_2O_2 \rightarrow Mn^{2+} + O_2 + 2H_2O$$

Notice that in this reaction, one mole of O_2 is produced by oxidation of *one* mole of H_2O_2 whereas in the thermal and catalytic decompositions discussed above, one mole of O_2 was produced from *two* mole of H_2O_2. The 'volume strength' of a hydrogen peroxide solution refers to the thermal or catalytic decomposition, not to the oxidation reaction.

Some further reactions of hydrogen peroxide, in which it is reduced are given in the next two chapters.

Preparation of Oxygen from Potassium Chlorate

If potassium chlorate, $KClO_3$, is heated strongly in a test tube, it first melts and then evolves oxygen. Molten potassium chlorate is dangerous, and if the tube breaks, allowing the potassium chlorate to spill out, it will inflame any combustible substance it falls on.

It is therefore customary to use a catalyst so that the decomposition will occur at a lower temperature.

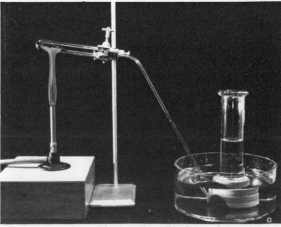

Fig. 30.2. Preparation and collection of oxygen: a mixture of potassium chlorate and manganese(IV) oxide is heated in the test tube.

Manganese(IV) oxide is the usual catalyst employed, although copper(II) oxide is also suitable.

A mixture of about one part of manganese(IV) oxide with two or three parts of potassium chlorate is prepared. The mixture is placed in a large hard-glass test tube, and heated gently as shown in fig. 30.2.

$$2KClO_3 \rightarrow 2KCl + 3O_2$$

The gas is collected over water.

This method is less convenient than the preparation from hydrogen peroxide. Moreover, it can be dangerous unless suitable precautions are taken. Points to be observed are as follows:

(a) The manganese(IV) oxide should be heated strongly on an iron tray before use. This is to burn out any combustible substances present. Commercial manganese(IV) oxide often contains small amounts of organic matter (carbon containing substances), and the presence of any relatively large amount of combustible material, could lead to a serious explosion when heated with potassium chlorate.

(b) The oxygen evolved tends to blow the fine powder into the delivery tube which can become blocked.

(c) The test tube should slope down slightly as shown. This is because moisture may be driven off, and after condensation, it could run back to the hot end of the test tube and crack it.

(d) The gas produced by this method always contains a little chlorine which can be smelt in the gas evolved.

Properties of Oxygen

Oxygen exists as diatomic molecules, O_2, although a more reactive form, ozone, O_3, can be prepared by an electrical discharge through oxygen.

PHYSICAL PROPERTIES

It can be observed during the preparation, that oxygen gas is colourless, odourless and not very soluble in water.

Even though the gas is not very soluble in water, its slight solubility is important as most aquatic life uses this dissolved oxygen for respiration.

CHEMICAL PROPERTIES

If a burning taper is thrust into a gas jar of oxygen, the taper flares up and burns brightly, but the gas itself does not burn.

● *Most substances burn vigorously in oxygen.*

To test the combustibility of a solid in oxygen gas, first heat the solid as shown in fig. 30.3 and then lower the hot substance into the gas.

A variety of substances can be burnt in oxygen in this way.

Fig. 30.3. Combustion of a substance in a gas. The substance is heated in a deflagrating spoon which is then placed in a jar full of the gas.

Carbon

Red hot carbon (charcoal) glows brightly in oxygen. Lime water turns milky if added to the jar, but does not turn milky in a jar of oxygen. The product of combustion is therefore carbon dioxide:

$$C + O_2 \rightarrow CO_2$$

Sulphur

Molten sulphur burns very strongly in oxygen with a mauve coloured flame. The residue is evidently an oxide of sulphur, and tests show that it is mainly sulphur dioxide with a little sulphur trioxide.

$$S + O_2 \rightarrow SO_2 \qquad \text{(considerable)}$$
$$2S + 3O_2 \rightarrow 2SO_3 \qquad \text{(very slight)}$$

Phosphorus

Molten phosphorus burns extremely vigorously in oxygen, with a blaze of yellowish light and the production of a thick white cloud. The product can be shown to be phosphorus pentoxide (P_4O_{10}) together with a little phosphorus trioxide (P_4O_6).

$$P_4 + 5O_2 \rightarrow P_4O_{10}$$
$$P_4 + 3O_2 \rightarrow P_4O_6$$

N.B. The names of P_4O_{10} and P_4O_6 are explained in chapter 29.

Sodium

Burning sodium flares up when lowered into oxygen. The flame is yellow and a white cloud is produced. The white cloud is sodium peroxide:

$$2Na + O_2 \rightarrow Na_2O_2$$

Magnesium

When lowered into oxygen, burning magnesium produces a dazzling white light and a white cloud. The light is dangerous to the eyes (see earlier). The product is magnesium oxide:

$$2Mg + O_2 \rightarrow 2MgO$$

Iron

If a bundle of hot, fine iron wires is lowered into oxygen it burns with a shower of sparks, producing a black residue of iron(II, III) oxide:

$$3Fe + 2O_2 \rightarrow Fe_3O_4$$

Summary of the Properties of Oxygen

PHYSICAL PROPERTIES OF OXYGEN

1. Colourless, odourless gas.
2. Slightly soluble in water (about 3 cm³ of oxygen dissolves in 100 cm³ of water at room temperature).
3. Slightly denser than air (the molecular mass of O_2 is about 32).
4. Difficult to liquefy (b.p. = $-182.7\ °C$).

CHEMICAL PROPERTIES OF OXYGEN.

1. Stable to heat.
2. Does not burn.

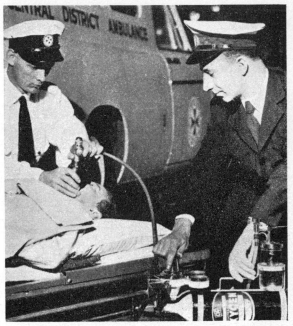

Portable resuscitation units employ oxygen drawn from cylinders. The picture shows oxygen being administered to the victim of a swimming accident.

3. Supports combustion very vigorously.
4. Its solution in water is neutral to litmus.
5. It is an oxidizer in almost all of its reactions.

TEST FOR OXYGEN

Oxygen is a colourless, odourless gas, which re-ignites a glowing string or splinter.

Uses of Oxygen

Many of the uses of oxygen are dependent on the fact that it is essential for life. A few very simple living organisms can exist without oxygen, but more complex living things cannot survive very long without it.

1. *In hospitals.* Pure oxygen, or a mixture of air and oxygen is sometimes given to patients suffering from respiratory complaints such as pneumonia. Oxygen is also given to people suffering from electric shock,

An oxy-acetylene cutting torch being used for scrap cutting. The scrap can then be returned to the furnaces to make new steel.

apparent drowning or suffocation with poisonous gases. The oxygen used is sometimes mixed with a small proportion of carbon dioxide, which stimulates breathing.

2. *In aviation.* At great altitudes the air is at very low pressure and does not supply sufficient oxygen for respiration. Commercial aircraft use pressurized cabins and other types of aircraft frequently carry an oxygen supply.

3. *In oxy-acetylene flames* used for cutting and welding metals. The gas acetylene is burnt in oxygen producing a very high temperature, sufficient to melt steel fairly easily.

The *oxy-hydrogen flame* is used for cutting metals under water.

4. *In explosives.* Liquid oxygen can be used, mixed with substances such as charcoal, as an industrial explosive.

5. *Liquid oxygen* is used in some rockets. It is vapourized and the oxygen is used for rapid combustion of the fuel.

STUDY QUESTIONS

1. What are the products of combustion in oxygen of: carbon, sulphur, phosphorus, sodium, magnesium, and iron. Label each product as one of the following types of oxides: acidic oxide, basic oxide, peroxide, compound oxide.

2. What explanation can be offered for the statement that a mixture of potassium chlorate and manganese(IV) oxide containing organic matter might explode on heating?

3. What volume of oxygen can be obtained at s.t.p. by catalytic decomposition of 50 cm³ of 20 volume hydrogen peroxide solution?

4. What volume of oxygen can be obtained at s.t.p. by complete electrolysis of 1 mole of water?

5. Calculate the volume of oxygen at s.t.p., evolved from 1 dm³ of 3.4% hydrogen peroxide solution by:
(a) catalytic decomposition;
(b) addition of excess acidified potassium permanganate solution.

$$(H = 1.0, \quad O = 16.0)$$

6. Copper(II) oxide may be used as a catalyst in the decomposition of potassium chlorate. Describe how you would establish experimentally that the copper(II) oxide is not used up during the reaction.

7. The formula of acetylene is C_2H_2.
(a) Write the structure for a molecule of acetylene.
(b) Write an equation for the combustion of acetylene in oxygen.

8. Liquid oxygen absorbed on charcoal is used as an industrial explosive. What reaction occurs when this explosive is detonated?

Answers to Numerical Problems

3. 1 dm³.
4. 11.2 dm³.
5. (a) 11.2 dm³. (b) 22.4 dm³.

31 : Sulphur and the Sulphides

Valency of Sulphur

Sulphur is in the same group of the periodic table as oxygen. It has six electrons in its outer shell and, like oxygen, it can have a covalency of 2 or an electro-valency of -2, e.g.

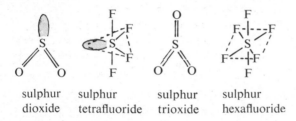

| hydrogen sulphide | carbon disulphide | sodium sulphide |

However, unlike oxygen, sulphur can have covalencies of 4 or 6, e.g.

| sulphur dioxide | sulphur tetrafluoride | sulphur trioxide | sulphur hexafluoride |

Also, the element usually consists of S_8 molecules in contrast to the usual form of oxygen, O_2 molecules.

Occurrence of Sulphur

The most important sources of sulphur are the American deposits in Texas and Louisiana. This sulphur is in the free state, but is deep underground and occurs in amongst other minerals such as gypsum and calcite. There are smaller deposits of free sulphur in the volcanic regions of Sicily, Mexico and Japan. Sulphur occurs widely in the combined state in mineral sulphides such as galena, PbS, zinc blende, ZnS, and cinnabar, HgS. A number of sulphates also occur naturally, an important one being gypsum, $CaSO_4,2H_2O$.

Extraction of Sulphur

Over 80% of the world's supply of sulphur is extracted from the American deposits by the *Frasch method*. The sulphur-bearing strata lie several hundred metres below ground level. Three concentric pipes are sunk into the sulphur-bearing layer. The pipes are 15 cm, 7.5 cm, and 2.5 cm in diameter respectively. Super-heated water (at about 170 °C and under pressure) is pumped down the 15 cm pipe and melts the sulphur in amongst the other minerals in the sulphur-bearing strata. The molten sulphur collects as a pool at the bottom of the pump. Hot compressed air is blown down the 2.5 cm pipe so that a light froth of molten sulphur and air rises up the 7.5 cm pipe. The molten sulphur flows into enormous bins and after the sulphur has solidified the sides of the bins are dismantled and the blocks of sulphur are broken up. The sulphur obtained by this process is about 99.5% pure and does not need purification for most commercial uses.

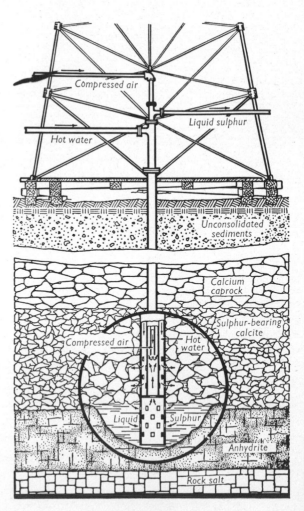

Fig. 31.1. A Frasch sulphur pump.

Sulphur as a Laboratory Chemical

Sulphur is supplied in the small quantities used in the laboratory in the form of 'roll sulphur' and 'flowers of sulphur'.

Roll sulphur is made by casting molten sulphur in suitable moulds. It is a yellow solid which is fairly hard but brittle, and it can be easily ground to a powder.

Flowers of sulphur is made by subliming sulphur vapour. It is an amorphous or non-crystalline allotrope of sulphur formed when the vapour is cooled rapidly. It is a pale yellow powder.

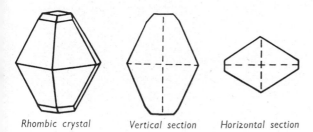

Rhombic crystal   Vertical section   Horizontal section

Fig. 31.2. Rhombic axes of symmetry: three axes at right angles to each other; the axes are of unequal length.

Crystalline Allotropes of Sulphur

RHOMBIC SULPHUR

If crystals of sulphur are allowed to form at any temperature below 96 °C, rhombic sulphur is formed (fig. 31.2). These crystals can be obtained by treating powdered roll sulphur with carbon disulphide. When the mixture is filtered, a clear yellow solution is obtained. If this solution is placed in a shallow dish and covered with a watch glass, the solvent evaporates slowly leaving the crystals of sulphur.

SPECIAL NOTE: Carbon disulphide must be used very carefully. The commercial liquid has a foul odour, it is poisonous and is very volatile. **Furthermore, the vapour is highly inflammable and serious fires and explosions may result if carbon disulphide is used near a flame.** *Carbon disulphide should only be used in an efficient fume cupboard and no flames should be allowed to remain alight in the vicinity, particularly if the odour can be detected in the laboratory.*

PRISMATIC SULPHUR

If crystals of sulphur are allowed to form at any temperature above 96 °C, prismatic sulphur is formed. These crystals can be readily obtained by melting sulphur in a deep dish. The sulphur is made just hot enough to become molten and then allowed to cool. As soon as a crust forms, it is pierced in two places and the remaining molten sulphur is poured off. The crust can then be cut out of the dish and inverted. The underside of the crust is a mass of needle-like crystals called prismatic (i.e. long and narrow) crystals, but they may be classified as monoclinic because of the arrangement of the axes of symmetry of the crystals (fig. 31.3).

At room temperature the crystals of monoclinic sulphur slowly become opaque and the hard mass of interlacing crystals becomes brittle.

This change is due to slow conversion of the monoclinic allotrope to a mixture of rhombic sulphur and non-crystalline forms because the temperature is below 96 °C.

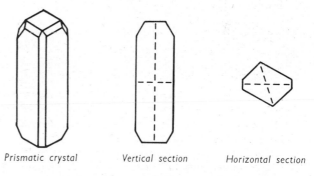

Prismatic crystal   Vertical section   Horizontal section

Fig. 31.3. Monoclinic axes of symmetry: two axes at right angles and one inclined; the axes are of unequal length.

TRANSITION TEMPERATURE

Rhombic sulphur is stable up to 96 °C, above which temperature monoclinic sulphur is the stable form. Thus any method of preparation which involves crystallization below this temperature (called the transition temperature) produces rhombic crystals, and crystallization at higher temperatures produces monoclinic crystals. The properties of these two crystalline allotropes are compared in the table (fig. 31.4).

PROPERTY	RHOMBIC SULPHUR	MONOCLINIC SULPHUR
Appearance	*Yellow crystalline*	*Yellow crystalline*
Melting Point	113 °C	119 °C
Density	2.07 g cm^{-3}	1.96 g cm^{-3}
Carbon Disulphide	*Soluble*	*Soluble*
Alternative Names	*α—sulphur*	*β—sulphur*

Fig. 31.4. Comparison of the crystalline allotropes of sulphur.

Non-crystalline Allotropes of Sulphur

If flowers of sulphur is treated with carbon disulphide, it is found that some of it is insoluble. This insoluble allotrope of sulphur is a very fine, pale yellow powder and is apparently an *amorphous* form of sulphur. When sulphur is precipitated from solution in chemical reactions, it forms as a finely divided or colloidal form of amorphous sulphur which may appear to be white and is called milk of sulphur.

Another allotrope of sulphur is obtained when boiling sulphur is poured into cold water. A number of changes are observable as the sulphur is heated to its boiling point, but a discussion of these will be deferred for the moment (see below). When boiling sulphur is quenched in cold water, a dark brown, rubbery, plastic mass is obtained which becomes hard in a short time. The rubbery mass is called *plastic sulphur* and is unstable, changing fairly quickly to a mixture of rhombic and amorphous sulphur. A clear amber form of plastic sulphur can be obtained by distilling roll sulphur from a retort (see fig. 31.5).

Changes in Liquid Sulphur on Heating

Just above the melting point, sulphur is a pale yellow, mobile liquid. As the temperature is increased, the liquid darkens and quite suddenly becomes viscous

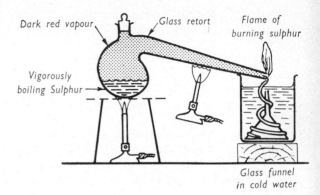

Fig. 31.5. Preparation of plastic sulphur.

at about 180 to 200 °C. The liquid at this temperature is so thick, that the vessel containing it can be inverted without causing the sulphur to pour out. At still higher temperatures, the molten sulphur again becomes mobile. The darkening is thought to be due to impurities because very carefully purified sulphur does not darken as it is heated.

The Structure of Sulphur

The many different forms of solid sulphur and liquid sulphur can be more readily understood in terms of the structure of sulphur molecules. X-ray methods show that *rhombic sulphur* consists of S_8 molecules which are shaped like puckered octagons (fig. 31.6).

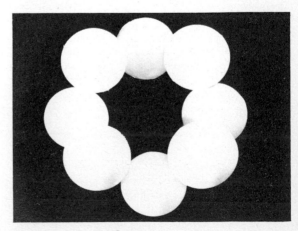

Fig. 31.6. A model representing the structure of an S_8 molecule showing the atoms as spheres in contact.

Monoclinic sulphur is composed of the same S_8 rings but the molecules are arranged differently to give a different crystalline form.

One difference between the arrangements of the S_8 molecules in rhombic and monoclinic sulphur can be inferred from their densities. The higher density of rhombic sulphur indicates a more compact arrangement.

Just above the melting point the *pale yellow mobile liquid* is probably composed of these same S_8 rings. The *increase in viscosity* at higher temperatures can be explained in the following way: the increased thermal motion of the S_8 rings causes them to open, and then they join together forming long chains of sulphur atoms which become entangled and hence unable to move freely. At still higher temperatures, the decrease in viscosity can be attributed to the increased thermal agitation which breaks the chains into smaller fragments, and also breaks up cross-linking between adjacent chains.

Sudden cooling of the boiling liquid gives *plastic sulphur*, which is thus a haphazard array of sulphur chains. In general, when a solid composed of long chains of atoms is distorted, the bonds between the atoms are strained and provide the force which tends to bring the solid back to its original shape, when the distorting force is removed. In this respect, plastic sulphur is like other elastic solids, such as rubber.

Sulphur vapour, just above the boiling point, has a density which indicates that most of the vapour is in the form of S_8 molecules, although a proportion of smaller molecules is present. At 900 °C, the density indicates that the vapour consists almost entirely of S_2 molecules. At still higher temperatures, the vapour contains single S atoms. It appears that as the temperature increases, the increased energy of the system causes the molecules to become split into progressively smaller units (fig. 31.7).

Uses of Sulphur

Sulphur is used mainly for the production of sulphuric acid (described in the next chapter). The element is used as a fungicide and in vulcanizing rubber. The vulcanizing of rubber consists of heating it with various substances so that it is toughened. Sulphur brings this about by forming bridges between adjacent long chain molecules in the rubber.

PHYSICAL STATE	TEMPERATURE	MOLECULAR STRUCTURE	APPEARANCE
Vapour	900 °C	$S_2 \rightleftharpoons 2S$	*Yellow vapour*
		$S_8 \rightleftharpoons 4S_2$	*Dark red vapour*
	444.4 °C (b.p.)		
Liquid	180-200 °C	*Short S chains* *Long S chains* S_8 *molecules*	*Mobile liquid* *Viscous liquid* *Mobile liquid*
	119 °C (m.p.)		
Solid		S_8 *molecules in rings*	*Monoclinic crystals*
	96 °C (*Transition temperature*)		
		S_8 *molecules in rings*	*Rhombic crystals*

Fig. 31.7. Change in structure of sulphur with temperature.

HYDROGEN SULPHIDE

Laboratory Preparation

Hydrogen sulphide can be prepared by the reaction of hydrochloric acid (equal parts of water and concentrated acid) with iron(II) sulphide:

$$FeS + 2H^+ \rightarrow H_2S + Fe^{2+}$$

The gas produced in this way is impure but is suitable for most laboratory purposes. Iron(II) sulphide is prepared by heating together iron and sulphur and the impurity in the hydrogen sulphide is mainly hydrogen liberated from the acid by unchanged iron.

● *Note: Hydrogen sulphide is extremely poisonous and should always be prepared in a fume cupboard.*

A Kipp's apparatus is most convenient for the preparation of the gas as it provides a ready supply. The gas should be bubbled through water to remove acid spray. The gas is best collected over water, although it is fairly soluble in water.

Physical Properties of Hydrogen Sulphide

The gas is colourless but has an offensive odour. It is often called 'rotten-egg gas', and is in fact formed during the decay of eggs. The gas is extremely poisonous. It is a reasonable guide to say that if the odour is offensive, it is not present in the air in dangerous concentrations. If the concentration becomes higher the gas tends to have a sickly sweet odour and causes headaches and nausea. Part of the danger of hydrogen sulphide lies in the fact that it paralyses the olfactory nerves. Even moderate concentrations of hydrogen sulphide may cause sudden unconsciousness or even death.

Hydrogen sulphide is slightly denser than air (the molecular mass of H_2S is about 34). It is fairly soluble in water (about $2\frac{1}{2}$ volumes of H_2S dissolve in one volume of water at room conditions) and it is fairly easily liquefied (b.p. = $-61\ ^\circ C$).

Chemical Properties of Hydrogen Sulphide

STABILITY

If hydrogen sulphide is passed through a heated glass tube a yellow deposit of sulphur can be seen, indicating that hydrogen sulphide is unstable to heat (fig. 31.8). The reverse reaction can also occur because a small

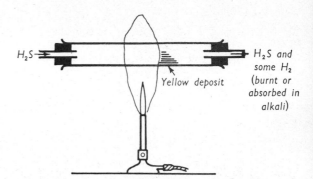

Fig. 31.8. Thermal instability of hydrogen sulphide.

proportion of hydrogen sulphide is formed when hydrogen is bubbled through molten sulphur:

$$H_2S \rightleftharpoons H_2 + S$$

COMBUSTION

If a lighted taper is plunged into a gas jar of hydrogen sulphide, the taper is extinguished but the gas burns with a blue flame and a yellow deposit is left on the walls of the jar:

$$2H_2S + O_2 \rightarrow 2H_2O + 2S$$

Hydrogen sulphide can be burnt at a jet but the experiment is dangerous unless suitable precautions are taken. The gas must be free from air before being lit and it is desirable to have a small wash bottle between the jet and the generator. The gas can be lit as it issues from a fine metallic jet, such as a simple blowpipe. This conducts the heat of the flame away sufficiently rapidly to prevent striking back. No deposit of sulphur is formed but the characteristic odour of sulphur dioxide can be detected:

$$2H_2S + 3O_2 \rightarrow 2H_2O + 2SO_2$$

● *Mixtures of hydrogen sulphide and air may explode violently when ignited.*

NATURE OF SOLUTION

A solution of hydrogen sulphide turns litmus purplish-red and so the solution is a weak acid:

$$H_2S + H_2O \rightleftharpoons H_3O^+ + HS^- \qquad \text{(slight)}$$
$$HS^- + H_2O \rightleftharpoons H_3O^+ + S^{2-} \qquad \text{(very slight)}$$

Hence, HS^- is amphiprotic and S^{2-} is a moderately strong base.

REDUCING PROPERTIES

1. The combustion of hydrogen sulphide in air shows that it can be oxidized by oxygen. A solution of hydrogen sulphide turns milky on standing in air due to formation of a precipitate of colloidal sulphur. This is evidently due to oxidation by oxygen in a similar way to the combustion in a limited supply of air:

$$2H_2S + O_2 \rightarrow 2H_2O + 2S$$

2. The reducing properties of hydrogen sulphide gas can also be shown by its reaction with *chlorine gas*. If a gas jar of chlorine is placed on top of a gas jar of hydrogen sulphide and the cover slips are removed, the yellow-green colour of the chlorine fades and a yellow deposit forms. The gas jars can be shown to contain hydrogen chloride.

$$H_2S + Cl_2 \rightarrow 2HCl + S$$

The HCl can be identified by its effect on litmus, and its reaction with ammonia.

3. Hydrogen sulphide acts as a reducer when bubbled into solutions of oxidizers. Thus, if concentrated nitric acid and acidified solutions of iron(III) chloride, potassium permanganate and potassium dichromate, are each saturated with hydrogen sulphide, a yellow precipitate forms in each solution. Thus the half reaction for the reducer may be represented as:

$$H_2S \rightarrow 2H^+ + S + 2e^-$$

The colour changes which can be observed in each solution enable the reduction products of the oxidizers to be identified. The balanced equation for each reaction can be obtained by addition of the appropriate half reactions so that the electrons are cancelled out:

$$H_2S + 2NO_3^- + 2H^+ \rightarrow S + 2NO_2 + 2H_2O$$
brown gas

$$H_2S + 2Fe^{3+} \rightarrow S + 2Fe^{2+} + 2H^+$$
yellow   green

$$5H_2S + 2MnO_4^- + 6H^+ \rightarrow 5S + 2Mn^{2+} + 8H_2O$$
purple   colourless

$$3H_2S + Cr_2O_7^{2-} + 8H^+ \rightarrow 3S + 2Cr^{3+} + 7H_2O$$
orange   green

Hydrogen sulphide is also oxidized by hydrogen peroxide to form a precipitate of sulphur but in this case there is no other colour change in the solution:

$$H_2S + H_2O_2 \rightarrow S + 2H_2O$$

4. Hydrogen sulphide reacts with sulphur dioxide. When gas jars of hydrogen sulphide and sulphur dioxide are allowed to mix, a yellow deposit forms on the walls of the gas jars. The hydrogen sulphide reduces the sulphur dioxide to sulphur and is itself oxidized to sulphur:

$$2H_2S + SO_2 \rightarrow 2H_2O + 3S$$

Test for Hydrogen Sulphide

Hydrogen sulphide gas may be identified as a colourless gas with a characteristic offensive odour. A piece of filter paper soaked in a solution of a silver or lead(II) salt, is turned black by hydrogen sulphide (explained below).

SULPHIDES

The precipitation of metallic sulphides from solutions of salts, by saturating the solutions with hydrogen sulphide, is important in the analysis of inorganic salts. A suitable method of doing this is shown in fig. 31.9. The precipitation is accompanied by the formation of an acid in the solution. This can be seen by using a solution of silver nitrate, which is neutral to litmus. With H_2S, a black precipitate of silver sulphide forms and the remaining solution becomes acidic to litmus:

$$H_2S + 2Ag^+ \rightarrow Ag_2S\,(s) + 2H^+$$

In this reaction, Ag_2S is precipitated from an aqueous solution and so it can be concluded that Ag_2S is insoluble in water. Also, since the precipitation results in the formation of an acid, it can be concluded that Ag_2S is insoluble in dilute solutions of acids.

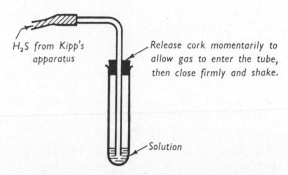

Fig. 31.9. Saturating a solution with H_2S gas.

Precipitation of Sulphides

If *acidified* solutions of various salts are saturated with hydrogen sulphide using the method shown in fig. 31.9, the following precipitations occur:

$$2Ag^+ + H_2S \rightarrow Ag_2S\,(s) + 2H^+$$
black

$$Hg^{2+} + H_2S \rightarrow HgS\,(s) + 2H^+$$
black

$$Cu^{2+} + H_2S \rightarrow CuS\,(s) + 2H^+$$
black

$$Pb^{2+} + H_2S \rightarrow PbS\,(s) + 2H^+$$
black

$$Sn^{2+} + H_2S \rightarrow SnS\,(s) + 2H^+$$
brown

$$Sn^{4+} + 2H_2S \rightarrow SnS_2\,(s) + 4H^+$$
yellow

$$2Fe^{3+} + H_2S \rightarrow 2Fe^{2+} + 2H^+ + S\,(s)$$
white or yellow

It may be noted that all of the metals below iron on the activity series form sulphides which are insoluble in water and insoluble in acid. Iron(III) salts oxidize hydrogen sulphide forming a precipitate of sulphur in acid solution.

To form precipitates of sulphides in the absence of acid, it is necessary to use some other sulphide than hydrogen sulphide. Ammonium sulphide is suitable because it can be easily prepared in the laboratory by saturating ammonia solution with hydrogen sulphide:

$$2NH_3 + H_2S \rightarrow 2NH_4^+ + S^{2-}$$

If solutions of various salts are tested with this solution, all of the above sulphides are precipitated, and in addition the following are also precipitated:

$$Fe^{2+} + S^{2-} \rightarrow FeS\,(s)$$
black

$$Zn^{2+} + S^{2-} \rightarrow ZnS\,(s)$$
white

Thus iron(II) sulphide and zinc sulphide are insoluble in water but soluble in acids.

The sulphides of the metals above zinc on the activity series are not precipitated by either of the methods described above and so are apparently soluble in water. Alternatively they may not exist at all in the presence of water. For example, aluminium sulphide reacts completely with water:

$$Al_2S_3 + 6H_2O \rightarrow 2Al(OH)_3 + 3H_2S$$

In this reaction, the base S^{2-} accepts protons from the acid H_2O, and the Al^{3+} and OH^- ions combine because $Al(OH)_3$ is insoluble in water.

Therefore it cannot be prepared in the presence of water. It can be prepared by direct combination of the elements.

SPECIAL NOTE: **Because of the unpleasant and poisonous nature of hydrogen sulphide,** *another method of precipitating sulphides is becoming increasingly popular. Some thioacetamide is warmed with an acidified solution of the salt. Hydrogen sulphide is released:*

$$CH_3-\overset{S}{\underset{\|}{C}}-NH_2 + H_2O \rightarrow CH_3-\overset{O}{\underset{\|}{C}}-NH_2 + H_2S$$
Thioacetamide · Acetamide

The hydrogen sulphide is used immediately in forming a precipitate with the cations of the salt, and little escapes into the atmosphere.

If a balloon is fitted over the mouth of the tube, none of the gas can escape.

Identification of Cations

One scheme for identification of cations is based on precipitation of sulphides. The procedure begins with an aqueous solution which is acidified with dilute HCl. If present, Ag^+ is completely precipitated as AgCl but Pb^{2+} is only partially precipated as $PbCl_2$. If precipitation occurs here (or in later tests), the mixture is filtered.

The filtrate is saturated with H_2S to form precipitates of HgS, CuS, PbS, SnS and SnS_2 from the corresponding cations, or a precipitate of sulphur by reaction with Fe^{3+}, if these cations are present.

The filtrate is then boiled to free it of dissolved H_2S and also boiled with HNO_3 to oxidize any Fe^{2+} to Fe^{3+}. This solution is cooled and made alkaline with NH_3 solution. Precipitates of $Al(OH)_3$ and $Fe(OH)_3$ form if the corresponding cations are present.

Once again, the filtrate is saturated with H_2S and, if Zn^{2+} is present, ZnS precipitates. After filtration and boiling off the H_2S, $(NH_4)_2CO_3$ solution is added to form precipitates of $CaCO_3$ and $MgCO_3$ from the respective cations.

The precipitates can be subjected to further tests to identify the cations. NH_4^+, K^+ and Na^+ do not form any precipitates and must be identified in other ways.

STUDY QUESTIONS

1. Explain how samples of each of the following allotropes of sulphur may be prepared:
   (a) rhombic sulphur;
   (b) amorphous sulphur.

2. Explain why two of the crystalline forms of sulphur are called rhombic and monoclinic.

3. Devise a method by which it could be proved that rhombic sulphur and amorphous sulphur are allotropes of the same element.

4. What crystalline form of sulphur is formed when molten sulphur is allowed to solidify? What change occurs as the sulphur, so formed, is allowed to stand at room temperature?

5. Describe the changes which can be seen if sulphur is heated steadily in a test tube until it boils and the boiling liquid is then poured into cold water.
   Describe the changes in structure which occur during the heating and sudden cooling.

6. Write equations to illustrate the reactions of hydrogen sulphide when:
   (a) the gas is heated strongly;
   (b) the gas is burnt in a limited supply of air;
   (c) the gas is burnt in excess air;
   (d) the gas is mixed with chlorine gas.

7. Evaluate the advantages and disadvantages of collecting hydrogen sulphide by:
   (a) displacement of water;
   (b) upward displacement of air.

8. What volume of sulphur dioxide is required for complete reaction with 100 cm³ of hydrogen sulphide? (Both volumes are measured at the same temperature and pressure.)

9. Draw diagrams to illustrate the valency of sulphur in H_2S, S_8, SF_6 and MgS.

10. Compare and contrast oxygen and sulphur with respect to:
    (a) physical properties;
    (b) valency;
    (c) acidic character of the hydrides H_2O and H_2S.

11. Describe the changes which can be observed if:
    (a) hydrogen peroxide is reacted with hydrogen sulphide gas;
    (b) hydrogen peroxide solution is added to acidified potassium permanganate solution.
    Describe why these observations can be regarded as evidence to support the statement that hydrogen peroxide can act as either an oxidizer or a reducer.

12. A solution contains a mixture of copper(II) sulphate and zinc sulphate. How may each of the cations be completely precipitated and obtained as separate samples of the sulphides?

13. An acidified solution of iron(III) chloride is saturated with H_2S gas. The resulting solution is then neutralized with ammonia solution, and excess ammonium sulphide solution is added. Write the equations for the reactions which occur. Name the precipitates formed.

14. A salt is thought to be lead nitrate. How could it be established that the salt contains lead cations?

15. A solution contains the following cations: silver, copper(II) and zinc. How could three precipitates be obtained, each of which contains only one of the cations?

16. A mineral prospector has found a white rock which is thought to be dolomite, $CaCO_3.MgCO_3$. The rock is insoluble in water. How could it be verified that the mineral contains compounds of calcium and magnesium?

Answer to Numerical Problem

8. 50 cm³.

32 : Oxy-Compounds of Sulphur

SULPHUR DIOXIDE, SULPHUROUS ACID AND SULPHITES

Preparation of Sulphur Dioxide

1. INDUSTRIAL PREPARATION

On an industrial scale, sulphur dioxide is usually prepared by burning sulphur or sulphides in air or oxygen. The combustion of sulphur produces mainly sulphur dioxide:

$$S + O_2 \rightarrow SO_2$$

However, 3 to 4% of the sulphur is oxidized to sulphur trioxide, and if air is used, nitrogen and the other gases from the air are also present.

Zinc blende, ZnS, and pyrites, FeS_2, are two sulphides which are used as a source of sulphur dioxide:

$$2ZnS + 3O_2 \rightarrow 2ZnO + 2SO_2$$
$$4FeS_2 + 11O_2 \rightarrow 2Fe_2O_3 + 8SO_2$$

2. LABORATORY PREPARATION

In the laboratory, sulphur dioxide can be conveniently prepared by the reaction of dilute sulphuric acid with sodium sulphite.

H_3O^+ ions donate protons to SO_3^{2-} ions because the sulphite ion, SO_3^{2-}, is a strong base, i.e. a good proton acceptor.

$$2H^+ + SO_3^{2-} \rightarrow H_2SO_3$$
$$H_2SO_3 \rightarrow H_2O + SO_2$$

The gas is evolved more readily if the mixture is warmed, because this helps to decompose the unstable sulphurous acid. The gas may be dried by bubbling it through concentrated sulphuric acid and collected by upward displacement of air (fig. 32.1).

Almost dry sulphur dioxide can be prepared by reduction of hot concentrated sulphuric acid with copper:

$$Cu + H_2SO_4 + 2H^+ \rightarrow Cu^{2+} + 2H_2O + SO_2$$

A number of side reactions also occur resulting in the formation of a black residue of copper(I) and copper(II) sulphides. However, sulphur dioxide is the only gaseous product.

Physical Properties of Sulphur Dioxide

It can be observed during the preparation, that sulphur dioxide is a colourless gas, although it may form white fumes in moist air. It has an extremely irritating,

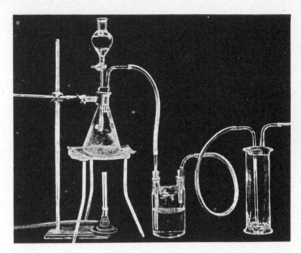

Fig. 32·1. Laboratory preparation and collection of sulphur dioxide: dilute sulphuric acid is added to sodium sulphite and the gas is dried by concentrated sulphuric acid.

choking odour which is characteristic of sulphur dioxide. The gas is more than twice as dense as air (the molecular mass of SO_2 is about 64).

The solubility of sulphur dioxide in water can be observed by opening a gas jar of the gas under water. If the jar is full of sulphur dioxide, the water will rise rapidly and fill the whole of the jar. This shows that sulphur dioxide is very soluble in water. Quantitative measurements show that water will dissolve about 40 times its volume of sulphur dioxide at ordinary temperatures and pressures.

Sulphur dioxide is one of the easiest of gases to liquefy. Its boiling point is $-10\,°C$ at 1 atmosphere but it can be easily liquefied by compression to about 3 atmosphere at room temperature (18-20 °C). Sulphur dioxide is usually transported as a liquid in metal cylinders.

Chemical Properties of Sulphur Dioxide

STABILITY

Sulphur dioxide can be prepared by the combustion of sulphur or sulphides at fairly high temperatures. It is thus stable to heat.

COMBUSTION

If a lighted taper is plunged into a gas jar of sulphur dioxide, the taper is extinguished. The gas neither burns nor supports ordinary combustion. However, if burning magnesium is plunged into a gas jar of sulphur dioxide, it continues to burn forming a white cloud. Some areas of a yellow deposit may also form on the walls of the jar close to the burning magnesium. The white solid is magnesium oxide and the yellow solid is sulphur (although side reactions result in the formation of some magnesium sulphite, $MgSO_3$, and magnesium thiosulphate, MgS_2O_3).

$$2Mg + SO_2 \rightarrow 2MgO + S$$

NATURE OF SOLUTION

A solution of sulphur dioxide in water is acidic to litmus and therefore sulphur dioxide is an acidic oxide. The solution formed is only weakly acidic and is probably similar to carbon dioxide solution in that it contains molecules of the gas which are dissolved but have not reacted with the water. The reactions occurring in the solution may be represented by the equations:

$$SO_2 + H_2O \rightleftharpoons H_2SO_3$$
$$H_2SO_3 + H_2O \rightleftharpoons H_3O^+ + HSO_3^- \qquad \text{(slight)}$$

Further ionization of the HSO_3^- ions will occur but only to a very slight extent:

$$HSO_3^- + H_2O \rightleftharpoons H_3O^+ + SO_3^{2-} \qquad \text{(very slight)}$$

This solution is called sulphurous acid and because it contains several different molecules and ions it cannot be simply represented by any one formula. However, as a matter of convenience, the formula H_2SO_3 is usually used to represent the solution.

REACTION AS AN OXIDIZER

Sulphur dioxide can act as an oxidizer as was shown by its reaction with burning magnesium (above) and its reaction with hydrogen sulphide (previous chapter). In both of these reactions the sulphur dioxide is reduced to sulphur, e.g.:

$$SO_2 + 2H_2S \rightarrow 3S + 2H_2O$$

REACTION AS A REDUCER

Although sulphur dioxide can act as an oxidizer, its more important reactions are those in which it behaves as a reducer. Thus the oxidation of sulphur dioxide by oxygen to form sulphur trioxide is the basis for the production of sulphuric acid. This reaction is discussed fully, later in this chapter. In the

laboratory, sulphur dioxide can be used in a number of redox reactions.

If a gas jar of moist chlorine is placed on top of a gas jar of moist sulphur dioxide, the yellow-green colour of the chlorine fades and suitable tests show that hydrogen chloride gas is formed. Thus the chlorine is reduced. If the gases are dry, no reaction takes place when they are mixed and therefore moisture is necessary. The reducer is usually taken to be sulphurous acid and is represented as H_2SO_3 as explained above. The products of the reaction between moist chlorine and moist sulphur dioxide can be identified by adding water to the gas jars when positive tests can be obtained for sulphate ions and chloride ions. The half reaction for the reducer may be represented as:

$$H_2SO_3 + H_2O \rightarrow SO_4^{2-} + 4H^+ + 2e^-$$

Addition to the half reaction for the oxidizer:

$$Cl_2 + 2e^- \rightarrow 2Cl^-$$

gives the total equation for the redox reaction:

$$H_2SO_3 + H_2O + Cl_2 \rightarrow 4H^+ + SO_4^{2-} + 2Cl^-$$

The other redox reactions to be discussed also involve water and are reactions of sulphurous acid.

Reducing Properties of Sulphurous Acid

Further redox reactions of sulphur dioxide solution can be observed if concentrated nitric acid and acidified solutions of iron (III) chloride, potassium permanganate and potassium dichromate are each saturated with sulphur dioxide. The colour changes indicate that reduction takes places:

$$H_2SO_3 + 2NO_3^- \rightarrow SO_4^{2-} + \underset{\text{brown gas}}{2NO_2} + H_2O$$

$$\underset{\text{yellow}}{H_2SO_3 + 2Fe^{3+}} + H_2O$$
$$\rightarrow SO_4^{2-} + \underset{\text{green}}{2Fe^{2+}} + 4H^+$$

$$5H_2SO_3 + \underset{\text{purple}}{2MnO_4^-}$$
$$\rightarrow 5SO_4^{2-} + \underset{\text{colourless}}{2Mn^{2+}} + 4H^+ + 3H_2O$$

$$3H_2SO_3 + \underset{\text{orange}}{Cr_2O_7^{2-}} + 2H^+$$
$$\rightarrow 3SO_4^{2-} + \underset{\text{green}}{2Cr^{3+}} + 4H_2O$$

The presence of sulphate ions and the other ions can be verified in each case by applying appropriate tests to the solution.

Sulphurous acid is also oxidized by hydrogen peroxide, but in this case there are no colour changes:

$$H_2SO_3 + H_2O_2 \rightarrow SO_4^{2-} + 2H^+ + H_2O$$

Oxidation also occurs if sulphurous acid is exposed to air:

$$2H_2SO_3 + O_2 \rightarrow 4H^+ + 2SO_4^{2-}$$

Because of this reaction, sulphurous acid cannot be kept for very long.

Sulphur dioxide can bleach some coloured substances such as wool or straw. If a coloured flower, e.g., a purple iris, is placed in sulphur dioxide it is bleached white. The bleaching action is due to reduction by sulphurous acid.

Uses of Sulphur Dioxide

Sulphur dioxide is used as a source of sulphur trioxide and hence sulphuric acid. It is also widely used as a bleaching agent for wool, silk and straw. The bleaching does not harm the fibres but is not permanent and the natural colour of the fibre returns very slowly. Sulphur dioxide is used as a fumigant to destroy bacteria and moulds and both its fumigant and bleaching properties are applied for the preservation of dried fruits.

Liquid sulphur dioxide is used in the purification of the kerosene and light oil fractions of petroleum. It was once used in refrigeration but has now been largely replaced by non-toxic substances, particularly for domestic refrigerators.

Tests for Sulphur Dioxide

A gas can be identified as sulphur dioxide if it is a colourless gas with a characteristic choking, irritating odour and it turns moist blue litmus paper red and also turns filter paper soaked in potassium dichromate from pale orange to a blue-green colour.

It should be noted that hydrogen sulphide gas causes the same colour change in potassium dichromate but the odour and the effect on $Pb(NO_3)_2$ solution will clearly distinguish the two gases.

Preparation of Sodium Sulphite and Sodium Hydrogen Sulphite

If caustic soda solution is saturated with sulphur dioxide, the acid salt is formed because of the presence of excess sulphurous acid:

$$SO_2 + H_2O \rightarrow H_2SO_3$$
$$H_2SO_3 + OH^- \rightarrow HSO_3^- + H_2O$$

The overall reaction may be represented by one equation:

$$SO_2 + Na^+ + OH^- \rightarrow Na^+ + HSO_3^-$$
(sodium hydrogen sulphite)

This reaction replaces each OH^- ion by one HSO_3^- ion. Thus, if the same volume of caustic soda solution as was used before, is now added to the hydrogen sulphite solution, one OH^- ion will be available to react with each HSO_3^- ion:

$$HSO_3^- + OH^- \rightarrow SO_3^{2-} + H_2O$$

The overall reaction may be represented:

$$Na^+ + HSO_3^- + Na^+ + OH^-$$
$$\rightarrow 2Na^+ + SO_3^{2-} + H_2O$$
(sodium sulphite)

Sodium sulphite can also be prepared from sodium carbonate and sulphur dioxide:

$$2Na^+ + CO_3^{2-} + SO_2 \rightarrow 2Na^+ + SO_3^{2-} + CO_2$$

Uses of Sulphites

Sulphites and hydrogen sulphites are used like sulphur dioxide for bleaching, disinfecting and preserving, and also for removing traces of chlorine from bleached cotton and linen.

Hydrogen sulphites are also used in the paper industry. When wood is digested with calcium hydrogen sulphite, $Ca(HSO_3)_2$, the lignin which binds the cellulose fibres together is dissolved and the cellulose fibers are separated and bleached. The product is called paper pulp.

Tests for Sulphites

Soluble sulphites form a white precipitate with barium chloride solution:

$$Ba^{2+} + SO_3^{2-} \rightarrow BaSO_3 \text{ (s)}$$

Sulphates also form a white precipitate, but the sulphite can be distinguished by the solubility of the precipitate in hot dilute hydrochloric acid:

$$BaSO_3 + 2H^+ \rightarrow Ba^{2+} + H_2O + SO_2$$

Barium sulphite is soluble in hot dilute hydrochloric acid but barium sulphate is insoluble.

The reason for this difference is that SO_3^{2-} is a moderately strong base and readily accepts protons, but SO_4^{2-} is a very weak base. The removal of SO_3^{2-} ion causes the $BaSO_3$ to dissolve.

SULPHUR TRIOXIDE, SULPHURIC ACID AND SULPHATES

Sulphur Trioxide

In contrast to all the reactions involving oxidation of sulphur dioxide discussed so far, the formation of sulphur trioxide can be brought about by using *dry* sulphur dioxide and *dry* oxygen.

$$2SO_2 + O_2 \rightleftharpoons 2SO_3$$

The extent of this reaction decreases as the temperature is increased and so it would seem desirable to prepare sulphur trioxide at a relatively low temperature. However, if the temperature is too low the reaction becomes very slow. By using a suitable catalyst, the sulphur dioxide and oxygen can be made to combine very rapidly at a moderate temperature (400 °C), and the conversion is almost complete (97-98%).

Platinum was formerly used as a catalyst but vanadium pentoxide and its related compounds are now widely used. The reaction takes place on the surface of the catalyst and thus the catalyst is usually treated so as to expose a very large surface area. If platinum is used as the catalyst it is spread over the surface of an inert carrier such as asbestos. This can be done on a small scale by dropping a solution of hexachloroplatinic acid, H_2PtCl_6 (sometimes labelled 'platinum chloride'), on asbestos fibre and then heating the asbestos until no more fumes are evolved. The product is called platinized asbestos or platinum black because the finely divided platinum is black. If vanadium compounds are used, they are mixed with inert substances such as zeolites or silica gel.

Laboratory Preparation of Sulphur Trioxide

A dry mixture of sulphur dioxide and oxygen is passed through a heated tube containing platinized asbestos (fig. 32.2).

The sulphur trioxide collects as a white fume which can be condensed to a white solid by passing it through a vessel cooled by ice. It is very difficult to dissolve sulphur trioxide completely in water but if water is

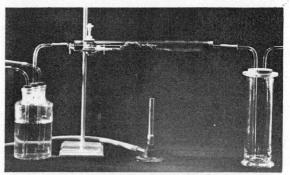

Fig. 32.2. Laboratory preparation of sulphur trioxide: oxygen and sulphur dioxide are dried by concentrated sulphuric acid and the mixture of gases is passed over heated platinized asbestos.

added, it will be found that a strongly acidic solution is formed.

Properties of Sulphur Trioxide

Sulphur trioxide is a volatile substance which decomposes on heating. Very little decomposition takes place up to 400-450 °C and this is the temperature range used in the preparation of sulphur trioxide. However, at higher temperatures, the extent of decomposition increases and is almost complete at 1000 °C.:

$$2SO_3 \rightarrow 2SO_2 + O_2$$

Sulphur trioxide is an acidic oxide, forming sulphuric acid with water:

$$SO_3 + H_2O \rightarrow H_2SO_4$$

The reaction is so vigorous that a mist of acid droplets forms rather than a solution. The mist remains suspended in the gas and it is almost impossible to absorb the sulphur trioxide completely once the mist forms. This difficulty can be avoided by dissolving the sulphur trioxide in fairly concentrated sulphuric acid instead of water. As the sulphur trioxide is absorbed the concentration of the acid increases. The acid may then be diluted and some of it used to absorb more sulphur trioxide.

Contact Process for Sulphuric Acid

The contact process is a widely used commercial method for the production of sulphuric acid and is designed to use contact catalysts (solids), e.g., vanadium pentoxide. A convenient method of obtaining

Fig. 32.3. Production of sulphuric acid.

The catalyst chamber of the contact plant at Yarraville, Victoria. Sulphur trioxide is produced by combination of sulphur dioxide and oxygen.

the sulphur dioxide is to burn sulphur in excess air which has been previously dried.

The combustion of the sulphur produces a great deal of heat and the gases from the burner must be cooled down to about 400 °C before passing to the catalyst. The hot gases are cooled by being used to generate steam, which provides energy for the operation of ancillary equipment.

In the catalyst chamber 97-98% conversion to sulphur trioxide is achieved and more heat is evolved:

$$2SO_2 + O_2 \rightleftharpoons 2SO_3 \quad \text{(exothermic)}$$

The gases leave the catalyst chamber at 400-450 °C and are cooled before passing to the absorption tower, where the sulphur trioxide reacts to form sulphuric acid:

$$H_2O + SO_3 \rightarrow H_2SO_4$$

If excess sulphur trioxide is absorbed, pyro-sulphuric acid is formed:

$$H_2SO_4 + SO_3 \rightarrow H_2S_2O_7$$

The technical names for this acid are fuming sulphuric acid or oleum.

Physical Properties of Sulphuric Acid

Sulphuric acid is usually sold as a 98% solution which is an oily liquid having a density of 1·84 g cm^{-3} and a boiling point of nearly 340 °C. 100% H_2SO_4 is not very stable and when heated it loses sulphur trioxide to form the 98% solution which is the azeotropic mixture.

Chemical Properties of Sulphuric Acid

IONIZATION

The 100% H_2SO_4 has a slight conductivity which indicates some ionization:

$$2H_2SO_4 \rightleftharpoons H_3SO_4^+ + HSO_4^-$$

It is also likely that ions such as H_3O^+ and $HS_2O_7^-$ are also present.

REACTION WITH WATER

When sulphuric acid reacts with water it evolves a considerable amount of heat. Because of this, the acid is always added to the water. If water is added to concentrated sulphuric acid the water may be vaporized to steam and spit out corrosive droplets of the acid. A dilute sulphuric acid solution is a good electrolytic conductor and thus sulphuric acid is a strong acid. Its reaction with water may be represented as:

$$H_2SO_4 + H_2O \rightarrow H_3O^+ + HSO_4^-$$
$$HSO_4^- + H_2O \rightleftharpoons H_3O^+ + SO_4^{2-}$$

DEHYDRATING PROPERTIES

Sulphuric acid is a strong absorber of water and is often used to dry gases. It reacts with substances which contain water, e.g., hydrated copper(II) sulphate. The blue crystals turn white in concentrated sulphuric acid but if the acid is drained off and then water is added (caution is needed here in case any acid is still present), the white residue turns blue. These reactions can be represented as:

$$\underset{\text{blue}}{CuSO_4,5H_2O} \rightleftharpoons \underset{\text{white}}{CuSO_4} + 5H_2O$$

The ability of concentrated sulphuric acid to dehydrate substances can also be seen by warming carbohydrates such as sugar or cellulose with the acid. A black residue of carbon is formed and so the elements of water are removed from the carbohydrate:

$$C_{12}H_{22}O_{11} \rightarrow 12C + 11H_2O$$
cane sugar

Formic acid and oxalic acid are also dehydrated by concentrated sulphuric acid (chapter 25).

The water which is removed from all of these substances is absorbed by the concentrated sulphuric acid:

$$H_2SO_4 + H_2O \rightarrow H_3O^+ + HSO_4^-$$

ACIDIC PROPERTIES

Because of its high boiling point, concentrated sulphuric acid can be used to prepare volatile acids such as hydrogen chloride and hydrogen nitrate from the corresponding salts:

$$H_2SO_4 + Cl^- \rightarrow HCl + HSO_4^-$$
$$H_2SO_4 + NO_3^- \rightarrow HNO_3 + HSO_4^-$$

In each of these reactions the H_2SO_4 donates a proton, H^+, to a Cl^- or NO_3^- ion. The H_2SO_4 is therefore functioning as an acid in these reactions.

The reactions occur, not because of any difference in acid strength, but because the products, HCl and HNO_3, are volatile.

OXIDIZING PROPERTIES

When copper is heated with concentrated sulphuric acid, sulphur dioxide can be identified as the gaseous product by its odour and effect on potassium dichromate:

$$Cu + H_2SO_4 + 2H^+ \rightarrow Cu^{2+} + SO_2 + 2H_2O$$

Thus concentrated sulphuric acid can act as an oxidizer.

Uses of Sulphuric Acid

Sulphuric acid is one of the most important chemicals manufactured. The wide use of sulphuric acid is due to the fact that it is the cheapest and most readily available strong acid.

Sulphuric acid is used in the manufacture of fertilizers such as superphosphate and ammonium sulphate and also in the manufacture of other chemicals including acids, dyes, drugs and explosives.

Some other uses include the refining of petroleum products, cleaning scale from steel which is to be drawn into wire and in electrolytic cells and lead storage batteries.

Preparation of Sodium Sulphate and Sodium Hydrogen Sulphate

Sodium sulphate can be prepared by adding dilute sulphuric acid to caustic soda solution until the resulting solution is neutral. This can be ascertained by using an indicator such as methyl orange:

$$2H^+ + SO_4^{2-} + 2Na^+ + 2OH^-$$
$$\rightarrow 2Na^+ + SO_4^{2-} + 2H_2O$$

Sodium hydrogen sulphate can be prepared by using twice the volume of acid and the same volume of alkali solution used in the preparation of sodium sulphate. Evaporation of these solutions yields sodium sulphate, Na_2SO_4, and sodium hydrogen sulphate, $NaHSO_4$. If the sodium sulphate solution is evaporated at a temperature above 32 °C, the anhydrous salt forms, but evaporation at temperatures below 32 °C yields the decahydrate, $Na_2SO_4,10H_2O$, which is known as Glauber's salt. Sodium hydrogen sulphate is obtained commercially as a by-product from the manufacture of nitric acid.

Reactions of Sulphates and Hydrogen Sulphates

The usual test for a sulphate or hydrogen sulphate is the addition of barium chloride solution which gives a white precipitate of barium sulphate:

$$Ba^{2+} + SO_4^{2-} \rightarrow BaSO_4$$

Both sodium sulphate and sodium hydrogen sulphate give this white precipitate with barium chloride.

Sodium sulphate and sodium hydrogen sulphate can be distinguished by testing the solutions with litmus. The sulphate solution is neutral but the hydrogen sulphate ion makes a solution of sodium hydrogen sulphate acidic.

Uses of Sulphates

Sodium sulphate is used as a substitute for some of the sodium carbonate in the manufacture of glass and *sodium hydrogen sulphate* can be used as a substitute for sulphuric acid in the manufacture of hydrochloric acid and for cleaning steel.

Selenite, a clear crystalline form of gypsum.

Calcium sulphate in the form of the mineral gypsum, $CaSO_4,2H_2O$, is used to manufacture plaster of Paris. The gypsum is heated at about 120-130 °C to partially dehydrate it, forming a residue in which the degree of hydration is $\frac{1}{2}$. This is plaster of Paris and may be represented as $CaSO_4,\frac{1}{2}H_2O$ or $(CaSO_4)_2,H_2O$. When mixed to a paste with water it changes back to $CaSO_4,2H_2O$ and sets to a solid which consists of a mass of interlacing prismatic crystals and is fairly hard. It expands slightly as it hardens and so is used for making casts.

If gypsum is heated at 200 °C, it loses all of its water of crystallization and the anhydrous salt formed will set only very slowly if mixed with water.

Aluminium sulphate crystallizes from solution as $Al_2(SO_4)_3,18H_2O$. However, it is more commonly crystallized with potassium sulphate as a double salt called potassium alum, $K_2SO_4.Al_2(SO_4)_3,24H_2O$ or $KAl(SO_4)_2,12H_2O$. This is only one of a series of double sulphates having similar empirical formulae which are known generally as alums.

Iron(II) sulphate, $FeSO_4,7H_2O$, is used as a disinfectant and a weed killer. It is used with tannin, gum and a blue dye to make blue-black writing ink. Iron(III) oxide, which is used as an abrasive (jeweller's rouge), is obtained by thermal decomposition of iron(II) sulphate:

$$FeSO_4,7H_2O \rightarrow FeSO_4 + 7H_2O$$
$$2FeSO_4 \rightarrow Fe_2O_3 + SO_2 + SO_3$$

A double salt obtained by crystallizing a mixture of iron(II) sulphate and ammonium sulphate, $(NH_4)_2SO_4, FeSO_4,6H_2O$, known as Mohr's salt, is often used in preference to iron(II) sulphate because it is not so readily oxidized by air.

Copper(II) sulphate, $CuSO_4,5H_2O$, sometimes called blue stone or blue vitriol, is used as a disinfectant and a fungicide.

Magnesium sulphate, $MgSO_4,7H_2O$, known as Epsom salts, is used as purgative.

STUDY QUESTIONS

1. Sulphur dioxide is obtained as a by-product when galena is burnt to litharge. Write the equation for this reaction.

2. Discuss reactions which show that sulphurous acid is unstable, and is a reducer.

3. Discuss reactions which show that sulphuric acid is a dehydrator, an oxidizer and an acid.

4. Write an equation to represent the reaction involved in the setting of plaster of Paris.

5. Four white solids are available. They are known to be sodium sulphite, sodium sulphate, sodium hydrogen sulphate and ammonium sulphate, but they are in unlabelled jars. Describe and explain with the aid of equations how each could be identified.

6. Calculate the volume of sulphur dioxide, expressed at s.t.p., formed by:
   (a) burning 8.0 g of sulphur;
   (b) treating 2.10 g of Na_2SO_3 with excess dilute sulphuric acid and warming.
   (Na = 23.0,   S = 32.1,   O = 16.0.)

7. Calculate the mass of sulphuric acid which could theoretically be produced from one kg of sulphur.
   (H = 1.0,   S = 32.1,   O = 16.0.)

8. Explain why sulphur dioxide and oxygen are reacted at about 400—450 °C in the industrial production of H_2SO_4.

Answers to Numerical Problems
   6. (a) 5.6 dm³. (b) 373 cm³.
   7. 3.1 kg.

33 : Chlorine and Chlorides

Valency of Chlorine

Chlorine is in group VII of the periodic table and each Cl atom has 7 electrons in its outer shell. Thus chlorine may have an electrovalency of -1, as in the ionic compound Na^+Cl^-, or a covalency of one, as in the molecular compound $H-Cl$.

However, chlorine is not restricted to a covalency of one. It can have covalencies of 1, 3, 5 or 7 as in the acids:

hypochlorous   chlorous   chloric   perchloric
   acid   acid   acid.   acid

Occurrence of Chlorine

About 0.2% by mass of the earth's crust, including the oceans, consists of chlorine. The element is much too reactive to occur in the free or uncombined state and it occurs mainly as sodium chloride.

The oceans contain about 2.7% of dissolved sodium chloride by mass, and vast underground deposits of solid sodium chloride, called rock salt, also occur.

Other chlorides including potassium chloride and magnesium chloride occur in sea water and these salts also occur in a few places as mineral deposits, e.g., in Germany.

Extraction of Common Salt

Enormous quantities of common salt are required annually. In Australia it is extracted by solar evaporation of sea water, whereas in some other countries the principal source of salt is rock salt. In Europe, rock salt is usually obtained by mining, whereas in America, rock salt is often dissolved in water and the resulting solution (brine) is pumped to the surface for subsequent evaporation.

Uses of Common Salt

Common salt is an essential item in man's diet and it is also used for preservation of food. However, industrial uses consume most of the sodium chloride produced. It is the source of chlorine and hydrochloric acid, and almost all sodium compounds. In addition, it is used in preserving hides, in glazing pottery and in many other fields.

Recovery of salt from sea water by solar evaporation.

Right: aerial view of evaporation areas near Laverton, Victoria.
Below: mechanical harvesting of salt at Dry Creek, South Australia.

Crystals of common salt, magnified 100 times.

The use of sodium chloride in the production of chlorine gas and sodium metal has been discussed already. Another important use involves the electrolysis of a concentrated aqueous solution of common salt to produce chlorine, caustic soda and hydrogen, all of which are useful products. The electrolysis can be carried out in a diaphragm type cell, e.g., the Nelson Cell.

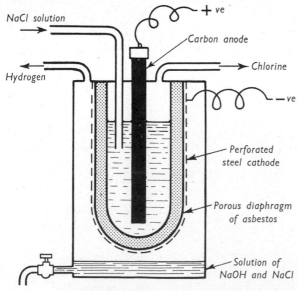

Fig. 33.1. The Nelson cell.

The Nelson Cell

A nearly saturated solution of purified sodium chloride is electrolyzed in a cell containing a partition or diaphragm separating the electrodes—see the diagram, fig. 33.1. The diaphragm prevents mixing of the products at the two electrodes.

Chloride ions migrate to the anode and are discharged:

ANODE REACTION:

$$2Cl^- \rightarrow Cl_2 + 2e^-$$

The anode is made of carbon because this is not attacked by the chlorine.

Sodium ions migrate to the cathode but they are not discharged. Instead hydrogen is evolved at the cathode.

CATHODE REACTION:

$$2H_2O + 2e^- \rightarrow H_2 + 2OH^-$$

The hydroxide ions produced, are partnered by the sodium ions arriving by migration. The liquid dripping from the cathode is therefore a solution of sodium ions and hydroxide ions, i.e., caustic soda solution, still containing about 12% sodium chloride.

Because the chlorine is produced in the inner compartment, it does not react with either the sodium hydroxide solution or the hydrogen gas produced in the outer compartment.

If the concentration of sodium chloride falls below 12%, oxygen is evolved at the anode as well as chlorine:

$$2H_2O \rightarrow O_2 + 4H^+ + 4e^-$$

This is avoided because it causes the chlorine gas to be contaminated with oxygen.

The solution of sodium hydroxide and sodium chloride obtained from the cell is concentrated by evaporation and most of the sodium chloride crystallizes out. The concentrated solution of sodium hydroxide is finally evaporated to dryness, still containing about 2% sodium chloride.

The solid sodium hydroxide is melted, cast into sticks, pellets, or flakes and sealed in containers.

The chlorine is liquefied by cooling it under pressure and is sold as the liquid in cylinders. Some of the hydrogen and chlorine are burnt together to give hydrogen chloride gas, which is dissolved in water to give hydrochloric acid of about 33% concentration.

HYDROGEN CHLORIDE

Laboratory Preparation and Collection

Hydrogen chloride gas is prepared by the reaction of concentrated sulphuric acid with sodium chloride in the cold. The acid, H_2SO_4, donates a proton to the very weak base, Cl^-.

$$H_2SO_4 + Cl^- \rightarrow HCl + HSO_4^-$$

The gas is collected by upward displacement of air—see fig. 33.2.

The residue is sodium hydrogen sulphate, $NaHSO_4$, and this will react with more sodium chloride at red heat. However, this is unsuitable for laboratory preparations, although it is used commercially:

$$HSO_4^- + Cl^- \rightarrow HCl + SO_4^{2-}$$

Properties of Hydrogen Chloride

PHYSICAL PROPERTIES

Hydrogen chloride is a colourless gas although it fumes in moist air. It has a sharp penetrating odour and a choking effect. The gas is denser than air. (The molecular mass of HCl is about 36.5.)

If a gas jar of hydrogen chloride is inverted into water and the cover is removed, water rushes into the gas jar with considerable violence. The gas is extremely soluble in water and may be used in the fountain experiment—see chapter 28. At room conditions, 1 volume of water dissolves about 440 volumes of hydrogen chloride.

The gas is fairly easily liquefied (b.p. $= -84\,°C$ at 1 atmosphere pressure).

CHEMICAL PROPERTIES

Hydrogen chloride is a stable gas and is hardly decomposed at all even at temperatures as high as 1000 °C.

If a burning wax taper is introduced into the gas, the taper goes out and the gas does not burn.

A solution of hydrogen chloride is acidic and is a good electrolytic conductor. Thus hydrochloric acid is a strong acid:

$$HCl + H_2O \rightarrow H_3O^+ + Cl^-$$

The fuming of hydrogen chloride gas in moist air is due to the production of minute droplets of hydrochloric acid and the choking effect of the gas is due to the production of the acid in the respiratory tract.

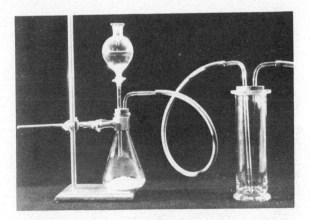

Fig. 33.2. Laboratory preparation and collection of hydrogen chloride: concentrated sulphuric acid is added to sodium chloride.

Hydrogen chloride gas forms a white cloud with ammonia gas. The acid, HCl, donates a proton to the base, NH_3, forming the conjugates NH_4^+ and Cl^-. These ions condense to form a cloud of white crystalline particles of $NH_4Cl\,(s)$.

$$HCl + NH_3 \rightleftharpoons NH_4^+\,Cl^-(s)$$

Tests for Hydrogen Chloride

The gas is usually identified by odour and effect on litmus. The formation of white clouds with ammonia does not provide clear identification of the gas, because other gases such as hydrogen bromide and hydrogen iodide react in a similar way. However, the formation of a white curdy precipitate with silver nitrate solution is a good test:

$$Ag^+ + Cl^- \rightarrow AgCl\,(s)$$

This precipitate turns purple in sunlight, and is soluble in ammonia solution—see under chlorides.

HYDROCHLORIC ACID

Production

The commercial production of hydrochloric acid involves first the formation of hydrogen chloride gas and then its dissolution in water.

1. *Pure hydrochloric acid* is obtained by burning hydrogen and chlorine together and dissolving the gas in water in small towers which are packed with stoneware or some other inert substance.

2. *Commercial grades* of hydrochloric acid are produced by heating sodium hydrogen sulphate (obtained from nitric acid manufacture) with sodium chloride at temperatures of about 650 °C. Alternatively, sulphuric acid and common salt are heated in stages:

$$H_2SO_4 + Cl^- \rightarrow HCl + HSO_4^-$$
$$HSO_4^- + Cl^- \rightarrow HCl + SO_4^{2-}$$

The gas produced is again dissolved in water in small towers. The hydrochloric acid produced is pale yellow due to the presence of iron compounds.

Uses of Hydrochloric Acid

The acid is used industrially for de-scaling steel sheet prior to galvanizing and also for the production of ammonium chloride. In addition, it is a very important laboratory reagent.

Properties of Hydrochloric Acid

Hydrochloric acid is a colourless fuming liquid and, like nitric acid and sulphuric acid, it forms a constant-boiling or azeotropic mixture—see fig. 33.3. If an aqueous solution of hydrochloric acid is boiled, the residue gradually approaches a concentration of 20% HCl by mass, and a boiling point of 110 °C, irrespective of whether the original solution is dilute or concentrated. A saturated solution of HCl in water at room conditions contains about 40% HCl by mass.

Hydrochloric acid is a strong monoprotic acid. Its reactions with metals, basic oxides, bases and carbonates have been discussed in earlier chapters.

It is oxidized by powerful oxidizers such as manganese(IV) oxide and potassium permanangate to form chlorine, and is therefore a reducer.

CHLORINE

Chlorine is produced commercially by the electrolysis of a concentrated aqueous solution of sodium chloride. It is also produced by the electrolysis of fused chlorides.

Fig. 33.4. Laboratory preparation and collection of chlorine: concentrated hydrochloric acid is heated with manganese(IV) oxide and the gas is bubbled through water to remove hydrogen chloride.

Laboratory Preparation and Collection

Chlorine is often prepared in the laboratory by the method used by Scheele, who discovered chlorine in 1774. The method consists of heating concentrated hydrochloric acid with manganese(IV) oxide. The chlorine produced is contaminated with hydrogen chloride gas and is purified by passing it through water—see fig. 33.4. The equation for the reaction is:

$$MnO_2 + 4H^+ + 2e^- \rightarrow Mn^{2+} + 2H_2O$$
$$2Cl^- \rightarrow Cl_2 + 2e^-$$
$$\overline{MnO_2 + 4H^+ + 2Cl^- \rightarrow Mn^{2+} + 2H_2O + Cl_2}$$

If dry chlorine is required, the gas can be bubbled through concentrated sulphuric acid after being washed free of acid spray.

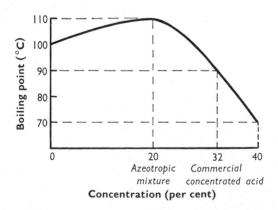

Fig. 33.3. Boiling point curve of aqueous solutions of hydrochloric acid at one atmosphere pressure.

A somewhat more convenient method of preparing chlorine is to drop concentrated hydrochloric acid on solid potassium permanganate, which oxidizes the hydrochloric acid at room temperature.

● *CAUTION: It is imperative to make sure that concentrated sulphuric acid is not reacted with potassium permanganate by mistake, for this could lead to a dangerous explosion.*

Physical Properties of Chlorine

Chlorine is a greenish-yellow, pungent smelling gas which has a corrosive effect on the respiratory tract. It is dangerous even in small amounts. The gas is about $2\frac{1}{2}$ times as dense as air (the molecular mass of Cl_2 is about 71), and it is fairly soluble in water, 1 volume of which dissolves about 2.3 volumes of the gas at room conditions. Chlorine is easily liquefied and its boiling point is -34 °C at atmospheric pressure. It can be liquefied by the application of 6.6 atmosphere pressure at a temperature of 20 °C.

Chemical Properties of Chlorine

Chlorine is a thermally stable gas, being only very slightly dissociated into atoms even at 1000 °C.

SUPPORT OF COMBUSTION

The gas does not burn in air, but it supports the combustion of a wide variety of substances including many elements. However, it does not react directly with carbon, oxygen or nitrogen. Chlorine supports the combustion of:

1. *Hydrogen and Hydrogen Compounds.*

(a) Hydrogen: If hydrogen burning at a jet is lowered into a jar of chlorine, the hydrogen continues to burn with an enlarged flame, producing clouds of white fumes—see fig. 33.5.

$$H_2 + Cl_2 \rightarrow 2HCl$$

(See chapter 22 for precautions when lighting hydrogen.)

(b) Wax tapers: A burning wax taper continues to burn in chlorine with a dull red smoky flame and the production of white fumes. The white fumes form a white cloud with ammonia.

The wax of candles and tapers usually contains long chain acids such as stearic acid, $C_{17}H_{35}COOH$. The wax reacts with chlorine to form mainly hydrogen chloride and carbon.

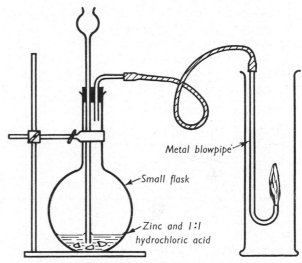

Fig. 33.5. Combustion of hydrogen in chlorine.

Metal blowpipe

Small flask

Zinc and 1:1 hydrochloric acid

(c) Turpentine: If glass wool or filter paper is soaked in warm turpentine and then dropped in chlorine, the turpentine ignites spontaneously. Turpentine is a mixture of hydrocarbons and it burns in chlorine to form hydrogen chloride and carbon.

2. *Metals.*

(a) Burning sodium continues to burn in chlorine with a yellow flame and the production of a white cloud of sodium chloride:

$$2Na + Cl_2 \rightarrow 2NaCl$$

(b) Burning magnesium continues to burn in chlorine with a spluttering sound and the production of white clouds:

$$Mg + Cl_2 \rightarrow MgCl_2$$

(c) Red hot iron continues to glow in chlorine producing clouds of dark brown fumes. The anhydrous iron(III) chloride formed, has the molecular formula Fe_2Cl_6 in the vapour state at moderate temperatures:

$$2Fe + 3Cl_2 \rightarrow Fe_2Cl_6$$

(d) Finely powdered antimony sparks spontaneously when sprinkled into chlorine and it produces clouds of white fumes consisting of antimony pentachloride:

$$2Sb + 5Cl_2 \rightarrow 2SbCl_5$$

The white fumes are poisonous.

3. *Non-metals.*

(a) White phosphorus takes fire spontaneously in chlorine, producing white fumes and some colourless droplets. The colourless droplets are phosphorus trichloride, PCl_3, and the white solid is phosphorus pentachloride, PCl_5:

$$P_4 + 6Cl_2 \rightarrow 4PCl_3$$
$$P_4 + 10Cl_2 \rightarrow 4PCl_5$$

(b) Burning sulphur continues to burn in chlorine but with difficulty. The product is a yellow liquid containing a mixture of chlorides of sulphur, mainly disulphur dichloride (also called sulphur monochloride), S_2Cl_2:

$$2S + Cl_2 \rightarrow S_2Cl_2$$

With excess chlorine some sulphur dichloride forms:

$$S + Cl_2 \rightarrow SCl_2$$

NATURE OF SOLUTION OF CHLORINE IN WATER

A solution of chlorine in water is greenish-yellow in colour and is called chlorine water. This solution is strongly acidic to litmus on account of fairly considerable reaction:

$$Cl_2 + H_2O \rightleftharpoons H^+ + Cl^- + HOCl$$

The substance HOCl is hypochlorous acid, and because it is relatively weak, the acid is written in the molecular form.

Solutions of hypochlorous acid are unstable, especially in sunlight, and gradually evolve oxygen:

$$2HOCl \rightarrow 2H^+ + 2Cl^- + O_2$$

Thus chlorine water slowly loses its colour on standing. Eventually it becomes a solution of hydrochloric acid.

CHLORINE AS AN OXIDIZER AND BLEACHER

Chlorine is a powerful oxidizing agent. It acts as an oxidizer in all the reactions of combustion in chlorine, and it oxidizes many other substances, e.g., hydrogen sulphide:

$$H_2S + Cl_2 \rightarrow 2HCl + S$$

Moist chlorine is also a powerful bleaching agent and bleaches by oxidation. It bleaches litmus, cotton, linen and other vegetable fibres. The bleaching is probably due to the powerful oxidizing action of hypochlorous acid:

$$HOCl + H^+ + 2e^- \rightarrow H_2O + Cl^-$$

Chlorine is not used for bleaching the more delicate animal fibres such as wool and silk.

Excess chlorine must be removed from the fibres after bleaching, because it would oxidize the dye to be added and furthermore it slowly rots the fibres.

Test for Chlorine

Chlorine can be identified as a greenish-yellow gas with a choking odour. The gas bleaches moist litmus.

Uses of Chlorine

Chlorine is used widely as a bleaching agent, both as chlorine gas and as compounds of chlorine, largely hypochlorites. It is also used for sterilizing water in swimming pools, and in some countries it is used for sterilizing drinking water.

Large quantities of chlorine are used in the production of hydrochloric acid, many dyes and drugs, and many solvents.

CHLORIDES

The salts of hydrochloric acid are called chlorides. Most chlorides are soluble in water but silver chloride, lead(II) chloride, mercury(I) chloride and copper(I) chloride can be precipitated from solution.

Soluble chlorides may be identified by adding silver nitrate solution, which precipitates the chloride as white silver chloride:

$$Ag^+ + Cl^- \rightarrow AgCl \ (s)$$

This substance is insoluble in dilute nitric acid, turns purple in sunlight and is soluble in ammonia solution:

$$AgCl + 2NH_3 \rightarrow Ag(NH_3)_2^+ + Cl^-$$

The ion $Ag(NH_3)_2^+$ is the diamminesilver(I) cation.

A further typical reaction of chlorides is their reaction to produce chlorine when warmed with manganese(IV) oxide and concentrated sulphuric acid:

$$MnO_2 + 4H^+ + 2e^- \rightarrow Mn^{2+} + 2H_2O$$
$$2Cl^- \rightarrow Cl_2 + 2e^-$$
$$\overline{MnO_2 + 4H^+ + 2Cl^- \rightarrow Mn^{2+} + 2H_2O + Cl_2}$$

The chlorine is detected as described above.

CHLORINATED HYDROCARBONS

Many valuable solvents for oils, greases, waxes and rubber are made by reacting chlorine with hydrocarbons.

Methane, CH_4 (see fig. 33.6), is the simplest member of the alkanes, a series of hydrocarbons of general formula C_nH_{2n+2}. The members of this series are said to be saturated hydrocarbons because there are no double bonds in the molecules.

Methane reacts with chlorine in diffused light forming a mixture of products:

$$CH_4 + Cl_2 \rightarrow CH_3Cl + HCl$$
methyl chloride

$$CH_3Cl + Cl_2 \rightarrow CH_2Cl_2 + HCl$$
methylene chloride

$$CH_2Cl_2 + Cl_2 \rightarrow CHCl_3 + HCl$$
chloroform

$$CHCl_3 + Cl_2 \rightarrow CCl_4 + HCl$$
carbon tetrachloride

Chlorination of saturated hydrocarbons is used industrially for the production of mixtures of chlorinated hydrocarbons, for use as solvents. The reactions are called substitution reactions.

● *CAUTION: A mixture of methane and chlorine* **explodes violently** *on exposure to* **direct sunlight** *and the reaction forms a deposit of carbon:*

$$CH_4 + 2Cl_2 \rightarrow C + 4HCl$$

Ethylene, C_2H_4, is a gas which is obtained during petroleum refining. The structure of ethylene is shown in the diagrams, fig. 33.7. It is said to be an unsaturated hydrocarbon because of the presence of the double bond. Ethylene reacts with chlorine to form ethylene dichloride.

$$H_2C=CH_2 + Cl_2 \rightarrow ClH_2C—CH_2Cl$$

Such a reaction is called an addition reaction. The product is an important substance, both as a solvent and as a starting point for further chemical syntheses.

Ethylene is the first member of the alkenes, a series of hydrocarbons of general formula C_nH_{2n} The alkenes all have a double bond.

Acetylene, C_2H_2, is a gas which is obtained by adding water to calcium carbide (chapter 24). It, too, is an unsaturated hydrocarbon, and its structure is rep-

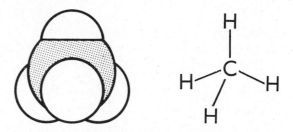

Fig. 33.6. Structure of methane.

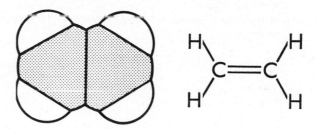

Fig. 33.7. Structure of ethylene.

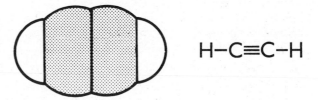

Fig. 33.8. Structure of acetylene.

resented in fig. 33.8. Acetylene is the first member of the alkyne series of hydrocarbons of general formula C_nH_{2n-2}

Acetylene reacts with chlorine in following way:

$$HC{\equiv}CH + 2Cl_2 \rightarrow Cl_2HC—CHCl_2$$

The product, called tetrachlorethane, is an excellent solvent. If tetrachlorethane, is treated with calcium hydroxide, it loses one molecule of HCl:

$$Cl_2HC—CHCl_2 \rightarrow ClHC=CCl_2 + HCl$$

The product is trichlorethylene, which is a widely used solvent for degreasing metal parts.

● *CAUTION: The vapours of many chlorinated hydrocarbons are poisonous.*

STUDY QUESTIONS

1. Write the equations for the electrode reactions which occur when a solution of sodium chloride is electrolyzed using inert electrodes, if:
(a) the solution is dilute;
(b) the solution is concentrated.
What explanation can be offered for the fact that the anode reactions are different?

2. Draw diagrams to represent the valency of chlorine in Cl_2, HCl, NaCl and PCl_5.

3. Explain what is meant by saying that aqueous HCl forms an azeotropic mixture.

4. Write the equation for the reaction of sodium chloride crystals with:
(a) cold concentrated sulphuric acid;
(b) sodium hydrogen sulphate at red heat.

5. Write the equation for the reaction of chlorine with:
(a) magnesium;
(b) phosphorus;
(c) cold water.

6. Write the equation for the reaction which occurs if concentrated hydrochloric acid is dropped on solid potassium permanganate.

7. If chlorine gas is passed into a solution of potassium bromide, bromine is produced. What explanation can be offered for this fact? Write the equation for the reaction which occurs.

8. If 20 cm³ of chlorine is mixed with 50 cm³ of hydrogen gas, what volume of hydrogen chloride gas is formed? (All volumes measured at the same temperature and pressure.)

9. What volume of chlorine, expressed at s.t.p., is produced by reacting 10.0 g of manganese(IV) oxide with excess concentrated hydrochloric acid?
(Mn = 54.9,   O = 16.0.)

10. What mass of silver chloride is formed by adding excess silver nitrate solution to 0.953 g of magnesium chloride in solution?
(Mg = 24.3,   Cl = 35.5,   Ag = 107.9.)

11. Compare and contrast oxygen and chlorine with respect to:
(a) molecular structure;
(b) reactivity;
(c) valency of the elements in their compounds;
(d) acidic nature of their hydrides H_2O and HCl.

12. If chlorine gas is passed over hot iron or hot tin, the higher valency chlorides are formed.
What explanation can be offered for this?

13. How would you endeavour to confirm experimentally, that the reaction of dry chlorine with hot aluminium foil produces a substance of molecular formula Al_2Cl_6?

14. Draw diagrams to represent the structures of the molecules CH_4, CH_3Cl, CH_2Cl_2, $CHCl_3$ and CCl_4.

15. By using methane and ethylene as examples, illustrate the difference in the reactions of saturated and unsaturated hydrocarbons with chlorine.

Answers to Numerical Problems

8. 40 cm³.
9. 2.58 dm³.
10. 2.87 g.

34 : The Halogens and their Hydrides

THE HALOGENS

The elements comprising group VII of the periodic table are called the halogens—a name which means 'salt producer'. This name is given because the halogens readily form anions and hence react with metals to form ionic salts, e.g. Na^+Cl^-.

Similarities between elements in the same group are most marked for groups at the extreme right and extreme left of the periodic table. The halogens are quite similar to one another, and in addition, they show fairly regular gradations from one element to the next. Gradations in properties are also quite strongly noticeable in the compounds of these elements.

The group VII elements are fluorine, chlorine, bromine, iodine and astatine. Astatine has been omitted from the following discussion because it has been isolated only in minute quantities. All the halogen atoms have seven electrons in their outer shells (see fig. 34.1) and therefore form singly-charged negative ions: fluoride, F^-; chloride, Cl^-; bromide Br^-; iodide, I^-.

The halogens also form molecular compounds. The covalency of fluorine is limited to a value of one, e.g.

| fluorine | hydrogen fluoride | oxygen difluoride |

Chlorine, bromine and iodine differ from fluorine in that they can have covalencies of 1, 3, and 5. Chlorine and iodine also show valencies of 7. For example, the acids of chlorine given in the previous chapter show these valencies.

Physical Properties of the Halogens

Some of the physical properties of the halogens are summarized in the table, fig. 34.2. The figures illustrate that the melting points, boiling points and densities increase from fluorine to iodine.

ELEMENT	SYMBOL	ATOMIC MASS	ATOMIC NUMBER	ELECTRON ARRANGEMENT K L M N O				
fluorine	F	18.998	9	2	7			
chlorine	Cl	35.453	17	2	8	7		
bromine	Br	79.904	35	2	8	18	7	
iodine	I	126.904	53	2	8	18	18	7

Fig. 34.1. The halogens.

HALOGEN	APPEARANCE at Room Temperature and Pressure	MELTING POINT (°C)	BOILING POINT (°C)	DENSITY (g cm⁻³)
fluorine	*pale yellow gas*	-223	-188	1.11 (at -188 °C)
chlorine	*greenish-yellow gas*	-103	-35	1.57 (at -35 °C)
bromine	*dark red-brown liquid*	-7	59	3.14
iodine	*lustrous purplish-black solid*	114	184	4.93

Fig. 34.2. Physical properties of the halogens.

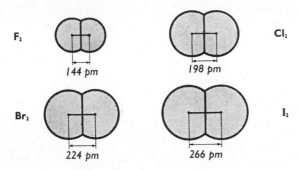

Fig. 34.3. The halogen molecules.

The Halogen Molecules

All the halogens form diatomic molecules: F_2, Cl_2, Br_2 and I_2. The relative sizes of the halogen molecules are shown in the diagrams, fig. 34.3.

The relative sizes of the molecules are helpful in explaining many of the physical and chemical properties of the halogens. With increasing size, the outer electrons are less tightly held in the molecule because they are further from the nuclei. These outer electrons therefore become freer to interact with electrons in adjacent molecules, causing an increase in the melting and boiling points. Furthermore, the difference in size between F_2 and Cl_2 is greater than the difference between Cl_2 and Br_2 or Br_2 and I_2 and these differences are paralleled by the differences between the melting points and also the boiling points of the halogens.

Occurrence of the Halogens

All the halogens occur in the form of compounds. The principal mineral of fluorine is fluorspar, CaF_2, which occurs in fairly large deposits. Chlorides are the most abundant halogen compounds and occur, together with bromides, in sea water and rock salt. A major source of iodine is sodium iodate which is present in small quantities in Chile saltpetre deposits. A certain amount of iodine is also obtained by burning kelp (a seaweed). The ash obtained in this way contains a small percentage of iodides. Astatine occurs naturally in barely detectable amounts; it was first recognized when trace amounts were made artificially. It has been omitted from the following discussions.

Preparation of the Halogens

Because the halogens occur naturally as the singly charged negative ions (halide ions), it is evident that the preparation of the elements involves the oxidation of the halide ion:

$$2X^- \rightarrow X_2 + 2e^-$$

The standard electrode potentials for a number of half-reactions are given in the table, fig. 34.4.

HALF-REACTION	STANDARD ELECTRODE POTENTIAL (VOLT)
$F_2 + 2e^- \rightleftharpoons 2F^-$	$+2.87$
$MnO_4^- + 8H^+ + 5e^- \rightleftharpoons Mn^{2+} + 4H_2O$	$+1.52$
$Cl_2 + 2e^- \rightleftharpoons 2Cl^-$	$+1.36$
$MnO_2 + 4H^+ + 2e^- \rightleftharpoons Mn^{2+} + 2H_2O$	$+1.28$
$Br_2 + 2e^- \rightleftharpoons 2Br^-$	$+1.07$
$Fe^{3+} + e^- \rightleftharpoons Fe^{2+}$	$+0.77$
$I_2 + 2e^- \rightleftharpoons 2I^-$	$+0.54$

Fig. 34.4. Standard electrode potentials for some half-reactions.

Fluorine is the most powerful of the chemical oxidizers, and it can be seen that the oxidizing strengths of the halogens decrease in the order F_2, Cl_2, Br_2, I_2. Conversely, the iodide ion is the strongest reducer of the halide ions and their reducing strengths decrease in the order I^-, Br^-, Cl^-, F^-. Thus the iodide ion can be oxidized by many oxidizers, e.g., Fe^{3+}, Br_2, Cl_2, etc., whereas the fluoride ion cannot be oxidized by any chemical oxidizer.

PRODUCTION BY ELECTROLYSIS

Fluorine is produced industrially by the electrolytic (anodic) oxidation of fluorides. If an aqueous solution of a fluoride is electrolyzed, oxygen is evolved at the anode, and hence to obtain fluorine by electrolysis, anhydrous fluorides must be used. A suitable electrolyte is potassium fluoride dissolved in anhydrous hydrogen fluoride. A graphite anode is used because graphite is not attacked by fluorine at moderate temperatures.

Chlorine is also produced industrially by electrolytic oxidation of chloride ions, although in this case it is not necessary to use anhydrous electrolytes. *Iodine* is also produced commercially, to a limited extent, by electrolytic oxidation of solutions of iodides.

PRODUCTION BY CHEMICAL OXIDATION

Chemical oxidizers are able to oxidize the ions Cl^-, Br^- and I^-. Thus the laboratory preparation of chlorine involves oxidation of concentrated hydrochloric acid with potassium permanganate or manganese(IV) oxide. The commercial production of bromine from sea water involves oxidation of bromide ions with chlorine:

$$2Br^- + Cl_2 \rightarrow Br_2 + 2Cl^-$$

Chemical oxidation of iodides present in kelp ash is often brought about by heating the ash with a mixture of manganese(IV) oxide and concentrated sulphuric acid:

$$MnO_2 + 4H^+ + 2I^- \rightarrow I_2 + Mn^{2+} + 2H_2O$$

Thermal Stability of the Halogens

When strongly heated, the diatomic molecules of the halogens dissociate into atoms:

$$X_2 \rightleftharpoons 2X$$
$$\text{or } :\overset{..}{X}:\overset{..}{X}: \rightleftharpoons :\overset{..}{X} + \cdot\overset{..}{X}:$$

The percentage dissociation of the halogens at 1000 K is shown in the table, fig. 34.5. The same trends are observable in the heats of dissociation shown in the table, fig. 34.6. The value of 155 for fluorine indicates that 155 000 joule are required to dissociate 1 mole of the gas into its atoms.

HALOGEN	PERCENTAGE DISSOCIATION AT 1000 K
fluorine	4.3
chlorine	0.035
bromine	0.23
iodine	2.8

Fig. 34.5. Percentage dissociation of the halogens at 1000 K.

HALOGEN	DISSOCIATION ENERGY (kJ mol^{-1})
fluorine	155
chlorine	239
bromine	190
iodine	149

Fig. 34.6. Heats of dissociation of the halogens.

The figures for percentage dissociation and heats of dissociation show that for chlorine, bromine, and iodine, the molecules become easier to dissociate as the size of the molecule increases. This is to be expected, because with increasing molecular size, the electrons comprising the bond are further from the nuclei and are less firmly held, due to the screening effect of the inner shells of electrons. The relatively high percentage dissociation and low heat of dissociation of fluorine, suggest that the bond between the two fluorine atoms is relatively weak and this partly explains the extraordinary reactivity of fluorine.

Chemical Reactions of the Halogens

REACTION WITH HYDROGEN

Hydrogen reacts with all the halogens forming covalent hydrides:

$$H_2 + X_2 \rightarrow 2HX$$
$$\text{or } H:H + :\overset{..}{X}:\overset{..}{X}: \rightarrow 2H:\overset{..}{X}:$$

Fluorine reacts explosively with hydrogen even at very low temperatures. Chlorine reacts explosively with hydrogen in direct sunlight, but more slowly in diffused light. Bromine reacts slowly in sunlight but more rapidly if heated and the reaction between hydrogen and iodine is slow and partial even on heating. Thus there is a gradation in the reactivity of the halogens towards hydrogen, fluorine being the most reactive and iodine the least reactive.

REACTION WITH METALS

The halogens are very reactive towards metals. The gradation in reactivity is shown by the fact that fluorine combines directly with all metals but iodine does not, e.g., iodine does not combine with platinum. The order of activity of the halogens is also illustrated by their combination with metallic sodium. The reactions involve the formation of ions:

$$2Na + X_2 \rightarrow 2Na^+X^-$$

The combination of a halogen with sodium generates heat and the quantity of heat evolved with each of the halogens is shown in fig. 34.7.

SODIUM HALIDE	HEAT OF FORMATION ($kJ\ mol^{-1}$)
NaF	569
NaCl	411
NaBr	360
NaI	288

Fig. 34.7. Heat evolved in formation of sodium halides.

It can be seen that the heat evolved in the formation of 1 mole of the salt Na^+X^- is greatest for NaF and least for NaI.

REACTION WITH WATER

Fluorine reacts very vigorously with water and the other halogens show a decreasing reactivity. Fluorine reacts in the following way:

$$2F_2 + 2H_2O \rightarrow 4HF + O_2$$

This reaction occurs completely and almost instantaneously.

The reactions of the other halogens with water can be represented:

$$X_2 + H_2O \rightleftharpoons H^+ + X^- + HOX$$

This reaction proceeds partially for both chlorine and bromine, but hardly at all for iodine.

REACTION WITH ALKALIS

The reactions of the halogens with alkalis form the ions corresponding to the acids which they form with water. Thus fluorine reacts with alkalis forming fluoride ions and oxygen gas:

$$2F_2 + 4OH^- \rightarrow 4F^- + O_2 + 2H_2O$$

However, with very dilute (2%) cold alkali, fluorine produces gaseous oxygen difluoride, OF_2:

$$2F_2 + 2OH^- \rightarrow OF_2 + 2F^- + H_2O$$

The other halogens react with cold alkalis as follows:

$$X_2 + 2OH^- \rightarrow X^- + XO^- + H_2O$$

(ClO^- ion = hypochlorite, BrO^- ion = hypobromite, IO^- ion = hypoiodite.)

With hot alkalis, chlorine, bromine and iodine react as follows:

$$3X_2 + 6OH^- \rightarrow XO_3^- + 5X^- + 3H_2O$$

(ClO_3^- ion = chlorate, BrO_3^- ion = bromate, IO_3^- ion = iodate.)

Uses of Halogens and their Compounds

Until quite recently elemental *fluorine* has been available only in small quantities and therefore most fluorine compounds are prepared from hydrogen fluoride. An example of this is the preparation of dichloro-difluoro-methane, CCl_2F_2, which is prepared by reaction of carbon tetrachloride with hydrogen fluoride in the presence of a catalyst:

$$CCl_4 + 2HF \rightarrow CCl_2F_2 + 2HCl$$

The substance CCl_2F_2, commonly called freon, is non-corrosive, non-toxic and non-inflammable. It is used as a refrigerant in domestic refrigerators and as a propellant in spray cans of insecticide.

Much of the *bromine* produced industrially is used in the preparation of ethylene dibromide, $Br-CH_2-CH_2-Br$. This substance is added to petrols to help remove lead formed in engine cylinders, resulting from the addition of tetraethyl lead to the petrol as an anti-knock additive.

Iodine is used as a solution in methylated spirits, called tincture of iodine. This solution has antiseptic properties and is used for the treatment of cuts. Many compounds of iodine are used in medicine because the compounds also have antiseptic properties. Iodine is also used in the manufacture of some dyes.

The uses of *chlorine* have been described already.

THE HALOGEN HYDRIDES

Preparation of the Halogen Hydrides

All four halogen hydrides are volatile and may be prepared by treating a metallic halide with a non-volatile acid.

Concentrated sulphuric acid may be used to prepare hydrogen fluoride and hydrogen chloride by reaction with calcium fluoride and sodium chloride respectively:

$$X^- + H_2SO_4 \rightarrow HX + HSO_4^-$$

Hydrogen chloride evolves readily, but hydrogen fluoride must be warmed to vaporize it because its boiling point is about 20 °C.

Bromide and iodide ions are oxidized to the halogen by concentrated sulphuric acid and hence a non-oxidizing and non-volatile acid must be used. Ortho-phosphoric acid has suitable properties and so may be used in preparing hydrogen bromide or hydrogen iodide:

$$X^- + H_3PO_4 \rightarrow HX + H_2PO_4^-$$

An alternative method of producing hydrogen bromide or hydrogen iodide is to hydrolyze the corresponding phosphorus trihalide with water:

$$PX_3 + 3HOH \rightarrow 3HX + H_3PO_3$$

Physical Properties of the Halogen Hydrides

The physical properties of the halogen hydrides are summarized in the table, fig. 34.8. The halogen

HALOGEN HYDRIDE	MELTING POINT (°C)	BOILING POINT (°C)
HF	−83	19.5
HCl	−115	−85
HBr	−87	−67
HI	−51	−35

Fig. 34.8. Physical properties of the halogen hydrides.

hydrides are colourless gases at room temperature except for hydrogen fluoride which is a colourless liquid below 19.5 °C. All fume in moist air and have pungent, choking odours. Hydrogen fluoride is particularly poisonous and is irritating to the skin.

Structure of the Halogen Hydrides

Anhydrous liquid halogen hydrides are very poor conductors of electricity. Their poor conductivity together with their low melting points indicate that they are covalent substances. The distances between

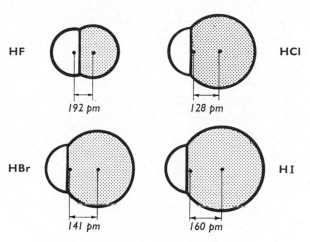

Fig. 34.9. The halogen hydride molecules.

the nuclei of the hydrogen and halogen atoms in the molecules of the gaseous halogen hydrides are shown in fig. 34.9.

At high temperatures and low pressures, gaseous hydrogen fluoride has a density which indicates a molecular mass corresponding to the molecular formula, HF, but at lower temperatures and higher pressures the density indicates that the molecular mass of the gas is considerably higher. This suggests that molecules of the gas combine together forming larger molecules, i.e., the gas molecules associate or polymerize. The unexpectedly high molecular mass of hydrogen fluoride explains the relatively high melting point and boiling point of the substance.

The association of molecules of hydrogen fluoride can be explained by a type of valency called *hydrogen bonding*. Hydrogen bonding occurs through unequal sharing of the electrons comprising the bond between the hydrogen and fluorine atoms in molecules of HF. The fluorine atoms gain a greater than equal share of the electrons, resulting in the fluorine atoms having a partial negative charge and the hydrogen atoms having a partial positive charge, causing electrostatic attractions between adjacent molecules. This is usually represented:

Hydrogen bonding also occurs in H_2O and NH_3.

Chemical Properties of the Halogen Hydrides

THERMAL STABILITY

The thermal stability of the halogen hydrides is illustrated by the figures in the table, fig. 34.10.

HALOGEN HYDRIDE	PERCENTAGE DISSOCIATION AT 1000 K
HF	—
HCl	0.0013
HBr	0.2
HI	28

Fig. 34.10. Percentage dissociation of the halogen hydrides at 1000 K.

It can be seen that hydrogen fluoride is the most stable and hydrogen iodide the least stable. The observed gradation is to be expected on the basis of the increasing distance between the centres of the hydrogen and halogen atoms in the molecules.

REACTIONS WITH AMMONIA

All the halogen hydride gases react with ammonia gas to form a white cloud of the solid ammonium halide:

$$HX + NH_3 \rightarrow NH_4^+ X^-$$

AQUEOUS SOLUTIONS

All the halogen hydrides are exceedingly soluble in water. Liquid hydrogen fluoride mixes with water in all proportions, i.e., it is completely miscible with water. The solubility of the gaseous hydrides HCl, HBr and HI are respectively about 500, 600 and 450 volumes of gas per volume of water at 0 °C.

Aqueous solutions of the halogen hydrides are acidic but hydrogen fluoride is a weak acid whereas hydrogen chloride, hydrogen bromide and hydrogen iodide are strong acids, the acid strength increasing from hydrogen fluoride (the weakest) to hydrogen iodide (the strongest). The increasing strength of the acids is to be expected on the basis of increasing distance between the centres of the hydrogen and halogen atoms in HF, HCl, HBr and HI. As the distance increases, the bond becomes weaker and

hence the ionization of the acid occurs more readily:

$$HX + H_2O \rightleftharpoons H_3O^+ + X^-$$

In dilute aqueous solutions, the substances HCl, HBr and HI are essentially fully ionized. The greater strength of hydrogen iodide as an acid is therefore easier to demonstrate in non-aqueous solvents. Thus in methanol, CH_3OH, hydrogen chloride is the most weakly ionized and hydrogen iodide the most strongly ionized of these three hydrides.

All of the halogen hydrides form azeotropic

ACID	PERCENTAGE OF ACID BY MASS	BOILING POINT (°C) AT 760 mmHg
HF	35.3	120
HCl	20.2	110
HBr	47.8	126
HI	57.0	127

Fig. 34.11. Azeotropic mixtures of the halogen hydrides.

mixtures with water. The table fig. 34.11 shows the composition and boiling points of these azeotropic mixtures.

REACTION OF HYDROFLUORIC ACID WITH SILICA

Hydrofluoric acid reacts with silica and with silicates present in glass:

$$SiO_2 + 4HF \rightarrow SiF_4 + 2H_2O$$

Silicon tetrafluoride is evolved on warming.

Hydrofluoric acid is therefore used to etch marks on glass and to make frosted glass. Parts of the glass which are to be protected from attack during etching, are covered with a layer of paraffin wax.

Hydrofluoric acid is available commercially as a 40% solution by mass and it is stored in polyethylene vessels.

● *CAUTION. Hydrofluoric acid is highly corrosive and can cause burns which are difficult and sometimes impossible to cure. Rubber gloves should be worn and extreme care should be taken to avoid inhaling any vapours.*

Salts of the Halogen Hydrides

The sodium halides are typical salts. The salts are difficult to melt, as seen in the table, fig. 34.12. Both the molten salts and their aqueous solutions are good electrolytic conductors.

SODIUM HALIDE	MELTING POINT (°C)
NaF	995
NaCl	801
NaBr	755
NaI	661

Fig. 34.12. Melting points of sodium halides.

The melting points of the salts show a regular gradation. Because the melting point is an indication of the strength of the cohesive force within the crystal lattice, it can be concluded that the inter-ionic force is strongest in sodium fluoride. This is to be expected on the basis of the ionic radii of the halide ions shown in the diagram, fig. 34.13. Thus in sodium fluoride, the centres of the charged particles are closest together, and the electrostatic attraction is greatest.

SOLUBILITY OF THE HALIDES

The solubility of the fluorides is somewhat different from the other halides. Thus calcium fluoride is very sparingly soluble in water, whereas calcium chloride, calcium bromide and calcium iodide are very soluble. Conversely silver chloride, bromide and iodide are sparingly soluble in water whereas silver fluoride is extremely soluble.

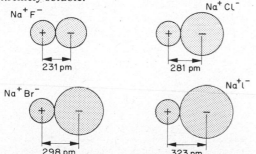

Fig. 34.13. The sodium halide ion-pairs.

Recent Developments

Fluorine is now produced on an industrial scale for the production of *uranium hexafluoride*, UF_6, which is the only gaseous compound of uranium. It is possible to separate the isotopes of uranium by using diffusion processes with uranium hexafluoride.

Fluorine reacts with hydrocarbons forming a series of compounds, called *fluorocarbons* of general formula C_nF_{2n+2}. The fluorocarbons are not inflammable and cannot be oxidized, and they are used as lubricants and solvents for special purposes.

'*Teflon*' is a recently developed plastic which contains only carbon and fluorine. It is extremely resistant to chemical attack and has found wide application in situations where resistance to oils or acids is necessary, e.g., in gears.

The last member of the halogens, *astatine*, is a radioactive element of very short half-life. Astatine occurs naturally in minute amounts and has recently been prepared artificially in very small quantities. It has been found that it resembles the other halogens in some ways although it is more metallic in character.

Hydrogen fluoride is used in the production of uranium metal. Ores containing as little as 0.1 to 0.3% uranium are processed to produce pure uranium dioxide which is then reacted with hydrogen fluoride at 250 °C—450 °C,

$$UO_2 + 4HF \rightarrow UF_4 + 2H_2O$$

The uranium tetrafluoride is mixed with magnesium and the mixture is heated to 950 °C to start the reaction. Magnesium reduces uranium tetrafluoride, because magnesium is high on the activity series of metals, and is a powerful reducer.

$$UF_4 + 2Mg \rightarrow U + 2MgF_2$$

The magnesium fluoride forms as a slag which is easily separated from the uranium metal.

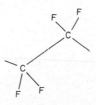

Fig. 34.14. 'Teflon' or polytetrafluoroethylene.

STUDY QUESTIONS

1. The halogens show fairly regular gradations in melting points, boiling points, and densities. Indicate the trends in these properties and provide an explanation of the trends.

2. Refer to the table, fig. 34.4, and offer an explanation why the reaction between hot concentrated hydrochloric acid and manganese(IV) oxide results in the evolution of chlorine, even though the standard electrode potentials appear to suggest that the reaction will not take place.

3. Write equations to represent the oxidation of iodide ions by:
   (a) chlorine;
   (b) an iron(III) salt;
   (c) a mixture of manganese(IV) oxide and concentrated sulphuric acid.

4. Write equations to illustrate the reactions between chlorine and:
   (a) water;
   (b) cold aqueous caustic soda;
   (c) an aqueous solution of hydrogen sulphide.

5. Write equations to illustrate the reactions of fluorine with:
   (a) hydrogen gas;
   (b) water;
   (c) sodium metal.

6. What explanation can be offered for the relatively high boiling point of hydrogen fluoride?

7. The melting points of sodium fluoride (995 °C), potassium fluoride (856 °C) and magnesium fluoride (1263 °C) are measures of the strengths of the inter-ionic forces in the compounds.
   What explanation can be offered, in terms of structure, for the observed differences in melting points?

8. The first member of a group in the periodic table frequently shows some properties which are different from other members of the group.
   (a) Illustrate this by reference to the properties of fluorine and its compounds.
   (b) What explanation can be offered for the exceptional character of fluorine?

9. Draw diagrams to illustrate the valency of:
   (a) fluorine in F_2 and OF_2;
   (b) chlorine in $HOCl$, $HClO_3$ and $HClO_4$;
   (c) bromine in BrF_5;
   (d) iodine in IF_7.

10. The reaction of the sodium halide (Na^+X^-) with concentrated H_2SO_4 is an effective method of preparing HF and HCl, but not HBr or HI. What explanation can be offered for this?
    What other acid could be used to prepare HBr and HI from the sodium salts?

11. Explain how the relative strengths of HCl, HBr and HI could be established.

12. What explanation can be offered for the resistance of teflon to chemical attack?

The many applications of nuclear science and radio-chemistry arose from the discovery of natural radioactivity in 1896, artificial nuclear transmutation between 1919 and 1934 and nuclear fission in 1939.

RADIOACTIVITY

The Discovery of X-rays

In 1895 world wide interest was aroused by the discovery of X-rays by W. Röntgen. Röntgen was studying electrical discharges through gases in a darkened laboratory, so that he could observe the glowing regions in the gas, when he noticed that a nearby object was glowing brightly. This chance observation was followed by a series of carefully devised experiments by which Röntgen was able to show that an invisible but very penetrating form of radiation is emitted by various substances when they are bombarded by cathode rays. He called the radiation X-rays and showed that the radiation could be detected because it caused various minerals, including those of uranium, to fluoresce. It also affected a photographic plate so that, on development, the emulsion became blackened. The penetrating rays are absorbed partially as they pass through substances and the degree of absorption varies with different substances. Röntgen used X-rays to obtain a photograph of the bones in his hand and when this was published it created great interest in the medical profession.

Natural Radioactivity

H. Becquerel had discovered that uranium minerals fluoresce after exposure to sunlight. This principle is now used in making fluorescent dials for watches, etc. In 1896 he set out to determine whether there is any connection between *this* fluorescence and the fluorescence caused by X-rays. For his experiments he decided to use photographic plates to detect any X-rays.

Becquerel placed samples of uranium compounds on photographic plates wrapped in black paper and exposed them to sunlight. On development, the plates showed black areas, but Becquerel did not jump to the conclusion that this proved that X-rays were being emitted. He carried out further investigations, in the course of which he happened to prepare a plate during a period of cloudy weather. After a few brief periods of sunlight, the sun clouded over and Becquerel put the plate away in a drawer with the uranium compound.

Some days later, he was developing the photographic plates and he included the one from the drawer thinking that it would be only faintly blackened, but instead, the blackening was intense. Thus Becquerel's careful and systematic experiments had led to a completely unexpected result. He had discovered that uranium compounds are capable of emitting a penetrating radiation even when they are not stimulated by sunlight.

In 1897, *E. Rutherford* was able to show that the invisible penetrating radiation discovered by Becquerel was composed of at least two components, which differed in their penetration of substances; he designated these as α (alpha) and β (beta). *P. Curie*, Madame Curie's husband, later found a third component which he called γ (gamma). At about this time *Madame Curie* gave the name radioactivity to Becquerel's discovery.

Classification of Radioactive Emissions

The classification of radioactivity into α-particles, β-particles and γ-rays is based on their penetrating properties (fig. 35.1) and on their behaviour in electrostatic and magnetic fields (fig. 35.2).

EMISSION	RANGE IN AIR	SHIELD TO STOP RADIATION
α-particle	*Few cm*	*Thin sheets e.g. paper.*
β-particle	*Few metre*	*Thin metal sheets e.g. a few cm of aluminium.*
γ-ray	*About 100 metre*	*Thick shields e.g. several cm of lead.*

Fig. 35.1. Penetration by radiations from radioactive sources.

(i) *α-particles* were identified as helium nuclei in 1909 by E. Rutherford and T. Royds. They placed an α-emitter in a very thin walled glass tube which was enclosed in an evacuated vessel. After several days, helium was detected in the vessel. An α-particle may be represented by the symbol $^4_2He^{2+}$, showing the atomic number, mass number and charge. α-particles

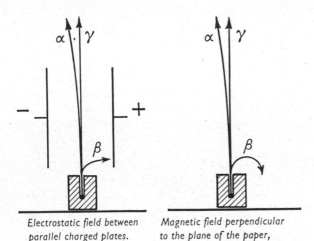

Electrostatic field between parallel charged plates.

Magnetic field perpendicular to the plane of the paper, north pole in front and south pole behind the plane of the paper.

Fig. 35.2. Deflection of radioactive emissions.

can be stopped by a sheet of paper and if this were the only kind of radioactive emission from uranium minerals, Becquerel's experiments would not have led to any positive results. In chapter 1 it was shown how Rutherford used the results of experiments on the penetrating properties of α-particles to postulate the nuclear theory of atomic structure.

(ii) A *β-particle* is identical with an electron and it may be represented by the symbol $_{-1}^{0}e^{-}$. The electrons emitted by radioactive substances usually have much higher speeds than the electrons in cathode rays.

(iii) *γ-rays* are electromagnetic energy waves similar to X-rays but with wave lengths of about 10^{-12} metre. This is about one-tenth of the wave lengths of X-rays used in medicine and about one-hundred-thousandth of the wave lengths of visible light. γ-rays are emitted at the same time as α- and β-particles from radioactive substances.

Detection of Radioactivity

SCINTILLATION COUNTERS
Radioactive emissions can be detected by their ability to cause certain minerals to fluoresce. Zinc sulphide glows when exposed to radiation, and when viewed under a microscope, this glow can be seen to be made up of a series of flashes called scintillations.

Each scintillation corresponds to the impact of an α- or β-particle and equipment has been devised to count the scintillations and hence measure the number of particles emitted by a radioactive source.

PHOTOGRAPHIC DETECTION
Radiation affects a photographic emulsion so that it becomes blackened on development. This is particularly useful in detecting the distribution of radioactive materials in a source. The source is placed against the photographic plate so that local areas of radioactivity affect the adjacent parts of the emulsion and the resulting pattern of darkened areas is called an *auto-radiograph*.

GEIGER-MÜLLER TUBES
Radiation ionizes gases and this property is used to detect radiation using a Geiger-Müller tube (fig. 35.3). Radioactive sources may be placed near the tube

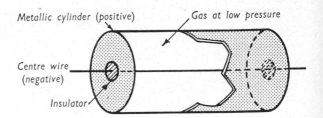

Fig. 35.3. The Geiger-Müller tube.

or inside the tube. Each emission knocks electrons from the molecules of the gas and results in the formation of ions and this in turn allows a pulse of current to pass between the electrodes. These current pulses may be recorded or counted and hence give a measure of the radiation.

CLOUD CHAMBERS
The cloud chamber is a device invented by C. T. R. Wilson. It consists of a shallow glass chamber filled with moist air and connected to a piston, so that withdrawal of the piston causes the air in the chamber to become supersaturated with water vapour (see fig. 35.4).

The passage of charged particles through the air forms positive ions due to electrons being knocked out of gas molecules, and water vapour condenses

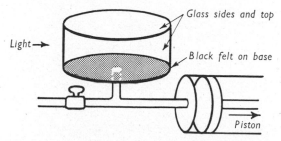

Fig. 35.4. The Wilson cloud chamber.

on these positive ions, forming a line of water droplets. Thus if a radioactive source is mounted inside the chamber, the tracks of α- and β-particles become clearly visible whenever the piston is withdrawn, and the tracks so produced can be photographed.

Rate of Radioactive Decay

It has been found that the rate of radioactive emission depends only on the mass of the radioactive substance present and is independent of all the factors which usually affect the rate of a chemical reaction such as temperature, pressure, etc. Because radioactive phenomena arise from the nuclei and chemical phenomena from the electronic structure of atoms, these are sometimes classified as nuclear and extra-nuclear properties respectively.

Consider a sample of pure uranium-238, $^{238}_{92}U$, which is an α-emitter. The α-particles are emitted as the nuclei of uranium disintegrate, and so the mass of uranium gradually decreases. This causes the rate of α-emission to fall off or decay.

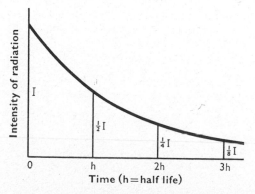

Fig. 35.5. Curve showing the decrease in the intensity of radiation during radioactive decay.

• *The rate of radioactive decay is usually expressed as a half-life, which is defined as the time taken for the rate of disintegration of the radioactive sample to fall to half its original value (see fig. 35.5).*

For $^{238}_{92}U$ the half-life is 4.5×10^9 year, which is one of the longest half-lives for natural radio-isotopes. As an example of the other extreme, polonium-214, $^{214}_{84}Po$, has a half-life of only 1.5×10^{-4} second.

Bubble chamber tracks: bubble chambers are like cloud chambers, but contain a nearly boiling liquid. The curved tracks are lines of bubbles formed along the paths of charged particles moving through a magnetic field in the chamber.

Example of Radioactive Series

The radioactive decay of $^{238}_{92}U$ can be represented by an equation. Since an ^4_2He particle is emitted from the nucleus, the atom left must have an atomic number of 90 and a mass number of 234, and since element number 90 is thorium:

$$^{238}_{92}U \rightarrow ^{234}_{90}Th + ^4_2He$$

It is not usual to show the charges on the particles because the charges are due to the extra-nuclear structure.

Thorium-234 is a β-emitter and so decays to an element with the same mass number, but an atomic number of 91. Its half-life is 24.5 day.

$$^{234}_{90}Th \rightarrow ^{234}_{91}Pa + ^0_{-1}e$$

Protactinium is also a β-emitter and has a half-life of 1.14 minute:

$$^{234}_{91}Pa \rightarrow ^{234}_{92}U + ^0_{-1}e$$

These reactions are the beginning of a series of changes

which ultimately lead to the formation of a non-radioactive isotope of lead, $^{206}_{82}Pb$ (see fig. 35.6), and here the series ends.

RADIOACTIVE DECAY SERIES FOR
URANIUM-238

Isotope	Emission	Half-Life
$^{238}_{92}U$	α	4.5×10^9 year
$^{234}_{90}Th$	β	24.5 day
$^{234}_{91}Pa$	β	1.14 minute
$^{234}_{92}U$	α	2.69×10^5 year
$^{230}_{90}Th$	α	8.3×10^4 year
$^{226}_{88}Ra$	α	1.59×10^3 year
$^{222}_{86}Rn$	α	3.8 day
$^{218}_{84}Po$	α	3.05 minute
$^{214}_{82}Pb$	β	26.8 minute
$^{214}_{83}Bi$	β	19.7 minute
$^{214}_{84}Po$	α	1.5×10^{-4} second
$^{210}_{82}Pb$	β	22 year
$^{210}_{83}Bi$	β	5.0 day
$^{210}_{84}Po$	α	140 day
$^{206}_{82}Pb$	Not radioactive	

A small amount of decay is also due to alternative sequences (see study question 1).

Fig. 35.6.

Induced Radioactivity

E. Rutherford was the first to show that if α-particles are moving fast enough, they may collide with the nuclei of a target element instead of being deflected away. The change observed by Rutherford was:

$$^{14}_{7}N + ^{4}_{2}He \rightarrow ^{17}_{8}O + ^{1}_{1}H$$

Many kinds of machines have been built to accelerate charged particles to very high speeds so that reactions of this type can be studied. Such machines are popularly known as atom-smashers.

However, the most significant discovery in the study of artificial transmutations came in 1934, when *E. Fermi* found that neutron bombardment could be used. Because the neutron carries no electrical charge, it does not have to overcome electrical repulsive forces from the nuclei of the target atoms, and it is therefore not necessary to accelerate it to high speeds. A typical neutron induced transmutation is:

$$^{27}_{13}Al + ^{1}_{0}n \rightarrow ^{24}_{11}Na + ^{4}_{2}He$$

The sodium-24 produced is a radioactive β-emitter with a half-life of 15 hour:

$$^{24}_{11}Na \rightarrow ^{24}_{12}Mg + ^{0}_{-1}e$$

Nuclear Fission

All of the natural radioactive changes mentioned so far result in emission of an α- or β-particle and a γ-ray, so the remaining nucleus is always an element with an atomic number close to the atomic number of the radio-isotope. In 1939 *O. Hahn* and *F. Strassmann* showed that the neutron bombardment of uranium ($Z = 92$) leads to the formation of barium ($Z = 56$) amongst a number of products. The uranium apparently breaks into fragments:

$$_{92}U + _{0}n \rightarrow _{56}Ba + _{36}Kr$$

Subsequent research has shown that this is only one of many possible fission reactions and that all elements between atomic numbers 30 and 63 are formed. The fission products most often formed are $^{96}_{40}Zr$ and $^{138}_{56}Ba$.

A further discovery of great significance is that *fission is accompanied by neutron emission*.

Thus for uranium:

$$_{92}U + _{0}n \rightarrow 2 \text{ fission} + 2 \text{ or } 3$$
$$\text{fragments} \quad \text{neutrons}$$

On the average about 2.6 neutrons are released by the fission of each uranium nucleus.

The importance of this discovery is due to two considerations. Firstly, there is the possibility of a self-sustaining chain of nuclear fissions in uranium. Secondly, there is the fact that *the total mass of the products of fission is less than the total mass before fission.* In 1905, *A. Einstein* postulated that the energy released by a change involving a decrease in mass is given by the expression:

$$E = mc^2$$

where: m is the mass lost;
c is the velocity of light;
and E is the energy released.

If the unit of mass is the kilogramme and the unit of velocity the metre second^{-1}, the unit of energy is the joule.

The energy released by fission reactions can be calculated, but, because of the many different reactions in the fission of uranium and because some energy is lost, it is better to use direct measurements. These show that the fission of uranium can produce over 10 000 times as much heat as the combustion of an equal mass of coal.

Nuclear Reactors

A typical nuclear reactor is shown in the diagram (fig. 35.7). The pile consists of pieces of uranium, in carefully designed shapes, separated by carbon blocks. The neutrons released by fission in the uranium are slowed down by collision with the carbon atoms in the graphite, without being absorbed.

Fig. 35.7. A graphite moderated nuclear reactor.

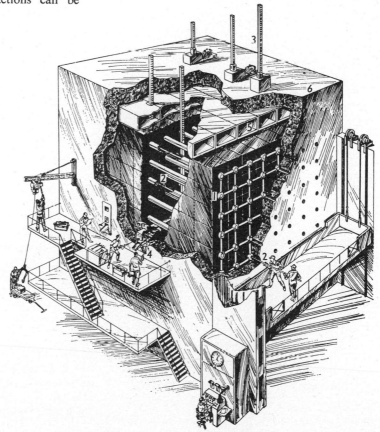

1 graphite
2 uranium rod
3 control rod of cadmium or boron
4 channel through which radioactive
   isotopes are obtained
5 air cooling
6 concrete screen

Natural uranium consists of about 99.3 % $^{238}_{92}$U and 0.7 % $^{235}_{92}$U. A very small percentage of $^{234}_{92}$U is also present but makes no significant contribution to the fission process. The two main isotopes of uranium differ in that $^{235}_{92}$U undergoes fission with neutrons but $^{238}_{92}$U can capture slow neutrons forming another isotope of uranium:

$$^{238}_{92}\text{U} + ^{1}_{0}\text{n} \rightarrow ^{239}_{92}\text{U}$$

Uranium-239 is a β-emitter and decays with a half-life of 23 minute. This produces element number 93 which does not occur naturally and has been called *neptunium*:

$$^{239}_{92}\text{U} \rightarrow ^{239}_{93}\text{Np} + ^{0}_{-1}\text{e}$$

The half-life of neptunium is only 2.3 day and it decays fairly rapidly with β-emission. This results in the production of another new element called *plutonium*:

$$^{239}_{93}\text{Np} \rightarrow ^{239}_{94}\text{Pu} + ^{0}_{-1}\text{e}$$

Plutonium is an α-emitter with a half-life of 2.4×10^4 year and so decays very slowly:

$$^{239}_{94}\text{Pu} \rightarrow ^{235}_{92}\text{U} + ^{4}_{2}\text{He}$$

In the operation of a reactor using natural uranium, fission products and plutonium are formed in about equal masses. After a time the fuel elements are removed from the pile and separated into the various products by chemical means.

Types of Nuclear Reactors

1. *Fast reactors* may use almost pure $^{235}_{92}$U, in which case virtually all collisions between neutrons and uranium nuclei result in fission. The term 'fast' refers to the speed of the neutrons, which is allowed to remain fairly high. The neutrons produced by fission may either lead to further fissions or may escape from the mass of uranium. The proportion which escapes depends on the size and shape of the uranium and it is possible to obtain a mass in which the reaction becomes self sustaining. This is referred to as the *critical mass*. In a reactor, neutrons are absorbed by the insertion of control rods made of neutron absorbing materials such as cadmium. The rods are withdrawn until the reactor just becomes critical.

An atomic bomb is essentially a fast reactor in which a greater than critical mass is allowed to undergo fission unchecked. The bomb contains $^{235}_{92}$U or $^{239}_{94}$Pu in a number of pieces which are each smaller than the critical mass and is detonated by bringing the pieces together very rapidly. One device is to surround one piece of the fissile material with other smaller pieces of fissile material which are attached to charges of chemical explosive. When the chemical charges are fired, the fissile material is shot into the central mass, which then becomes critical and detonates, releasing vast amounts of energy almost instantaneously.

2. *Thermal reactors* are so called because the speed of the neutrons is moderated to about the speed of of the thermal motion of molecules in a gas. It is possible to use natural uranium in such a reactor and generate electricity from the heat released by the fission process.

3. *Breeder reactors* are designed to produce heat and also enough $^{239}_{94}$Pu to replace the $^{235}_{92}$U used by the fission process. The plutonium may be used in both fast and thermal reactors or used with uranium to continue the operation of the breeder reactor. Another type of breeder reactor uses $^{232}_{90}$Th which is readily available. In the reactor:

$$^{232}_{90}\text{Th} + ^{1}_{0}\text{n} \rightarrow ^{233}_{90}\text{Th} + \gamma$$
$$^{233}_{90}\text{Th} \rightarrow ^{233}_{91}\text{Pa} + ^{0}_{-1}\text{e}$$
$$^{233}_{91}\text{Pa} \rightarrow ^{233}_{92}\text{U} + ^{0}_{-1}\text{e}$$

Uranium-233 may be used in the same way as plutonium-239 or uranium-235. The use of these reactions of uranium and thorium to form more fissile materials suggests the possibility of increasing the amount of heat obtainable from uranium by a thousand times, i.e., to 10 000 000 times the quantity of heat available from the same mass of coal.

Nuclear Fusion

At temperatures of several millions of degrees Celsius the nuclei of lighter elements combine to form heavier nuclei. This is called a nuclear fusion reaction, e.g., with the isotopes of hydrogen the reactions include:

$$^{2}_{1}\text{H} + ^{2}_{1}\text{H} \rightarrow ^{3}_{1}\text{H} + ^{1}_{1}\text{H}$$
$$^{2}_{1}\text{H} + ^{2}_{1}\text{H} \rightarrow ^{3}_{2}\text{He} + ^{1}_{0}\text{n}$$
$$^{2}_{1}\text{H} + ^{3}_{1}\text{H} \rightarrow ^{4}_{2}\text{He} + ^{1}_{0}\text{n}$$

All of these reactions involve a loss of mass and release more energy per unit mass than fission reactions.

The only fusion reactions used so far have been for hydrogen bombs because it has not yet been possible to devise a means of controlling thermonuclear reactions.

Hydrogen fusion is brought about by using a $^{235}_{92}U$ or $^{239}_{94}Pu$ fission bomb to attain the necessary temperature. Since fusion reactions are not limited by a critical mass in the way fission reactions are, it is theoretically possible to obtain an unlimited release of energy from fusion reactions.

ELEMENT	SYMBOL	HALF-LIFE
An α-emitter		
polonium	$^{210}_{84}$ Po	140 day
Some β-emitters		
tritium	$^{3}_{1}$ H	12.4 year
carbon	$^{14}_{6}$ C	5.6×10^3 year
phosphorus	$^{32}_{15}$ P	14.3 day
sulphur	$^{35}_{16}$ S	87.1 day
chlorine	$^{36}_{17}$ Cl	4.4×10^5 year
strontium	$^{90}_{38}$ Sr	20 year
Some γ-emitters (β particles also emitted)		
sodium	$^{24}_{11}$ Na	15 hour
iron	$^{59}_{26}$ Fe	45 day
cobalt	$^{60}_{27}$ Co	5.2 year
iodine	$^{131}_{53}$ I	8 day
gold	$^{198}_{79}$ Au	2.7 day

Fig. 35.8. Some radio-isotopes.

USES OF RADIO-ISOTOPES

The application of radioactive isotopes in science, medicine, and technology, may be classified into three main fields. These are based on:

(a) the effect of radiation on matter;

(b) the effect of matter on radiation;

(c) the labelling and tracing of atoms with radiation.

Some of the radio-isotopes at present available and in use are shown in the table (fig. 35.8).

Health Hazards

In making use of radioactivity it is important to take into account its effect on living tissues. The penetrating nature of radioactive emissions, together with their ionizing properties, may upset the complicated structure of a living cell and lead to malfunction or even destruction of the cell. Some of the units used to measure radioactivity are given below.

The Curie was originally taken to be the rate of radioactive disintegration in one gram of pure radium but is now defined more precisely as 3.7×10^{10} disintegrations per second. In practice, the millicurie and microcurie, 3.7×10^7 and 3.7×10^4 disintegrations per second respectively, are more often used.

The Röntgen is the unit used to measure a dose of radiation. A röntgen of radiation passing through one cm^3 of air at s.t.p. produces ions carrying one electrostatic unit of electrical charge. This is equivalent to about 2×10^9 singly charged positive or negative ions.

The dose rate of a radioactive source is expressed in röntgen per hour at one metre, abbreviated as rhm. Because of the much greater range of γ-rays the danger from a γ-source is much greater than from an α- or β-source and so more stringent precautions are necessary to shield the worker.

EXAMPLES OF RADIATION DOSES

Workers in radiation laboratories should not be exposed to more than 0.1 röntgen per week and this is checked by small dosimeters which they must carry. A sudden dose of about 500 röntgen is lethal to adults in about 50% of cases and a medical X-ray photograph requires exposure to about 0.2 röntgen.

The Effect of Radiation on Matter

MEDICAL USES

Penetrating γ-radiation can be used in the same way as X-rays in the treatment of malignant tumors and cancer. Cancerous tissue is more susceptible to radiation damage than normal healthy tissue because the cancer cells are dividing much more frequently. If a dividing cell is struck by ionizing radiation, the complex structure of the nucleus of the cell is disturbed and the cell will be destroyed. X-ray machines are often preferred to radio-isotopes because they can be switched off when not in use and do not require such elaborate shielding.

For the treatment of deep seated growths the very penetrating γ-radiation of $^{60}_{27}Co$ is particularly useful. The main problem with both X- and γ-ray treatment is that very large total doses are required (up to 5000 röntgen) and the surrounding healthy tissues may be extensively damaged. The use of radio-isotopes placed in contact with the malignant tissue can often overcome this. Thus small sealed capsules containing $^{60}_{27}Co$ or $^{198}_{79}Au$ may be placed surgically in the affected organ. Skin cancer can be treated by strapping a plastic sheet containing the β-emitters $^{32}_{15}P$ or $^{90}_{38}Sr$ on the diseased area.

BIOLOGICAL USES

Many useful results have been obtained by irradiating plant seeds and pollen. While in most cases, the seeds fail to germinate or develop into stunted or malformed plants, occasionally a useful change in the plant is induced. Improved strains of wheat and barley have been developed in this way. Success has also been claimed in the control of insect pests by irradiation. A substantial reduction in the number of screw-worm flies on the island of Curacao was achieved by sterilizing a large number of male screw-worm flies by irradiation, before releasing them.

Complete destruction of bacteria and enzymes is possible with very large doses of radiation. The problem is that even if 90% of a bacterial population is destroyed, the remaining 10% will continue to multiply and soon restore the original level of bacterial activity. For small objects such as surgical instruments, sterilizing equipment by this method is feasible, but requires doses of the order of 10^6 röntgen.

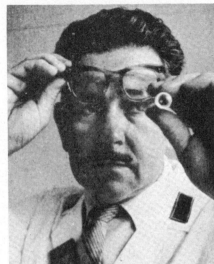

Above: Protection against radiation hazards.

All employees of the Australian Atomic Energy Commission who work with radioactive material are required to carry a dosimeter (left). This instrument records the radiation dosage received, and is read regularly (right) to ensure that this dosage does not exceed the safe level.

Below: Medical use of radiation.

A machine for treating cancer using the radiation from cobalt 60. This radioactive source is contained in the heavy shield mounted on the top of the counterbalanced arm. The position of the patient is carefully adjusted so that the region to be treated is at the centre as the arm rotates. This spreads out the dose over the surrounding healthy tissue.

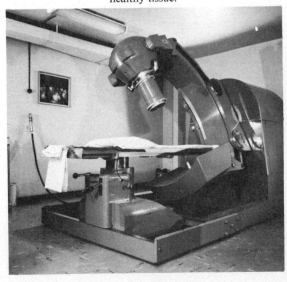

Handling radioactive material.
Skilled operators handle the master slave manipulators from outside the remote handling cells at Lucas Heights, while an A.A.E.C. scientist in the "umpire's chair" examines a specimen through a stereo viewer.

INDUSTRIAL USES

Static electricity, which develops on non-conducting materials subjected to friction, can be allowed to leak away if the surrounding air is ionized by using the radiation from a radio-isotope.

Specks of dust held on photographic negatives by electrostatic forces, can be easily brushed off after the negative has been passed very close to an α-source such as $^{210}_{84}Po$. The operation of high speed printing presses and textile looms has been improved by using β-sources such as $^{90}_{38}Sr$.

Ionizing effects are also utilized in making *luminescent signs* and *atomic batteries*. If a $^{90}_{38}Sr$ source is included in a mixture of zinc and cadmium sulphides, the β-particles excite electrons in the sulphides, and light is emitted as they fall back to lower energy levels. Such self-luminescent markers are useful as emergency

signs. Radioactive sources have been used to power small electric batteries. Although only very small currents can be generated in this way, quite high voltages can be developed and the battery can be made to have a very long life by using radio-isotopes with long half-lives.

The Effect of Matter on Radiation

THICKNESS GAUGES

Radiation is absorbed as it passes through a substance and this can be utilized in determining the thickness or density of a sample. A β- or γ-source with a long half-life can be used to control the *thickness of paper or plastic sheets* (see fig. 35.9).

Two identical sources are prepared simultaneously, so that the disintegration rates of the two sources remain equal as they decay. The radiation reaching the detectors is converted into electrical signals which are balanced against each other. Any inequality can be made to control the rollers and so control the thickness of the sheet. This method is reliable and involves no contact with the material and so there is no marking or tearing of the product. A similar method can be used to control the machines which pack tobacco into cigarettes although here the density of the product rather than its thickness is being controlled.

The mechanical filling of packets and tanks can be monitored by placing a source and a detector on either side of the container at the desired height. A sudden decrease in the transmitted radiation level, indicates that the contents have reached the predetermined depth in the container. Objects on a conveyor belt can be counted in similar way.

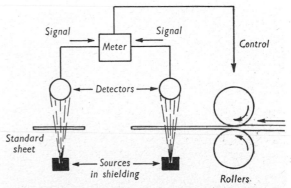

Fig. 35.9. Continuous control of thickness.

In the *petroleum industry*, one pipe may be used to convey a number of different hydrocarbon liquids one after the other and these may be checked by using a radioactive source. The density of a hydrocarbon liquid depends on the average size of the molecules in the mixture and this can be measured by the radiation absorbed.

Welds and metal castings can be checked by using the penetrating γ-rays of $^{60}_{27}Co$. For example, when two lengths of pipes are welded together, a 1 or 2 curie $^{60}_{27}Co$ source sealed in an aluminium sheath can be placed in the pipe. A photographic film enclosed in thin lead foil and black paper is then wrapped around the outside of the weld. The purpose of the lead foil is to intensify the effect on the photographic emulsion. γ-rays knock electrons from the lead atoms

Industrial use of radiation.

Above: The manhole duct on the right has been welded and is being checked for imperfections. A strip of special photographic film has been wrapped around the welded joint and the radio-isotope iridium 192 has been introduced into the duct from its portable safety shield by remote control.

Below: A gamma-radiograph of a revolver showing its interior mechanism. Iridium 192 was used as the source of gamma-rays.

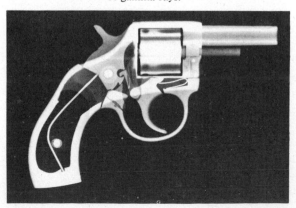

and these 'secondary electrons' enter the emulsion where they form many more ions than would be formed by the γ-rays. Any cracks in the weld show up as darker areas on development.

Industrial radiographs obtained as described above require fairly long exposure times, ranging from $\frac{1}{2}$ hour to several hours depending on the thickness of the casting. The exposure time could be shortened by using larger sources but this would present great difficulties in terms of safety and require elaborate sheathing when not in use. Much shorter exposures are needed to obtain radiographs through body tissue and it has been suggested that the use of $^{60}_{27}Co$ sources and self-developing films could be of great value in emergency work. For example, an accident victim could be checked for bone fractures in a few minutes without the need for elaborate equipment, power supplies or a dark room.

BACK-SCATTER GAUGES

Some of the radiation falling on a material is reflected back and the relative intensity of the reflected radiation depends on the composition, thickness and density of the material. Industrial back-scatter gauges use this principle to control the thickness of coatings on metals. The thickness of tin plating on steel can be controlled by using a β-source and balancing the signal against the reflection from a piece of the uncoated steel (fig. 35.10).

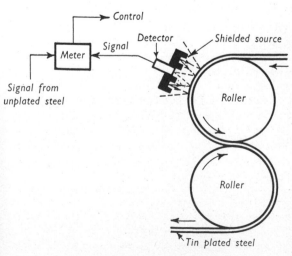

Fig. 35.10. A back-scatter gauge.

The Labelling and Tracing of Atoms with Radiation

INDUSTRIAL USES.

The efficiency of ventilating systems and mixing equipment can be followed by introducing a very small amount of radioactive material. Short lived isotopes such as $^{24}_{11}Na$ are preferred for this work so that any residual radioactivity will decay rapidly and be insignificant when the equipment is returned to normal use.

Underground water movements and leaks in underground pipes have been traced using soluble $^{131}_{53}I$ compounds. $^{198}_{79}Au$ has been used to trace silt movements in harbours and ports.

MEDICAL USES

Any element which is important to a particular metabolic process can be traced by using radio-isotopes. Two particularly important applications are the uptake of iron by the blood, traced by $^{59}_{26}Fe$, and uptake of iodine by the thyroid gland, traced by $^{131}_{53}I$.

CHEMICAL USES

Radio-isotopes have proved to be particularly useful to the chemist because they provide a method for following the course of particular atoms in a reaction.

The reaction of chlorine with water is represented as:

$$Cl_2 + H_2O \rightleftharpoons H^+ + Cl^- + HOCl$$

It has been suggested (chapter 33) that this reaction is an equilibrium system, i.e., both forward and reverse reactions are proceeding at the same rate. This has been verified by adding a soluble chloride containing $^{36}_{17}Cl$. After a time, evaporation of the solution yields Cl_2 gas containing $^{36}_{17}Cl$ and leaves less $^{36}_{17}Cl$ in the residue of chloride ions.

The structure of the $S_2O_3^{2-}$ ion can be better understood by using $^{35}_{16}S$. If a solution of a sulphite is boiled with sulphur containing this radio-isotope, the following reaction occurs:

$$SO_3^{2-} + {}^{35}S \rightarrow S_2O_3^{2-}$$

If the solution is acidified, it is found that all of the $^{35}_{16}S$ is precipitated and none remains in the solution:

$$S_2O_3^{2-} + 2H^+ \rightarrow H_2SO_3 + {}^{35}S$$

The structure of the thiosulphate ion can be based on

Tracing test using radio-isotopes.
A special counter is lowered into creek water to trace the movement of silt which has been "labelled" with radioactive gold-198.

the implication that the two sulphur atoms do not occupy equivalent or interchangeable positions, i.e., the reaction is consistent with the formulation:

It is evident that the $S_2O_3^{2-}$ ion should be named thiosulphate to indicate that it has the SO_4^{2-} structure but has a sulphur atom in place of an oxygen atom. The older name, hyposulphite, which is still used in photography, is not justified by the structure.

The use of isotopes has led to a much more complete understanding of the complex *process of photosynthesis*. The synthesis of carbon dioxide and water to a hexose (a sugar containing six carbon atoms) can be represented:

$$6CO_2 + 6H_2O \rightarrow C_6H_{12}O_6 + 6O_2$$

If water containing the non-radioactive isotope $^{18}_8O$

is used it is found that all of the O_2 gas comes from the water and not from the carbon dioxide. Hence:

$$6CO_2 + 12H_2O \rightarrow C_6H_{12}O_6 + 6O_2 + 6H_2O$$

Some of the intermediate steps in the process can be followed by using carbon dioxide containing $^{14}_{6}C$. Using the green alga, chlorella, in water containing radioactive carbon dioxide and analyzing the mixture after a few seconds exposure to light, the compound 3-phosphoglyceric acid (P.G.A.) has been isolated:

$^{14}COOH$

|

CHOH

|

$CH_2-O-PO(OH)_2$

The presence of the $^{14}_{6}C$ in the P.G.A. molecule indicates that the uptake of carbon dioxide gas is necessary for the formation of P.G.A. and that it is an intermediate in the photosynthetic reactions. The P.G.A. is reduced by hydrogen transferred from water by other complex intermediates and phosphoric acid and oxygen are released. Two reduced P.G.A. molecules then combine to form a six-carbon compound and ultimately a sugar.

Dating the Past with Radio-isotopes

THE AGE OF THE EARTH

The decay of $^{238}_{92}U$ suggests a possibility of dating rocks and minerals. If it is assumed that the mineral originally contained pure uranium, then the present ratio of $^{238}_{92}U$ to $^{206}_{82}Pb$ in the mineral can be used to calculate the time elapsed since the mineral was formed. Such calculations indicate ages varying from 50 to 4000 million years and this latter figure is often taken as the age of the earth.

DATING WITH CARBON-14

$^{14}_{6}C$ is being constantly formed in the upper atmosphere by reaction of nitrogen with neutrons knocked out of atoms by cosmic rays:

$$^{14}_{7}N + ^{1}_{0}n \rightarrow ^{14}_{6}C + ^{1}_{1}H$$

As a result of this transmutation, an almost constant concentration of carbon dioxide containing $^{14}_{6}C$ is present in the atmosphere. This is absorbed into plants as described above but, when the plant dies, no more carbon is absorbed and the radioactivity of the $^{14}_{6}C$ decays. By comparing the $^{14}_{6}C$ content of ancient organic remains with similar substances now growing, the age of the remains can be determined. Ages between 600 and 10 000 years can be determined quite accurately and older remains, up to about 40 000 years old, can be dated with less reliability.

DATING WITH TRITIUM

$^{3}_{1}H$ is also formed in the atmosphere and is present in water in fairly constant concentration. By concentrating the tritium in water by electrolysis, the level of radioactivity can be determined fairly accurately and compared with the level in present day samples of water. Because of the relatively short half-life of $^{3}_{1}H$, 12.4 year, this method is suitable for fairly recent samples. It has been checked and found reliable by using vintage wines as samples.

STUDY QUESTIONS

1. The radioactive series starting from $^{238}_{92}U$ given in table 35.6, shows only the main sequence, and does not include some alternative steps which account for a small proportion of the decay series. These are:
   (a) from $^{218}_{84}Po$, a β-particle and then an α-particle are emitted;
   (b) from $^{214}_{83}Bi$, first $^{210}_{81}Tl$ and then $^{210}_{82}Pb$ are formed.
Write equations to represent these steps of the decay sequence.

2. The main sequence in the decay series starting from $^{235}_{92}U$ and ending at $^{207}_{82}Pb$ involves the successive emission of the following particles in the order listed:
   $\alpha \ \beta \ \alpha \ \alpha \ \beta \ \alpha \ \alpha \ \alpha \ \beta \ \alpha \ \beta$.
Write the symbols of the elements which make up this decay series.

3. The leaves of a charged electroscope fall when a sample of pitch blende, a uranium mineral, is brought near the plate of the electroscope. Explain why this occurs.

4. Becquerel obtained an unexpected result from his experiments using uranium compounds which were placed on photographic plates wrapped in paper.

(a) Explain the blackening of the photographic plates observed by Becquerel.

(b) Were the uranium compounds pure? If not, what other elements may have been present?

(c) What modern applications use the effect of radiation on photographic emulsions?

5. Complete the following equations:

(a) $\quad ^{3}_{1}\text{H} \rightarrow \qquad + \quad ^{0}_{-1}\text{e}.$

(b) $\quad ^{14}_{6}\text{C} \rightarrow \qquad + \quad ^{0}_{-1}\text{e}.$

(c) $\quad ^{210}_{84}\text{Po} \rightarrow \qquad + \quad ^{4}_{2}\text{He}.$

(d) $\quad ^{9}_{4}\text{Be} + ^{4}_{2}\text{He} \rightarrow ^{12}_{6}\text{C} +$

(e) $\quad ^{16}_{8}\text{O} + ^{1}_{0}\text{n} \rightarrow \qquad + \quad ^{4}_{2}\text{He}.$

(f) $\quad ^{9}_{4}\text{Be} + \qquad \rightarrow ^{6}_{3}\text{Li} + ^{4}_{2}\text{He}.$

(g) $\quad ^{209}_{83}\text{Bi} + ^{2}_{1}\text{H} \rightarrow ^{210}_{83}\text{Bi} +$

6. A typical neutron induced fission of a $^{235}_{92}\text{U}$ atom yields $^{144}_{56}\text{Ba}$ and $^{89}_{36}\text{Kr}$.

(a) How many neutrons are released?

(b) Represent the change by an equation.

7. Calculate the energy in joule released by the fusion of two deuterons according to the reaction:

$$^{2}_{1}\text{H} + ^{2}_{1}\text{H} \rightarrow ^{3}_{2}\text{He} + ^{1}_{0}\text{n}$$

Given the mass of $^{2}_{1}\text{H} = 2.014\ 102$ on the ^{12}C scale

,, ,, ,, ,, $^{3}_{2}\text{He} = 3.016\ 049$,, ,, ,, ,,

,, ,, ,, ,, $^{1}_{0}\text{n} = 1.008\ 665$,, ,, ,, ,,

and that one unit on the ^{12}C scale $= 1.660 \times 10^{-27}$ kilogram and the velocity of light $= 2.998 \times 10^{8}$ metre second^{-1}.

8. A sample of a radioactive β-source is placed near a geiger counter and both are enclosed in a heavy lead shield. The counter registers 8870 counts per minute. Pieces of aluminium sheet are placed between the source and the

counter and the following results are obtained:

Thickness of Aluminium (in mg cm^{-2})	Radiation Measured (counts per minute)
25.4	7 390
50.7	6 330
76.7	5 340
102.1	4 540
132.2	3 750
157.6	3 190

Draw a graph from these results and hence estimate the thickness of aluminium necessary to reduce the radiation to half of its intensity. (The density of aluminium $- 2.70 \times 10^{3}$ mg cm^{-3}.)

9. A sample of an ancient piece of wood contains carbon in which the rate of β-emission from $^{14}_{6}\text{C}$ is only half of the rate from carbon in wood now growing. Estimate the age of the sample.

10. A vintage wine bottled 62 years ago is compared with recently collected rain water with respect to the tritium content. How should the rate of β-emission from the tritium in the two samples compare?

11. A reaction known as the crossed Cannizzaro reaction between formaldehyde, H—CHO, and benzaldehyde, C_6H_5—CHO, occurs in alkaline solution according to the stoichiometric equation:

C_6H_5—CHO + H—CHO + OH$^-$
benzaldehyde   formaldehyde

$\rightarrow C_6H_5$—CH$_2$OH + H—COO$^-$
benzyl alcohol   formate ion

If the carbon atom in the —CHO group of the benzaldehyde is labelled with $^{14}_{6}\text{C}$, it is found that all of the radio-isotope is obtained in the formate ion produced by the reaction. What information does this give concerning the course of the reaction?

12. If a sample of carbon dioxide containing $^{14}\text{CO}_2$ is heated in the presence of solid carbon, ^{14}CO is produced rapidly and then non-radioactive CO is slowly evolved. What information does this give concerning the reaction represented by the stoichiometric equation:

$CO_2 + C \rightarrow 2CO$?

Answers to Numerical Problems

7. 5.21×10^{-13} joule.

8. 106 mg cm^{-2} or 0.0392 cm.

9. 5600 year old.

10. Rates should be in the ratio 1:32.

APPENDIX ONE: The nomenclature of chemical compounds

The naming of chemical compounds poses many problems because several systems of nomenclature are in use. Many well known substances are known by traditional names such as water and ammonia. However, most chemical compounds are named on the basis of their structure according to either of two systems currently in use.

Naming of Ionic Compounds

In naming an ionic compound, *the cation (positive ion)* *is named first*. Most positive ions are related to metallic atoms and are given the same name, e.g.:

Na^+	sodium;
Ca^{2+}	calcium;
Zn^{2+}	zinc.

If the same metal may have more than one valency, the name may be derived from the Latin name of the metal with the ending *-ous* or *-ic to indicate the lower and higher valency states* respectively. The alternative system is to *state the valency by using Roman numerals*, e.g.:

Fe^{2+}	ferrous	iron(II)
Fe^{3+}	ferric	iron(III)
Cu^+	cuprous	copper(I)
Cu^{2+}	cupric	copper(II)

Most *negative ions* are related to non-metals and are named by using the name of the non-metal as a prefix and adding a suffix:

1. The ending *-ide* is used if the compound is composed of two elements (i.e. a binary compound), e.g.:

Na_2S	sodium sulphide	
$FeCl_2$	ferrous chloride	iron(II) chloride
$FeCl_3$	ferric chloride	iron(III) chloride.

2. The endings *-ite* and *-ate* are used to indicate a lower and higher oxidation state respectively, e.g.:

$CaSO_3$	calcium sulphite
$CaSO_4$	calcium sulphate.

These endings must be supplemented by prefixes where more than two oxidation states exist, e.g.:

$KClO$	potassium	hypochlorite
$KClO_2$	potassium	chlorite
$KClO_3$	potassium	chlorate
$KClO_4$	potassium	perchlorate.

In the naming of salts which contain hydrogen, the number of hydrogens should be indicated but "mono" is usually omitted if no ambiguity results, e.g.:

$NaHCO_3$	sodium hydrogen carbonate
$NaHSO_4$	sodium hydrogen sulphate
Na_2HPO_4	sodium monohydrogen orthophosphate
NaH_2PO_4	sodium dihydrogen orthophosphate

Naming of Molecular Compounds

Binary molecular compounds are named by using Greek prefixes to denote the number of atoms of each element in each molecule of the compound, e.g.:

mono	1	hex(a)	6
di	2	hept(a)	7
tri	3	oct(a)	8
tetr(a)	4	enne(a)	9
pent(a)	5	dec(a)	10

The prefix mono- is usually omitted if it is clear that only one atom of the element is contained in each molecule of the compound, e.g.:

HCl	hydrogen chloride
CO	carbon monoxide
CO_2	carbon dioxide.

The prefixes are necessary in naming the oxides of carbon to clearly distinguish between them.

The system is not always used rigorously because many substances have long established familiar names, e.g. water, ammonia. Systematic names for the oxides of nitrogen are compared with older names in the following set:

N_2O	dinitrogen monoxide	nitrous oxide
NO	nitrogen monoxide	nitric oxide
NO_2	nitrogen dioxide	nitrogen dioxide
N_2O_4	dinitrogen tetroxide	nitrogen tetroxide
N_2O_3	dinitrogen trioxide	nitrogen sesquioxide
N_2O_5	dinitrogen pentoxide	nitrogen pentoxide.

Naming of Acids

Acids are named in a similar way to the corresponding salts but the ending -ite is replaced by *-ous* and the ending -ate is replaced by *-ic*.

H_2SO_3	sulphurous acid
H_2SO_4	sulphuric acid
$HOCl$	hypochlorous acid
$HClO_2$	chlorous acid
$HClO_3$	chloric acid
$HClO_4$	perchloric acid.

APPENDIX TWO: The abundance of the elements

1. oxygen	49.5	9. hydrogen	0.9
2. silicon	25.7	10. titanium	0.6
3. aluminium	7.4	11. chlorine	0.2
4. iron	4.7	12. phosphorus	0.1
5. calcium	3.4	13. manganese	0.1
6. sodium	2.6	14. sulphur	0.1
7. potassium	2.4	15. carbon	0.1
8. magnesium	1.9	16. fluorine	0.1

This table shows the estimated relative abundance of elements in the earth's crust, calculated as a percentage by mass.

The remaining 76 naturally occurring elements amount to only 0.2% relative abundance.

APPENDIX THREE: Saturated vapour pressure of water

Temp. (°C)	Vapour Pressure (mmHg)	Temp. (°C)	Vapour Pressure (mmHg)
0	4.6	24	22.4
10	9.2	25	23.8
11	9.8	26	25.2
12	10.5	27	26.7
13	11.2	28	28.3
14	12.0	29	30.0
15	12.8	30	31.8
16	13.6	40	55.3
17	14.5	60	149.4
18	15.5	80	355.1
19	16.5	90	525.8
20	17.5	95	633.9
21	18.7	100	760.0
22	19.8	105	906.1
23	21.1	110	1075

APPENDIX FOUR: The shapes of molecules

PLANAR

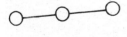

linear

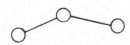

bent

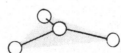

trigonal

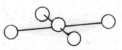

square

NON-PLANAR

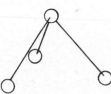
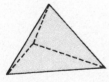

pyramidal

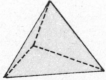

tetrahedral

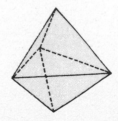

trigonal bipyramidal

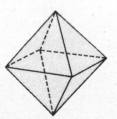
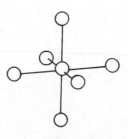

octahedral

265

APPENDIX FIVE: La Systeme Internationale d'Unites
The International System of Units
SI

The international authority on physical quantities is the International Organization for Standardization (ISO) which confers with organizations such as the International Union of Pure and Applied Chemistry (IUPAC) and the International Union of Pure and Applied Physics (IUPAP). The international authority on units is the General Conference on Weights and Measures (CGPM) and its International Committee on Weights and Measures (CIPM). These organizations are in the process of forming agreements on uniform usage of physical quantities and units.

The seven fundamental quantities and units

It has been agreed to adopt a system based on six quantities —length, mass, time, electric current, thermodynamic temperature and luminous intensity. In addition, chemists have agreed to adopt a seventh quantity called amount of substance, although this has not yet been formally adopted by CGPM. These seven quantities and their units are given in Table A.

PHYSICAL QUANTITY	SYMBOL OF QUANTITY	NAME OF UNIT	SYMBOL OF UNIT
length	l	metre	m
mass	m	kilogramme	kg
time	t	second	s
thermodynamic temperature	T	kelvin	K
amount of substance	n	mole	mol
electric current	I	ampere	A
luminous intensity	Iv	candela	cd

Table A.

Derived SI units

Special names are given to certain derived quantities. Some of these are given in Table B.

PHYSICAL QUANTITY	NAME OF UNIT	SYMBOL OF UNIT	DEFINITION OF UNIT
force	newton	N	$kg\,m\,s^{-2} = J\,m^{-1}$
energy	joule	J	$kg\,m^2\,s^{-2}$
electric charge	coulomb	C	$A\,s$
electric potential difference	volt	V	$kg\,m^2\,s^{-3}\,A^{-1}$ $= J\,A^{-1}\,s^{-1}$
frequency	hertz	Hz	s^{-1}

Table B.

Multiples and sub-multiples

The units given in Tables A and B are fundamental units but they may not be convenient in all circumstances. For example, densities may be conveniently measured in $g\,cm^{-3}$ or $g\,dm^{-3}$ instead of $kg\,m^{-3}$. For this reason, multiples and sub-multiples given in Table C, are provided for use with

10^1	deka	da	10^{-1}	deci	d
10^2	hecto	h	10^{-2}	centi	c
10^3	kilo	k	10^{-3}	milli	m
10^6	mega	M	10^{-6}	micro	μ
10^9	giga	G	10^{-9}	nano	n
10^{12}	tera	T	10^{-12}	pico	p

Table C.
Multiples, sub-multiples, prefixes and their symbols.

SI units. These prefixes are used where appropriate throughout this text. It is not always necessary to use the fundamental unit. Thus, volumes are generally quoted in cm^3 and dm^3 because these are the units in which volumes are conveniently measured.

The only circumstances in which SI units must be used are in calculations involving constants expressed in SI units. For example, in using the following equation, all quantities must be expressed in SI units.

$$\frac{pV}{T} = 8.314\,J\,K^{-1}\,mol^{-1}$$

Thus p must be expressed in $N\,m^{-2}$, v must be expressed in $m^3\,mol^{-1}$ and T must be expressed in K. If the numerical value of the constant is not used, non-SI units may be used. For example, any consistent units can be used for p and V in the following equation:

$$\frac{p_1\,V_1}{T_1} = \frac{p_2\,V_2}{T_2}$$

However, in this equation, T must be on the Kelvin scale.

Non-SI units

Some units which have not been adopted by CGPM are used in this text. Values of physical and chemical quantities for gases are generally quoted at 1 atm pressure, a non-SI unit, and in experimental work, the non-SI unit mmHg is commonly used.

In this text, atm and mmHg have been retained. Definitions of these non-SI units are given in Table D.

PHYSICAL QUANTITY	NAME OF UNIT	SYMBOL OF UNIT	DEFINITION OF UNIT
pressure	atmosphere	atm	$101\,325\,N\,m^{-2}$
pressure	millimetre of mercury	mmHg	$13.595\,1$ $\times\,9.806\,65\,N\,m^{-2}$

Table D.

ACKNOWLEDGEMENTS

We would like to thank the following for their assistance with the specific topics in this book and for permission to reproduce photographs:

The Australian Atomic Energy Commission for photographs on pp. 258-260.

Messrs Australian Carbon Black Pty Ltd for the photograph on the back cover.

Messrs B.P. Australia Ltd for the photograph on p. 172.

The British Phosphate Commission for fig. 29.1.

Messrs Broken Hill Pty Co Ltd for the photographs on pp. 129 and 133 (upper).

Messrs Carba Dry Ice (Australia) Pty Ltd for photographs on pp. 158 and 160 (lower).

Messrs Cheetham Salt Ltd for the photograph on p. 253 (right).

Mr Leon Comber for the photographs on p. 210.

Messrs CIG (Victoria) Pty Ltd for the photographs on pp. 133 (lower), 215, 218 and 219.

The Commonwealth Scientific and Industrial Research Organization for figs 1.2 and 4.3 and for photographs on pp. 4, 8 and 261.

Messrs David Mitchell Estate Ltd for photographs on pp. 164 and 165 (lower).

Messrs Dunklings the Jewellers Pty Ltd for the photograph on p. 155.

Messrs Electrolytic Zinc Co of Australasia Ltd for the photograph on p. 136.

Messrs Elsevier Uitgeverf Mij. N.V., Amsterdam, and Messrs John Murray Ltd for fig. 35.7 from *The Book of the Atom* by Leonard de Vries.

Messrs Herald and Weekly Times Ltd for the photograph on p. 127 (lower).

Mrs Pamela Hudson for the attractive cover design.

Messrs I.C.I.A.N.Z. Ltd for figs 31.1 and 32.3, and for photographs on pp. 192, 197, 213, 232 and 235.

The Lawrence Radiation Laboratory, Berkeley, California, for the photograph on p. 253.

Dr Ross Miller for the photograph on p. 38 (top right).

Messrs Mount Lyell Mining and Railway Co Ltd and the Visual Aids Branch, Department of Education, Hobart, for photographs on p. 92.

The Petroleum Information Bureau for the photograph on p. 173.

The United States Information Office for photographs on pp. 174 and 175.

The Vacuum Oil Co for the photograph on p. 172.

The Victorian Government Tourist Bureau for photograph on p. 169.

Messrs Woodall-Duckham (Australasia) Pty Ltd for fig. 26.5.

Messrs Wormald Bros (Victoria) Pty Ltd for the photograph on p. 160.

We would also like to thank the many other firms, including the Shell Co of Australia Ltd, The Commonwealth Aluminium Corporation Pty Ltd and The Australian Paper Manufacturers Ltd, who at our request supplied photographs and information on their operations.

TABLE OF LOGARITHMS

	0	1	2	3	4	5	6	7	8	9	1	2	3	4	5	6	7	8	9
10	0000	0043	0086	0128	0170	0212	0253	0294	0334	0374	4	8	12	17	21	25	29	33	37
11	0414	0453	0492	0531	0569	0607	0645	0682	0719	0755	4	8	11	15	19	23	26	30	34
12	0792	0828	0864	0899	0934	0969	1004	1038	1072	1106	3	7	10	14	17	21	24	28	31
13	1139	1173	1206	1239	1271	1303	1335	1367	1399	1430	3	6	10	13	16	19	23	26	29
14	1461	1492	1523	1553	1584	1614	1644	1673	1703	1732	3	6	9	12	15	18	21	24	27
15	1761	1790	1818	1847	1875	1903	1931	1959	1987	2014	3	6	8	11	14	17	20	22	25
16	2041	2068	2095	2122	2148	2175	2201	2227	2253	2279	3	5	8	11	13	16	18	21	24
17	2304	2330	2355	2380	2405	2430	2455	2480	2504	2529	2	5	7	10	12	15	17	20	22
18	2553	2577	2601	2625	2648	2672	2695	2718	2742	2765	2	5	7	9	12	14	16	19	21
19	2788	2810	2833	2856	2878	2900	2923	2945	2967	2989	2	4	7	9	11	13	16	18	20
20	3010	3032	3054	3075	3096	3118	3139	3160	3181	3201	2	4	6	8	11	13	15	17	19
21	3222	3243	3263	3284	3304	3324	3345	3365	3385	3404	2	4	6	8	10	12	14	16	18
22	3424	3444	3464	3483	3502	3522	3541	3560	3579	3598	2	4	6	8	10	12	14	15	17
23	3617	3636	3655	3674	3692	3711	3729	3747	3766	3784	2	4	6	7	9	11	13	15	17
24	3802	3820	3838	3856	3874	3892	3909	3927	3945	3962	2	4	5	7	9	11	12	14	16
25	3979	3997	4014	4031	4048	4065	4082	4099	4116	4133	2	3	5	7	9	10	12	14	15
26	4150	4166	4183	4200	4216	4232	4249	4265	4281	4298	2	3	5	7	8	10	11	13	15
27	4314	4330	4346	4362	4378	4393	4409	4425	4440	4456	2	3	5	6	8	9	11	13	14
28	4472	4487	4502	4518	4533	4548	4564	4579	4594	4609	2	3	5	6	8	9	11	12	14
29	4624	4639	4654	4669	4683	4698	4713	4728	4742	4757	1	3	4	6	7	9	10	12	13
30	4771	4786	4800	4814	4829	4843	4857	4871	4886	4900	1	3	4	6	7	9	10	11	13
31	4914	4928	4942	4955	4969	4983	4997	5011	5024	5038	1	3	4	6	7	8	10	11	12
32	5051	5065	5079	5092	5105	5119	5132	5145	5159	5172	1	3	4	5	7	8	9	11	12
33	5185	5198	5211	5224	5237	5250	5263	5276	5289	5302	1	3	4	5	6	8	9	10	12
34	5315	5328	5340	5353	5366	5378	5391	5403	5416	5428	1	3	4	5	6	8	9	10	11
35	5441	5453	5465	5478	5490	5502	5514	5527	5539	5551	1	2	4	5	6	7	9	10	11
36	5563	5575	5587	5599	5611	5623	5635	5647	5658	5670	1	2	4	5	6	7	8	10	11
37	5682	5694	5705	5717	5729	5740	5752	5763	5775	5786	1	2	3	5	6	7	8	9	10
38	5798	5809	5821	5832	5843	5855	5866	5877	5888	5899	1	2	3	5	6	7	8	9	10
39	5911	5922	5933	5944	5955	5966	5977	5988	5999	6010	1	2	3	4	5	7	8	9	10
40	6021	6031	6042	6053	6064	6075	6085	6096	6107	6117	1	2	3	4	5	6	8	9	10
41	6128	6138	6149	6160	6170	6180	6191	6201	6212	6222	1	2	3	4	5	6	7	8	9
42	6232	6243	6253	6263	6274	6284	6294	6304	6314	6325	1	2	3	4	5	6	7	8	9
43	6335	6345	6355	6365	6375	6385	6395	6405	6415	6425	1	2	3	4	5	6	7	8	9
44	6435	6444	6454	6464	6474	6484	6493	6503	6513	6522	1	2	3	4	5	6	7	8	9
45	6532	6542	6551	6561	6571	6580	6590	6599	6609	6618	1	2	3	4	5	6	7	8	9
46	6628	6637	6646	6656	6665	6675	6684	6693	6702	6712	1	2	3	4	5	6	7	7	8
47	6721	6730	6739	6749	6758	6767	6776	6785	6794	6803	1	2	3	4	5	5	6	7	8
48	6812	6821	6830	6839	6848	6857	6866	6875	6884	6893	1	2	3	4	4	5	6	7	8
49	6902	6911	6920	6928	6937	6946	6955	6964	6972	6981	1	2	3	4	4	5	6	7	8
50	6990	6998	7007	7016	7024	7033	7042	7050	7059	7067	1	2	3	3	4	5	6	7	8
51	7076	7084	7093	7101	7110	7118	7126	7135	7143	7152	1	2	3	3	4	5	6	7	8
52	7160	7168	7177	7185	7193	7202	7210	7218	7226	7235	1	2	2	3	4	5	6	7	7
53	7243	7251	7259	7267	7275	7284	7292	7300	7308	7316	1	2	2	3	4	5	6	6	7
54	7324	7332	7340	7348	7356	7364	7372	7380	7388	7396	1	2	2	3	4	5	6	6	7
	0	1	2	3	4	5	6	7	8	9	1	2	3	4	5	6	7	8	9

	0	1	2	3	4	5	6	7	8	9	1	2	3	4	5	6	7	8	9
55	7404	7412	7419	7427	7435	7443	7451	7459	7466	7474	1	2	2	3	4	5	5	6	7
56	7482	7490	7497	7505	7513	7520	7528	7536	7543	7551	1	2	2	3	4	5	5	6	7
57	7559	7566	7574	7582	7589	7597	7604	7612	7619	7627	1	2	2	3	4	5	5	6	7
58	7634	7642	7649	7657	7664	7672	7679	7686	7694	7701	1	1	2	3	4	4	5	6	7
59	7709	7716	7723	7731	7738	7745	7752	7760	7767	7774	1	1	2	3	4	4	5	6	7
60	7782	7789	7796	7803	7810	7818	7825	7832	7839	7846	1	1	2	3	4	4	5	6	6
61	7853	7860	7868	7875	7882	7889	7896	7903	7910	7917	1	1	2	3	4	4	5	6	6
62	7924	7931	7938	7945	7952	7959	7966	7973	7980	7987	1	1	2	3	3	4	5	6	6
63	7993	8000	8007	8014	8021	8028	8035	8041	8048	8055	1	1	2	3	3	4	5	5	6
64	8062	8069	8075	8082	8089	8096	8102	8109	8116	8122	1	1	2	3	3	4	5	5	6
65	8129	8136	8142	8149	8156	8162	8169	8176	8182	8189	1	1	2	3	3	4	5	5	6
66	8195	8202	8209	8215	8222	8228	8235	8241	8248	8254	1	1	2	3	3	4	5	5	6
67	8261	8267	8274	8280	8287	8293	8299	8306	8312	8319	1	1	2	3	3	4	5	5	6
68	8325	8331	8338	8344	8351	8357	8363	8370	8376	8382	1	1	2	3	3	4	4	5	6
69	8388	8395	8401	8407	8414	8420	8426	8432	8439	8445	1	1	2	2	3	4	4	5	6
70	8451	8457	8463	8470	8476	8482	8488	8494	8500	8506	1	1	2	2	3	4	4	5	6
71	8513	8519	8525	8531	8537	8543	8549	8555	8561	8567	1	1	2	2	3	4	4	5	5
72	8573	8579	8585	8591	8597	8603	8609	8615	8621	8627	1	1	2	2	3	4	4	5	5
73	8633	8639	8645	8651	8657	8663	8669	8675	8681	8686	1	1	2	2	3	4	4	5	5
74	8692	8698	8704	8710	8716	8722	8727	8733	8739	8745	1	1	2	2	3	4	4	5	5
75	8751	8756	8762	8768	8774	8779	8785	8791	8797	8802	1	1	2	2	3	3	4	5	5
76	8808	8814	8820	8825	8831	8837	8842	8848	8854	8859	1	1	2	2	3	3	4	5	5
77	8865	8871	8876	8882	8887	8893	8899	8904	8910	8915	1	1	2	2	3	3	4	4	5
78	8921	8927	8932	8938	8943	8949	8954	8960	8965	8971	1	1	2	2	3	3	4	4	5
79	8976	8982	8987	8993	8998	9004	9009	9015	9020	9025	1	1	2	2	3	3	4	4	5
80	9031	9036	9042	9047	9053	9058	9063	9069	9074	9079	1	1	2	2	3	3	4	4	5
81	9085	9090	9096	9101	9106	9112	9117	9122	9128	9133	1	1	2	2	3	3	4	4	5
82	9138	9143	9149	9154	9159	9165	9170	9175	9180	9186	1	1	2	2	3	3	4	4	5
83	9191	9196	9201	9206	9212	9217	9222	9227	9232	9238	1	1	2	2	3	3	4	4	5
84	9243	9248	9253	9258	9263	9269	9274	9279	9284	9289	1	1	2	2	3	3	4	4	5
85	9294	9299	9304	9309	9315	9320	9325	9330	9335	9340	1	1	2	2	3	3	4	4	5
86	9345	9350	9355	9360	9365	9370	9375	9380	9385	9390	1	1	2	2	3	3	4	4	5
87	9395	9400	9405	9410	9415	9420	9425	9430	9435	9440	0	1	1	2	2	3	3	4	4
88	9445	9450	9455	9460	9465	9469	9474	9479	9484	9489	0	1	1	2	2	3	3	4	4
89	9494	9499	9504	9509	9513	9518	9523	9528	9533	9538	0	1	1	2	2	3	3	4	4
90	9542	9547	9552	9557	9562	9566	9571	9576	9581	9586	0	1	1	2	2	3	3	4	4
91	9590	9595	9600	9605	9609	9614	9619	9624	9628	9633	0	1	1	2	2	3	3	4	4
92	9638	9643	9647	9652	9657	9661	9666	9671	9675	9680	0	1	1	2	2	3	3	4	4
93	9685	9689	9694	9699	9703	9708	9713	9717	9722	9727	0	1	1	2	2	3	3	4	4
94	9731	9736	9741	9745	9750	9754	9759	9763	9768	9773	0	1	1	2	2	3	3	4	4
95	9777	9782	9786	9791	9795	9800	9805	9809	9814	9818	0	1	1	2	2	3	3	4	4
96	9823	9827	9832	9836	9841	9845	9850	9854	9859	9863	0	1	1	2	2	3	3	4	4
97	9868	9872	9877	9881	9886	9890	9894	9899	9903	9908	0	1	1	2	2	3	3	4	4
98	9912	9917	9921	9926	9930	9934	9939	9943	9948	9952	0	1	1	2	2	3	3	4	4
99	9956	9961	9965	9969	9974	9978	9983	9987	9991	9996	0	1	1	2	2	3	3	3	4
	0	1	2	3	4	5	6	7	8	9	1	2	3	4	5	6	7	8	9

Index

PERIODIC CLASSIFICATION OF THE ELEMENTS

Period	Group I.	Group II.												Group III.	Group IV.	Group V.	Group VI.	Group VII.	Group O.
1	1 H																	→	2 He
2	3 Li	4 Be												5 B	6 C	7 N	8 O	9 F	10 Ne
3	11 Na	12 Mg												13 Al	14 Si	15 P	16 S	17 Cl	18 Ar
4	19 K	20 Ca	21 Sc	22 Ti	23 V	24 Cr	25 Mn	26 Fe	27 Co	28 Ni	29 Cu	30 Zn		31 Ga	32 Ge	33 As	34 Se	35 Br	36 Kr
5	37 Rb	38 Sr	39 Y	40 Zr	41 Nb	42 Mo	43 Tc	44 Ru	45 Rh	46 Pd	47 Ag	48 Cd		49 In	50 Sn	51 Sb	52 Te	53 I	54 Xe
6	55 Cs	56 Ba	57 La	72 Hf	73 Ta	74 W	75 Re	76 Os	77 Ir	78 Pt	79 Au	80 Hg		81 Tl	82 Pb	83 Bi	84 Po	85 At	86 Rn
7	87 Fr	88 Ra	89 Ac																

Transition Elements

Lanthanide Series	58 Ce	59 Pr	60 Nd	61 Pm	62 Sm	63 Eu	64 Gd	65 Tb	66 Dy	67 Ho	68 Er	69 Tm	70 Yb	71 Lu
Actinide Series	90 Th	91 Pa	92 U	93 Np	94 Pu	95 Am	96 Cm	97 Bk	98 Cf	99 Es	100 Fm	101 Md	102 No	103 Lr